GIScience and Geo-environmental Modelling

The "GIScience and Geo-environmental Modelling" book series seeks to publish a broad portfolio of scientific books addressing the interface between geography and the environment. The aim of the book series is to present geospatial technology approaches to data mining techniques, data analytics, modeling, risk assessment, visualization, and management strategies. The series includes peer-reviewed monographs, edited volumes, textbooks, and conference proceedings. The focus of Geo-environmental is on geospatial modelling in the frontier area of earth-environment related fields, such as urban and peri-urban environmental issues, ecology, hydrology, basin management, geohazards, estuarine-ecology, groundwater, agriculture, climate change, land-water, and forest resources, and related topics.

Geo-environmental modelling techniques have enjoyed an overwhelming interest in recent decades among the earth environmental and social sciences research communities for their powerful ability to solve and understand various complex problems and develop novel approaches toward sustainable earth and human society. Geo-environmental modelling using data mining, machine learning, and cloud computing technology is focused on spatiotemporal data analysis and modeling for sustainability in our environment.

Swapan Talukdar · Shahfahad ·
Swades Pal · Mohd Waseem Naikoo ·
Shakeel Ahmed · Atiqur Rahman
Editors

Water Resource Management in Climate Change Scenario

Innovations in Geospatial Techniques and Models

Springer

Editors
Swapan Talukdar
Department of Geography
Faculty of Sciences
Jamia Millia Islamia
New Delhi, Delhi, India

Department of Geography
Asutosh College
University of Calcutta
Kolkata, West Bengal, India

Swades Pal
Department of Geography
University of Gour Banga
Malda, West Bengal, India

Shakeel Ahmed
Centre for Disaster Risk Reduction
Islamic University of Science
and Technology
Awantipora, Jammu and Kashmir, India

Shahfahad
Department of Geography
Faculty of Sciences
Jamia Millia Islamia
New Delhi, Delhi, India

Mohd Waseem Naikoo
Department of Geography
Faculty of Sciences
Jamia Millia Islamia
New Delhi, Delhi, India

Department of Geography and Disaster
Management
University of Kashmir
Srinagar, Jammu and Kashmir, India

Atiqur Rahman
Department of Geography
Faculty of Sciences
Jamia Millia Islamia
New Delhi, Delhi, India

ISSN 2730-7506 ISSN 2730-7514 (electronic)
GIScience and Geo-environmental Modelling
ISBN 978-3-031-61120-9 ISBN 978-3-031-61121-6 (eBook)
https://doi.org/10.1007/978-3-031-61121-6

This Springer imprint is published by the registered company Springer Nature Switzerland AG
The registered company address is: Gewerbestrasse 11, 6330 Cham, Switzerland

Dedicated to Our Parents

Foreword

Amidst the evolving dynamics of global climate change, the imperative of water resource management intensifies, intersecting crucially with sustainability, development and technological innovation. The book volume entitled *Water Resource Management in Climate Change Scenario: Innovations in Geospatial Techniques and Models* offers a critical examination of innovative geospatial techniques and models, equipping researchers, policymakers, and practitioners with the knowledge and tools to address the pressing challenges of water resource management in a changing climate.

The editors of this volume, Swapan Talukdar, Shahfahad, Swades Pal, Mohd Waseem Naikoo, Shakeel Ahmed and Atiqur Rahman, have compiled a comprehensive collection of research that combines theoretical frameworks with practical applications, highlighting the multi-faceted approach required to tackle the complexities of managing water resources in the midst of climate change. Their collective expertise, deeply embedded in India's diverse and vibrant landscapes, weaves a rich tapestry of perspectives that enhances the depth and breadth of the discourse presented in this book. The clear focus on innovations in geospatial techniques and modelling makes this book an invaluable resource for anyone working on this important topic. I congratulate the editors of the book for their hard work in bringing out this book volume on such a important theme in the present days global water crisis.

The carefully compiled chapters navigate the latest advances in geospatial technology and artificial intelligence, offering ground-breaking methods for the effective monitoring and management of the water resources. From the intricate dynamics of groundwater discharge in mountain catchments to the innovative integration of machine learning and game theory in analysing soil erosion, the research summarised in this book reflects a profound understanding of the challenges at hand and the innovative solutions required to overcome them.

The commendable structure of the book allows for a deep dive into the topic. Part I explores advances in water resources modelling and mapping, and presents applications of stable isotopes, SWAT models and machine learning algorithms. Part II shifts the focus to water resources planning and management and addresses challenges and opportunities related to urban water bodies, groundwater resources and smart urban planning. Finally, Part III looks at surface water resource modelling and geospatial technologies and examines their role in understanding floodplains, wetland dynamics, glacial retreat and land use and land cover change.

The editors have assembled a team of highly qualified authors to ensure that the book reflects the expertise of leading researchers in the field of water resources. With chapters covering different geographical areas, the book offers insights that appeal to a global audience, enhancing its universal appeal and applicability.

As a reader deeply involved in water resource research and education, I consider that this book is an indispensable addition to the field of water resource management. It not only provides valuable knowledge and innovative methods, but also encourages proactive engagement with the challenges of climate change. By focusing on innovative approaches and comprehensively examining the contemporary challenges, *Water Resource Management in a Climate Change Scenario* is able to make a significant contribution to the field in the era of climate change.

Dr. S. P. Aggarwal
director@nesac.gov.in
Director, Department of Space
North Eastern Space Applications Centre
Government of India
Umiam, Meghalaya, India

Preface

The world is witnessing several problems related to various facets of natural resources, i.e. air, land, soil and water. Amongst all, water resource is one of the top most resources which is under the sever pressure mainly due to two factors, i.e. fast increase in population and faster economic growth globally. It has been observed that the withdrawal of groundwater has increased more than twice as fast as population growth. Therefore, there is urgent need to manage water resources with the increasing water demand. One of the shortcomings in the management of water is to focus on developing new sources rather than managing existing ones better and top-down sector approaches for better management of water, result in uncoordinated development and management of the resource. More and more development means a greater the demand of water and resultant impacts will be on the surrounding environment and climate change. Current concerns about climate variability and climate change demand improved management of water resources to cope with more intense floods and droughts which are taking place across the world more frequently. The process of planning, developing and managing water resources in terms of water quantity and quality, within and across water uses will be good for the benefit of humans, animals and ecosystems.

This book entitled *Water Resource Management in Climate Change Scenario: Innovations in Geospatial Techniques and Models* is an important resource that provides deep insights and innovative methods in the field of geospatial technology and its application in water resource management. In 16 comprehensive chapters, this book presents a complex web of research, case studies and theoretical frameworks. Each chapter addresses a unique aspect of water resource management, from the integration of water features into urban planning in a smart city to the pressing issues of wetland decline in West Bengal, vulnerability to flooding in Uttarakhand, and the impact of hydrological changes in different regions of South Asia.

A particular highlight is the study of very high-resolution optical satellite imagery for the extraction of urban water bodies, a study that emphasises the potential of GeoEye remote sensing data. Chapters also deal with the dynamics of glacier retreat and expansion in the Himalayas, groundwater analysis in the upper Yamuna basin, and the impact of flood inconsistencies and hydrological changes on wetland and floodplain morphology. The book also traverses the fields of machine learning and game theory and explores their integration for understanding soil erosion, wetland health and groundwater dynamics in mountain catchments. In addition, it addresses the impact of human activities on the carbon balance and the integration of urban water management with industrial corridor development.

This volume is a testament to the dedicated efforts of the authors, who have carefully crafted their contributions and thus made an important contribution to the field. It is a valuable resource for scientists, scholars and students across the disciplines like geographers, ecologists, geologists, climate scientists, water resource scientist and planners, researchers, environmentalists and policymakers concerned with water resource management. We extend our profound gratitude to the Springer publishing house, in particular Prof. Prabhat Kumar Shit, the editor of the Springer series, and Ms. Doris Bleier and her team for their professional collaboration in the rapid publication of this book. Their support has made a decisive contribution to the realization of this insightful and important work.

This book, with its diverse perspectives and innovative approaches, promises to be an indispensable guide for anyone wishing to address the complexities of managing water resources in the context of climate change. It is a beacon of knowledge that inspires innovative solutions and promotes a deeper understanding of the challenges and opportunities in this important area.

New Delhi, India — Swapan Talukdar
New Delhi, India — Shahfahad
Malda, India — Swades Pal
New Delhi, India — Mohd Waseem Naikoo
Awantipora, India — Shakeel Ahmed
New Delhi, India — Atiqur Rahman

Contents

Editors and Contributors

About the Editors

Dr. Swapan Talukdar is currently working as Assistant Professor of Geography in Department of Geography, Asutosh College, University of Calcutta, West Bengal. Dr. Talukdar has served as Dr. D. S. Kothari Post-doctoral fellow in the Department of Geography, Faculty of Sciences, Jamia Millia Islamia, New Delhi during 2021–2023. He completed his Ph.D. from the Department of Geography, University of Gour Banga, West Bengal, India, in 2020. He served as the junior research fellow (JRF) during 2015–2017 and senior research fellow (SRF) during 2017–2019 in the Department of Geography, University of Gour Banga. He has also completed several short- and long-term courses on remote sensing, machine learning and AI from different Scientific Institutes of India such as IITs and NITs. He has completed two international projects as Co-PI in collaboration with the King Khalid University, Saudi Arabia. He has published one edited book volume and more than 80 scientific research articles in the journals of international repute such as *Journal of Cleaner Production*, *Environmental Pollution*, *Scientific Reports*, *Ecological Indicators*, *Sustainable Cities and Society*, and *Remote Sensing*. His main areas of research are urban environment, urban wetlands, natural hazards, remote sensing and AI and their applications in environmental management. He has served as the reviewer for several reputed international journals. He was featured in the list of top 2% Global Elite Scientists list in a study by the University of Stanford, USA, published in the journal *PLOS Biology* in 2023.

Dr. Shahfahad is Assistant Professor of Geography at Department of Geography, Faculty of Sciences, Jamia Millia Islamia. He completed his B.A. (H) Geography in 2015, M.A. Geography in 2017 and Ph.D. in 2022 from the Department of Geography, Jamia Millia Islamia, New Delhi. He was junior research fellow (JRF) during 2017–2019 and senior research fellow (SRF) during 2019–2022 in the Department of Geography, Jamia Millia Islamia. His core areas of research are urban expansion, urban climate change, urban hydrology as well as applications of geospatial techniques and artificial intelligence in urban studies. He has extensively studied the climate pattern and change over the metropolitan cities of Indian subcontinent. He has published one edited book volume and more than 50 scientific research articles in the journals of international repute such as *Habitat International*, *Urban Climate*, *Journal of Environmental Management*, *Scientific Reports*, *Ecological Indicators*, *Remote Sensing*, *Geocarto International and Ecological Informatics*. He has served as the reviewer for several reputed international journals.

Prof. Swades Pal Professor of Geography, University of Gour Banga, has been working since last 13 years. His research interest is on river and fluvial geomorphology, groundwater and wetland hydrology as well as wetland ecology. Focusing on these issues, he has guided eleven Ph.D. research scholars and published more than 125 research articles from different peer-reviewed journals of international repute such as *Journal of Cleaner Production*, *Journal of Environmental Management*, *Environmental Pollution*, *Science of the Total Environment*, *Ecological Indicators* and *Sustainable Cities and Society*. He was featured in the top 2% scientist list in a study by the *University of Stanford, USA*, published in the journal *PLOS Biology* during 2020, 2021, 2022 and 2023.

Dr. Mohd Waseem Naikoo is Lecturer of Geography in the Department of Geography and Disaster Management, University of Kashmir. He completed his graduation in 2014, post-graduation in 2016 and doctorate in 2022 from the Department of Geography, Jamia Millia Islamia, New Delhi, India. He was junior research fellow (JRF) during 2017–2019 and senior research fellow (SRF) during 2019–2022 in the Department of Geography,

Jamia Millia Islamia. He has published more than 30 scientific research articles in the journals of international repute such as *Journal of Environmental Management*, *Applied Water Sciences*, *Journal of Cleaner Production*, *Urban Climate*, *Environmental Science and Pollution Research*, *Remote Sensing and Geocarto International*. His main areas of research are urban expansion and environment, water resource management, natural hazards, remote sensing and AI and their applications in environmental management. He has served as the reviewer for several reputed international journals.

Prof. Shakeel Ahmed (M.Sc. Tech. 1980 from BHU, India and Ph.D., 1987 from Mines Paris Tech, France) is Visiting Professor at Centre for Disaster Risk Reduction, Islamic University of Science and Technology, Awantipora, Jammu and Kashmir. Dr. Ahmed recently superannuated as professor from Maulana Azad National Urdu University, Hyderabad. An Earth Scientist worked extensively for Groundwater Hydrology in his 37 years' career at the CSIR's National Geophysical Research Institute, Hyderabad. Later he served as M. K. Gandhi Chair Professor in Jamia Millia Islamia, New Delhi. He led a large number of national and international projects on Earth Science including a mega programme of National Aquifer Mapping in India. With almost 200 SCI publications and three books, he supervised 41 doctoral theses. He is managing the Secretariat of the Asian GWADI, a UNESCO flagship programme as its Secretary. He has established an Indo-French Centre for Groundwater Research in Hyderabad and led it for about two decades as its founder head. His awards include National Geoscience Award, International Prize for Water Science, National Mineral Award, INC-IAH Excellence Award, NGRI-AHI National Lecture Award, IGU-Anni Talwani Memorial Prize, Young Scientist Award, etc. He is Elected Fellow of The World Academy of Sciences (TWAS), the National Academy of Sciences of India (NASI), National Environmental Science Academy (NESA), Indian Social Science Academy (ISSA), Geological Society of India (GSI), Telangana Academy of Sciences (TAS). He has been visiting Fellow at the Aligarh Muslim University and visiting professor to Kashmir University. He is the top publisher in the field of Groundwater Hydrology during the last three decades in Indian sub-continents as well as stands 25th in the National Ranking and 4673rd in the World Ranking in the World Best Earth Scientists (by Research.com).

Prof. Atiqur Rahman is a Professor of Geography at Department of Geography, Faculty of Sciences, Jamia Millia Islamia. He completed B.Sc. (Hons.), M.Sc., M.Phil. and Ph.D. from Aligarh Muslim University, Aligarh. His research interest is urban environmental management, water resources and use of remote sensing satellite data, GIS, and GPS. He was Post-Doctoral Fellow (PDF) at UFZ-Centre for Environmental Research, Leipzig, Germany (1999). He was Co-PI of Indo-Germany DST-DAAD major project. He is the recipient of prestigious Young Scientist Project Grant Award (2001–2004) from the Department of Science and Technology (DST), Government of India. He worked as one a member of Scientific Research Team of NASA-funded UEM project on *Urban Ecology and Sustainability* (2004–2007). He is the Co-PI Indo-Canadian major research projects on *Economic Transformation and Childhood Obesity* funded by ICMR (India) and CHI (Canada) during 2010–2015. He was Co-PI of a major research project funded by the Ministry of Environment and Forest, Government of India. He was the collaborating Scientist of another NASA-funded major project on *Impacts of Desert Urbanization on Climate using Remote Sensing and Numerical Modeling* (2012–2015). He has also done professional development initiative training on *Climate Change and Its Impacts: Resilience and Adaptation to Changes in Precipitation*, at Brown University USA (2010). He served as a Steering Committee (SC) Member (2009–2012) of Population and Environment Research Network (PERN), CISIN, Columbia University USA. He was the Advisor-Technical cum Consultant on Remote Sensing and GIS for the project (2013–2018) in the PHFI and AIIMS, and Emory University, USA project. He also served as a Member, Governing Body (GB), of some colleges of University of Delhi as a University Nominee. He was the Faculty Coordinator, Inspired Teachers In-Residence Programme organized by the Office of the President of India at Rashtrapati Bhavan, New Delhi, 6–12 June 2015. He was Honorary Deputy Director, (Academic) of Jamia Millia Islamia, (2015–2018). He is the Coordinator, UGC-DRS SAP-I, Special Assistance Programme (SAP-I), Department of Geography, JMI. Ten students have been awarded Ph.D. degrees under his supervision. He has published 15 books and 150 peer-reviewed research papers in the journals of international repute such as

Applied Geography, *Ecological Indicators*, *Urban Climate* and *Nature's Scientific Reports*. He has widely travelled across the globe. He is amongst the top 2% Global Elite Scientists in a study by the *University of Stanford, USA*, published in the journal *PLOS Biology* during 2020, 2021, 2022 and 2023.

Contributors

Syed Towseef Ahmad Department of Geography and Disaster Management, School of Earth and Environmental Sciences, University of Kashmir, Srinagar, India

Ishita Afreen Ahmed Department of Geography, Faculty of Sciences, Jamia Millia Islamia, New Delhi, India;
Department of Civil Engineering, Indian Institute of Technology Delhi, New Delhi, India

Pervez Ahmed Department of Geography and Disaster Management, School of Earth and Environmental Sciences, University of Kashmir, Srinagar, India

Rayees Ahmed Department of Geography and Disaster Management, School of Earth and Environmental Sciences, University of Kashmir, Srinagar, India

Shakeel Ahmed CSIR-National Geophysical Research Institute, Hyderabad, India; Centre for Disaster Risk Reduction, Islamic University of Science and Technology, Awantipora, Jammu and Kashmir, India

Mohammad Itahdur Ali Department of Geography, Faculty of Sciences, Jamia Millia Islamia, New Delhi, India

Md. Sarfaraz Asgher Department of Geography, University of Jammu, Jammu Tawi, Jammu & Kashmir, India

Asif Department of Geography, Faculty of Sciences, Jamia Millia Islamia, New Delhi, India

Mirza Razi Imam Baig Department of Geography, Faculty of Sciences, Jamia Millia Islamia, New Delhi, India

Yogesh Bhardwaj Department of Architecture, Jamia Millia Islamia, New Delhi, India

Somen Das Department of Geography, Kazi Nazrul University, Asansol, West Bengal, India

Tanmoy Das Department of Geography, Faculty of Sciences, Jamia Millia Islamia, New Delhi, India

Ratnadeep Dutta Centre for the Study of Regional Development, Jawaharlal Nehru University, New Delhi, India

Farhana Ferdos Department of Geography and Environmental Studies, University of Rajshahi, Rajshahi, Bangladesh

Asma Hamid Department of Architecture, Jamia Millia Islamia, New Delhi, India

Hoang Thi Hang Department of Geography, Faculty of Sciences, Jamia Millia Islamia, New Delhi, India

Zeeshan Ibrar Department of Architecture, Jamia Millia Islamia, New Delhi, India

Qamar Irshad Department of Architecture, Jamia Millia Islamia, New Delhi, India

Md. Rejaul Islam Department of Geography, Faculty of Sciences, Jamia Millia Islamia, New Delhi, India

Basir Ali Karikar Department of Geography, Faculty of Sciences, Jamia Millia Islamia, New Delhi, India

Nisar Khan Department of Architecture, Jamia Millia Islamia, New Delhi, India

Rumki Khatun Department of Geography, Kazi Nazrul University, Asansol, West Bengal, India

Manabendra Let Department of Geography, University of Gour Banga, Malda, West Bengal, India

Susanta Mahato Special Centre for Disaster Research (SCDR), Jawaharlal Nehru University, New Delhi, India

Manirul Mia Department of Geography, Faculty of Sciences, Jamia Millia Islamia, New Delhi, India

Riyaz Ahmad Mir Geological Survey of India, UT: Jammu and Kashmir, Srinagar, India

Sk Mohibul Department of Geography, Faculty of Sciences, Jamia Millia Islamia, New Delhi, India

Monidip Mondal Department of Geography, Faculty of Sciences, Jamia Millia Islamia, New Delhi, India

Mohd Waseem Naikoo Department of Geography, Faculty of Sciences, Jamia Millia Islamia, New Delhi, India; Department of Geography and Disaster Management, University of Kashmir, Hazratbal, Srinagar, Jammu and Kashmir, India

Swades Pal Department of Geography, University of Gour Banga, Malda, West Bengal, India

Neha Parveen Department of Geography, Faculty of Sciences, Jamia Millia Islamia, New Delhi, India

Suvamoy Pramanik Centre for the Study of Regional Development, Jawaharlal Nehru University, New Delhi, India; Assistant Professor, Department of Geography, Shaheed Bhagat Singh Evening College, University of Delhi, New Delhi, India

Milap Punia Centre for the Study of Regional Development, Jawaharlal Nehru University, New Delhi, India

Atiqur Rahman Department of Geography, Faculty of Sciences, Jamia Millia Islamia, New Delhi, India

Md. Rejaur Rahman Department of Geography and Environmental Studies, University of Rajshahi, Rajshahi, Bangladesh

Abid Farooq Rather Department of Geography and Disaster Management, School of Earth and Environmental Sciences, University of Kashmir, Srinagar, India

Nitin Rathi Centre for the Study of Regional Development (CSRD), Jawaharlal Nehru University, New Delhi, India

Mohd Rihan Department of Geography, Faculty of Sciences, Jamia Millia Islamia, New Delhi, India

Akthab Roshan Department of Geography, Faculty of Sciences, Jamia Millia Islamia, New Delhi, India

Tamal Kanti Saha Department of Geography, Faculty of Sciences, Jamia Millia Islamia, New Delhi, India

Sarah Sarah Deparment of Earth Sciences, University of Kashmir, Srinagar, Jammu and Kashmir, India

Rajesh Sarda Department of Geography, Delhi School of Economics, New Delhi, India

Md Nawaj Sarif Department of Geography, Faculty of Sciences, Jamia Millia Islamia, New Delhi, India

Shahfahad Department of Geography, Faculty of Sciences, Jamia Millia Islamia, New Delhi, India

Taha Shamim Department of Geography and Disaster Management, School of Earth and Environmental Sciences, University of Kashmir, Srinagar, India

Sanjeev Sharma Centre for the Study of Regional Development (CSRD), Jawaharlal Nehru University, New Delhi, India

Ahmad Mujtaba Siddiqui Department of Geography, Faculty of Sciences, Aligarh Muslim University, Aligarh, Uttar Pradesh, India

Lubna Siddiqui Department of Geography, Faculty of Sciences, Jamia Millia Islamia, New Delhi, India

Jasmine Sultana Amity Institute of Geoinformatics and Remote Sensing, Amity University, Noida, Uttar Pradesh, India

Swapan Talukdar Department of Geography, Faculty of Sciences, Jamia Millia Islamia, New Delhi, India; Department of Geography, Asutosh College, University of Calcutta, Kolkata, West Bengal, India

Gowhar Farooq Wani Department of Geography and Disaster Management, School of Earth and Environmental Sciences, University of Kashmir, Srinagar, India

Shah Waseem Deparment of Earth Sciences, University of Kashmir, Srinagar, Jammu and Kashmir, India

Recent Trends in Application of Geospatial Technologies and AI for Monitoring and Management of Water Resources

1

Swapan Talukdar, Shahfahad, Swades Pal, Mohd Waseem Naikoo, Shakeel Ahmed, and Atiqur Rahman

Abstract

In response to escalating global water crises, there has been a notable advancement in the methods and approaches used for monitoring and management of water resources. From traditional data-driven models to sophisticated and advanced geospatial techniques and artificial intelligence (AI) models, the studies on water resources have witnessed significant advancement in the past few decades. This article outlines the progress in water resource studies and recent trends that signify a paradigm shift in the field, emphasizing advancements in methodologies. Geospatial technologies, encompassing satellite-based remote sensing and Geographic Information Systems (GIS), are increasingly employed for precise and large-scale data collection. The AI algorithms such as machine and deep learning models are being deployed to analyze this data, offering predictive insights, and enhancing decision-making processes. The geospatial techniques and artificial intelligence (AI) have provided several advantages in water resource studies over traditional models by offering high-resolution spatial data, facilitating real-time monitoring, predictive modelling, complicated pattern detection, and climate change adaptation for more sustainable and informed management. The integration of geospatial techniques and AI has prompted innovation in hydrological modelling, enabling more accurate assessments of water contamination, groundwater recharge, and flow dynamics. These advancements have enabled hydrologists and policymakers to make effective policies and planning for water resource management which is important to meet the changing needs brought on by climate change.

Keywords

Water resources · Data-driven techniques · Geostatistical techniques · Geospatial techniques · Artificial intelligence

S. Talukdar · Shahfahad · M. W. Naikoo · A. Rahman (✉)
Department of Geography, Faculty of Sciences, Jamia Millia Islamia, New Delhi 110025, India
e-mail: arahman2@jmi.ac.in

S. Pal
Department of Geography, University of Gour Banga, NH12, Malda 732101, India

M. W. Naikoo
Department of Geography and Disaster Management, University of Kashmir, Hazratbal, Srinagar, Jammu and Kashmir 190006, India

S. Ahmed
CSIR-National Geophysical Research Institute, Hyderabad 500007, India

S. Talukdar
Department of Geography, Asutosh College, University of Calcutta, West Bengal, Kolkata 700026, India

S. Talukdar et al. (eds.), *Water Resource Management in Climate Change Scenario*, GIScience and Geo-environmental Modelling, https://doi.org/10.1007/978-3-031-61121-6_1

1.1 Introduction

Water is unequivocally the most vital natural resource, serving as a cornerstone for life on Earth. Water is indispensable for agriculture, industrial processes, and ecological balance, playing a crucial role in global food security, economic development, biodiversity support, and climate regulation (Pastor et al. 2019; Shu et al. 2021; Iyiola et al. 2022). However, the escalating global population, combined with the impact of climate change, poses significant threats to both the quantity and quality of water resources (Yoon et al. 2021; Shivanna et al. 2022). The current trend of depletion of water resources coupled with the changing climatic conditions is expected to result in the imminent occurrence of "Day Zero" in numerous regions worldwide, when water taps will run dry, and usage will be constrained solely to essential services (Ahmadi et al. 2020; Savelli et al. 2023). This multifaceted crisis may arise due to population growth, climate change, and unsustainable water uses and management practices (Dolan et al. 2021). In such a scenario, there is an urgent need to assess the stress of population pressure and climate change in water resources as well as identify the best practices to deal with water scarcity and contamination.

Studies are being done to analyze the water resources availability and contamination as well as to propose suitable measures to deal with the problems related to water scarcity and pollution in different parts of the world (Cosgrove and Loucks 2015). Among various issues related to water resources in the world, the availability of the fresh water and water pollution are the most common (Salehi 2022). Further, the depletion of surface and groundwater resources is another important issue which is faced by several parts of the world (Wada et al. 2010; Dangar et al. 2021). For assessing the availability and scarcity of the fresh water resources, researchers have studied the groundwater potentiality in different parts of the world (Thanh et al. 2022). Further, for assessing and predicting the quality of water, researchers have used water quality index (WQI) which describes overall water quality on the basis of several parameters (Uddin et al. 2021). This involves comprehensive analysis of physical, chemical, and biological characteristics to assess its suitability for specific purposes (Alshehri and Rahman 2023). Moreover, for assessing the depletion of surface water, researchers quantify the changes in area under wetlands and lakes (Huu Nguyen et al. 2016), while for examining the groundwater depletion, the analysis of changes in water table is most commonly used approach (Joshi et al. 2021; Liu et al. 2022).

Traditionally, the statistical models such as frequency ratio, fuzzy logic, weight of evidence, and logistic regression have been widely used for examining the surface and groundwater availability and quality (Kenda et al. 2020; Uddin et al. 2021). During first two decades of twenty-first century, remote sensing data has played pivotal role in advancing surface and groundwater studies, offering a comprehensive and efficient means to gather valuable information for hydrological research (Jiang and Wang 2019; Adams et al. 2022). Remote sensing provides the possibility of uncovering new insights into hydrological dynamics as it evolves, thereby contributing in sustainable water resource management and decision-making processes for the benefit of ecosystems and human societies (Sheffield et al. 2018). Geospatial techniques, which include geographical information system (GIS), remote sensing satellite images, and aerial photographs, provide a comprehensive overview of the broader areas, allowing researchers to track and analyze changes in land use land cover (LULC), surface characteristics, and water bodies in unprecedented detail (Zhu et al. 2022a; Bill et al. 2022). The capacity to collect data over remote or difficult terrain expands study in terms of geographical coverage, contributing to a more comprehensive knowledge of hydrological processes (Cui et al. 2018). Furthermore, remote sensing allows for the monitoring of temporal fluctuations, which assists in the assessment of long-term trends and the effects of climate change on surface water bodies and groundwater resources (Adjovu et al. 2023). The benefits extend to the cost-effectiveness and speed of data capture, allowing for frequent and consistent monitoring.

Recently, the use of artificial intelligence (AI)-based machine and deep learning algorithms has transformed water resource research, ushering in a new era of efficiency and precision (Ghobadi and Kang 2023). Machine learning algorithms and data analytics, for example, have been essential in improving our understanding of water systems by analyzing large datasets with remarkable speed and precision (Talukdar et al. 2020). Furthermore, AI has proven beneficial in optimizing water management practices, allowing for more informed decision-making (Doorn 2021). These improvements have made it possible to create complex models for predicting water supply, quality, and usage trends (Shams et al. 2023). Furthermore, the combination of AI with geospatial approaches has emerged as a potent synergy in the field of water resource studies (Dogo et al. 2019). Researchers have achieved a complete understanding of spatial pattern and dynamics within water ecosystems by integrating geospatial technologies with AI algorithms (Bill et al. 2022; Gonzales-Inca et al. 2022). The combination of AI with GIS approaches not only improves predictive model accuracy, but also allows for real-time monitoring and control of water resources (Zhu et al. 2022b; Fang et al. 2023). This chapter aims to investigate the evolutionary trajectory of water resource studies, transitioning from conventional data-driven models to the integration of geospatial techniques and AI, and to explore the prospective contributions of geospatial techniques and AI in forthcoming research endeavours in the realm of water resources.

1.2 Progress in Water Resource Modelling

Traditionally, the researches on water resources were based on conventional hydrological models, characterized by their reliance on empirical relationships and foundational adherence to simplistic physical principles (Singh 2018). These conventional methodologies invariably necessitated manual data collection, wherein hydrologists undertook field measurements of crucial parameters such as groundwater, rainfall, and stream flow (Jiang and Wang 2019). But, the efficacy of these traditional methods was circumscribed by inherent constraints in spatial and temporal resolutions, thereby obstructing a comprehensive elucidation of the intricate dynamics inherent to complex water systems. Simultaneously, data-driven techniques, notably statistical analyses, were routinely employed, although, the inadequacy in capturing the nuanced intricacies of hydrological processes remained evident in these methods (Hauswirth et al. 2021; Dong et al. 2023). Despite these limitations, traditional approaches stand as the indisputable progenitors of our initial understanding of water resources. They provided invaluable insights into local hydrological cycles, facilitating the formulation of rudimentary frameworks for water management (Loucks and van Beek 2017).

With the advancement in geospatial techniques, which include satellite-based remote sensing and aerial photography, a new era of enhanced observational precision begun (Sheffield et al. 2018). This advancement in technology makes it possible to conduct thorough assessments of changes in earth surface features, track LULC changes, and keep a close eye on surface water bodies (Talukdar et al. 2021; Shahfahad et al. 2022; Zhu et al. 2022a). Furthermore, the introduction of AI methods, specifically machine learning algorithms like support vector machines (SVM), random forests (RF), and artificial neural networks (ANN), has advanced the field by enabling models to analyze large datasets, identify complex patterns in hydrological processes, and generate predictions with increased precision (Zhu et al. 2022b; Mia et al. 2023). Recently, advanced AI techniques such as ensemble machine learning and deep learning techniques have been developed for studying water resources (Ewuzie et al.

2022; Talukdar et al. 2023). These AI models have a significant impact on the development of optimal decision support systems, early warning systems, and real-time monitoring for the management of water resources (Kamyab et al. 2023). AI models provide a detailed understanding of the interplay between natural and anthropogenic variables affecting water supplies by dynamically adjusting to changing conditions and iteratively learning from data (Tung and Yaseen 2020). Despite all these advancements, the problems like maintaining the integrity and coherence of data from multiple sources still exist.

The working mechanisms of AI models are complex to understand that generate significant obstacles in achieving interpretability of their decision-making processes (Ali et al. 2023; Rahman et al. 2023). To tackle these obstacles, strong validation techniques, constant research, and an open line of communication are required (Xu et al. 2021). However, it is crucial that capacity development efforts can be ongoing in order to give water resource experts the knowledge and abilities they need to successfully use and manage these cutting-edge technologies. Moreover, a more comprehensive, flexible, and successful strategy for managing sustainable water resources is promised by the complex interactions between conventional knowledge and cutting-edge technologies. Combining historical knowledge with modern inventions puts us in a position to manage the intricate water systems and build resilience against changing social and environmental issues (Mishra et al. 2021). The groundwater flow and transport models too have been improved with the use of GIS platform and application of remote sensing in supplying the numerous inputs hitherto required to fabricate, calibrate, and validate the models. With the availability of huge dataset, it is feasible to work in stochastic framework and predict the probabilistic output (Ahmed et al 2008, 2010; Sarah et al. 2014). The stochastic prediction provides protection in decision-making.

1.3 Integration of Geostatistical and Geospatial Techniques for Water Resource Monitoring

In beginning, the geostatistical techniques were utilized in water resource monitoring for spatial analysis and mapping (Chakma et al. 2022). Geostatistical methods, such as kriging and variography, enable the interpolation of measured data points to create continuous spatial maps (Atkinson and Lloyd 2021). Geostatistical techniques are widely employed in groundwater monitoring, providing insights into the spatial variability of groundwater quality and quantity (Ahmed and de Marsily 1993; Ahmed 2001, 2004; Ahmed et al 2007a, b; Ahmed and Devi 2007 etc.). Kriging, for example, can be used to estimate groundwater levels at unmonitored locations based on the observed values from neighbouring wells (Haaf et al. 2023). By incorporating spatial correlation into water quality data analysis, geostatistical models were utilized for estimating the pollutant concentrations in different parts of the world (Yang and Jin 2010). This aids in identifying areas susceptible to pollution, understanding the dynamics of pollutant dispersion, and developing effective strategies for water quality improvement.

The transition from geostatistical to geospatial techniques represents a paradigm shift in water resource monitoring, offering enhanced capabilities for understanding, managing, and conserving this vital resource (Machiwal et al. 2018). Geostatistical techniques, such as kriging and variography, have long been instrumental in assessing the spatial variability of water quality parameters, groundwater levels, and surface water flow including monitoring network optimization (Ahmed 2002). However, the effectiveness is often constrained by the availability and density of monitoring data (Dabas et al. 2022). Geospatial techniques address this limitation by incorporating a wealth of spatial information derived from various sources (Sheffield et al. 2018). Satellite imagery, for example, allows for a synoptic view of larger areas, enabling the identification of changes in LULC pattern, vegetation cover, and water bodies over the period of time. This information enriches the spatial context for water resource monitoring, providing valuable insights into the broader environmental factors influencing water dynamics (Zhu et al. 2022a, b).

The integration of geostatistical and geospatial techniques marks a synergistic approach that significantly enhances the precision and breadth of water resource monitoring. Geostatistical techniques, with their focus on statistical analysis of spatially correlated data, provide a robust foundation for understanding the variability of water quality parameters and groundwater levels (Machiwal et al. 2018; Pavlides et al. 2023). On the other hand, geospatial techniques bring a broader spatial context by incorporating satellite imagery, GIS data, and remote sensing technologies (Bill et al. 2022). By combining these approaches, researchers and water resource managers can achieve a comprehensive understanding of water dynamics, capturing both the fine-scale spatial variations and broader environmental influences (Mallick et al. 2015). This integration facilitates the creation of accurate and dynamic spatial models, allowing for real-time monitoring, early anomaly detection, and informed decision-making (Buenemann et al. 2011; Mallick et al. 2015). The synergy between geostatistical and geospatial techniques is particularly valuable for addressing complex water management challenges, such as assessing the impact of LULC changes, tracking temporal variations, and implementing effective conservation strategies in the face of climate change and growing water demand.

1.4 Application of Geospatial Techniques and AI for Surface and Groundwater Modelling

The integration of geospatial techniques and AI has revolutionized water resource studies, particularly in the assessment of surface and groundwater quality, surface water monitoring, and groundwater potential mapping (Bhunia et al. 2021). Geospatial techniques, encompassing GIS, remote sensing, and global positioning system (GPS), provide invaluable tools for spatial analysis and visualization of water-related data (Mallick et al. 2021; Pandey et al. 2022). AI, on the other hand, introduces advanced algorithms that enhance the processing and interpretation of complex datasets, enabling more accurate predictions and decision-making (Xu et al. 2021; Rahman et al. 2023). The combined use of geospatial techniques and AI in water resource studies allows for a comprehensive understanding of the dynamic interactions within hydrological systems (Gonzales-Inca et al. 2022).

In the surface water quality assessment, geospatial techniques facilitate the mapping of pollutant concentrations and their spatial distribution, enabling the identification of pollution sources and the monitoring of water quality over time (Alshehri and Rahman 2023; Talukdar et al. 2023). AI algorithms can analyze large datasets to identify pattern and trend, contribute to early detection of potential water quality issues (Kamyab et al. 2023). This integrated approach enhances the efficiency and precision of water quality management strategies. Surface water monitoring benefits significantly from the synergy of geospatial techniques and AI (Rana et al. 2023). Real-time data from remote sensing platforms can be processed using AI algorithms to detect changes in surface water bodies, monitor water levels, and identify potential flood or drought events (Sun and Scanlon 2019). The integration of these technologies allows for timely responses to water-related emergencies and the development of effective mitigation measures.

In the realm of water resource studies, the exploration of groundwater potential is a vital facet, and it is significantly advanced through the integration of geospatial techniques and AI. GIS provides a robust platform for the creation of groundwater potential maps, facilitating the demarcation of recharge zones, detection of potential contamination threats, and evaluation of the sustainable yield of aquifers (Rao et al.

2022). By employing GIS-based mapping, researchers can not only visualize but also analyze the spatial distribution of groundwater resources, enabling informed decision-making for sustainable water management practices (Jha et al. 2007; Roy et al. 2022). Moreover, the incorporation of AI into this process adds a dynamic dimension. AI models, when trained on historical groundwater data, demonstrate the capacity to forecast future trends in groundwater levels (Tao et al. 2022). This predictive capability proves invaluable in anticipating potential challenges related to water scarcity or contamination, thereby contributing to the formulation of proactive and effective strategies for the conservation and sustainable utilization of this indispensable natural resource. In essence, the convergence of geospatial techniques and AI in groundwater potential mapping stands as a powerful synergy, empowering researchers, and resource managers with the tools necessary for informed decision-making and long-term water resource sustainability (Pal et al. 2020).

1.5 Current Status of Water Resource Studies in Developed and Developing Countries

There exist notable differences in the approaches and applications in water resource studies in the developed and developing countries. In developed countries, these studies are often characterized by sophisticated technologies, extensive data collection networks, and well-established institutional frameworks (Cosgrove and Loucks 2015; Loucks and van Beek 2017). These nations possess the financial and technological capabilities to invest in advanced monitoring systems, remote sensing technologies, and modelling tools that provide accurate and real-time information about water availability, quality, and distribution (O'Grady et al. 2021). The emphasis is on efficient water management and sustainability. Advanced hydrological models help in predicting water availability, optimizing reservoir operations, and managing water distribution systems (Hughes et al. 2021).

In order to provide a dependable and secure supply of water for residential, commercial, and agricultural uses, the developed countries make huge investments in water infrastructure, such as dams, pipelines, and wastewater treatment facilities (Loucks and van Beek 2017). Utilizing state-of-the-art technologies, such as satellite images and GIS mapping, enables thorough geographical analysis, which helps identify possible water sources and evaluate environmental implications (Chabuk et al. 2020). Furthermore, environmental preservation and water conservation are top priorities in developed world (Yang et al. 2020). Water resources and ecosystems are protected through the implementation of policies and laws. A culture of conscientious water usage is further fostered by educational initiatives and public awareness campaigns (Hunt and Shahab 2021). Furthermore, research in these countries frequently concentrates on new problems like the effects of climate change on water resources, guaranteeing a proactive approach to upcoming difficulties (Dolan et al. 2021; van Vliet et al. 2021).

On the other hand, water resource studies in developing countries face unique challenges. Limited financial resources, inadequate infrastructure, and data gaps pose significant hurdles (Abbass et al. 2022). Many developing nations struggle with outdated or insufficient water monitoring systems, hindering the collection of accurate and timely data. As a result, decision-making processes related to water management often lack the necessary information, leading to inefficiencies and potential crises (Jury and Vaux Jr. 2005; Cosgrove and Loucks 2015). In developing countries, the primary focus of water resource studies is often on addressing immediate water-related challenges, such as ensuring access to clean drinking water and sanitation facilities (Mishra et al. 2021; Aneesh 2021). Basic water infrastructure development takes precedence over sophisticated modelling and monitoring systems (Cosgrove and Loucks 2015). Community-based approaches and local knowledge play a crucial role in understanding water dynamics, and efforts are directed towards establishing sustainable water supply systems that cater to the needs of both urban and rural populations (Whaley and Cleaver 2017). Furthermore, developing countries are particularly vulnerable to the impacts of climate change, including erratic rainfall patterns, prolonged droughts, and increased frequency of extreme weather events (Solecki et al. 2019; Abbass et al. 2022; Shahfahad et al. 2023). Water resource studies in these regions increasingly involve climate resilience assessments and the development of adaptive strategies to mitigate the adverse effects on water availability and quality.

1.6 Way Forward

Scarcity of water resources is a serious issue worldwide that endangers human health, food security, and ecosystems. The demand for water rises with the population growth, industrialization, faster economic growth, and climate change acceleration, aggravating the scarcity situation. A thorough understanding of water availability, usage trends, and sustainable management techniques is necessary to address these issues. Innovative approaches have been explored in recent research, encompassing everything from precision agriculture and water-use efficiency methods to cutting-edge water purification technologies. These studies highlight how crucial it is to manage water resources holistically, considering both quantity and quality factors. To effectively implement initiatives to reduce water scarcity, communities, governments, and scientists must work together. Even if there has been progress, more research is necessary to create robust and adaptable solutions that will lessen the effects of water scarcity and guarantee a sustainable future for everybody. The development of cutting-edge technology for effective water management, such as sophisticated sensors, machine learning algorithms, and remote sensing technologies to monitor and evaluate water availability in real-time, should be given top priority in the future researches on water resources. Understanding the intricate relationships between human activity, climate change, and water resources requires interdisciplinary collaborations involving hydrologists, engineers, ecologists, and social scientists. The climate change has modified the dynamics of water impacting all the components of the water cycle. Thus, most of the techniques need to be revisited to incorporate the impact of climate change and the extreme events. Although the applications of satellite data and remote sensing generative numerous data and AI and ML are capable of using them adequately, groundwater has become crucial for ensuring sustainability and information on groundwater needs geoscientific knowledge (Chandra et al. 2016). Furthermore, studies should concentrate on investigating natural alternatives, including restoring watersheds, rain water harvesting, development of ground water recharge sites, building green infrastructure etc., to improve the availability and quality of water while building resilience to shifting environmental circumstances. Future research projects aimed at resolving the global water scarcity issue will depend critically on the persistence of emphasis on community involvement and the fusion of traditional knowledge with modern breakthroughs.

References

Abbass K, Qasim MZ, Song H, Murshed M, Mahmood H, Younis I (2022) A review of the global climate change impacts, adaptation, and sustainable mitigation measures. Environ Sci Pollut Res 29(28):42539–42559

Adams KH, Reager JT, Rosen P, Wiese DN, Farr TG, Rao S et al (2022) Remote sensing of groundwater: current capabilities and future directions. Water Resour Res 58(10):e2022WR032219

Adjovu GE, Stephen H, James D, Ahmad S (2023) Overview of the application of remote sensing in effective monitoring of water quality parameters. Remote Sens 15(7):1938

Ahmadi MS, Sušnik J, Veerbeek W, Zevenbergen C (2020) Towards a global day zero? Assessment of current and future water supply and demand in 12 rapidly developing megacities. Sustain Cities Soc 61:102295

Ahmed S (2001) Regionalization of aquifer parameters for groundwater modeling including monitoring network design. In: Elango L, Jayakumar R (eds) Modeling in hydrogeology. Allied Publishers Limited, India, pp 39–57

Ahmed S (2002) Groundwater monitoring network design: applications of geostatistics with a few case studies from a granitic aquifer from semi-arid region, in a semi-arid region. In: Sherif MM et al (eds) Groundwater hydrology, vol 2. A.A. Balkema Publishers, pp 37–57

Ahmed S (2004) Application of geostatistics: parameter estimation to predictive aquifer modelling. In: Rai SN (ed) Role of mathematical modeling in groundwater resource management. NGRI Publication, pp 357–381

Ahmed S, de Marsily G (1993) Cokriged estimation of aquifer transmissivity as an indirect solution of inverse problem: a practical approach. Water Resour Res 29(2):521–530

Ahmed S, Devi K (2007) Kriging for estimating hydrogeological parameters. In: Ahmed S, Jayakumar R, Salih A (eds) Groundwater dynamics in hard rock aquifers. Capital Pub. Co., New Delhi, pp 172–178

Ahmed S, Nabi A, Owais S, Kumar D (2007a) Optimization of groundwater monitoring networks: application of geostatistics with case studies from a granitic aquifer in a semi-arid region, chap XV. In: Chery L, de Marsily G (eds) Aquifer systems management: Darcy's legacy in a world of impending water shortage. Taylor and Francis, London, pp 527–540

Ahmed S, Kumar D, Bhat AN (2007b) Application of geostatistics in optimal groundwater monitoring network design. In: Ahmed S, Jayakumar R, Salih A (eds) Groundwater dynamics in hard rock aquifers. Capital Pub. Co., Springer, New Delhi, pp 179–190

Ahmed S, Maréchal JC, Ledoux E, de Marsily G (2008) Groundwater flow modelling in hard-rock terrain in semi-arid areas: experience from India, chap XI. In: Wheater H, Sooroshian S, Sharma KD (eds) Hydrological modelling in arid and semi-arid areas. Cambridge University Press, pp 157–190

Ahmed S, Sarah S, Nabi A, Owais S (2010) Performing unbiased groundwater modelling: application of the theory of regionalized variables, chap 5. In: Wheater H, Mathias S, Li X (eds) Groundwater modeling for arid and semi-arid areas. Cambridge University Press, pp 63–74

Ali S, Abuhmed T, El-Sappagh S, Muhammad K, Alonso-Moral JM, Confalonieri R et al (2023) Explainable artificial intelligence (XAI): what we know and what is left to attain trustworthy artificial intelligence. Inf Fusion 99:101805

Alshehri F, Rahman A (2023) Coupling machine and deep learning with explainable artificial intelligence for improving prediction of groundwater quality and decision-making in arid region, Saudi Arabia. Water 15(12):2298

Aneesh MR (2021) Quality of drinking water and sanitation in India. Indian J Hum Dev 15(1):138–152

Atkinson PM, Lloyd CD (2021) Geostatistical models and spatial interpolation. In: Handbook of regional science. Springer Berlin Heidelberg, Berlin, Heidelberg, pp 1813–1827

Bhunia GS, Shit PK, Adhikary PP (2021) Concept of artificial intelligence and its applications in groundwater spatial studies. In: Geostatistics and geospatial technologies for groundwater resources in India, pp 41–57

Bill R, Blankenbach J, Breunig M, Haunert JH, Heipke C, Herle S et al (2022) Geospatial information research: state of the art, case studies and future perspectives. PFG J Photogramm Remote Sens Geoinform Sci 90(4):349–389

Buenemann M, Martius C, Jones JW, Herrmann SM, Klein D, Mulligan M et al (2011) Integrative geospatial approaches for the comprehensive monitoring and assessment of land management sustainability: rationale, potentials, and characteristics. Land Degrad Dev 22(2):226–239

Chabuk A, Al-Madhlom Q, Al-Maliki A, Al-Ansari N, Hussain HM, Laue J (2020) Water quality assessment along Tigris River (Iraq) using water quality index (WQI) and GIS software. Arab J Geosci 13:1–23

Chakma A, Bhowmik T, Mallik S, Mishra U (2022) Application of GIS and geostatistical interpolation method for groundwater mapping. In: Advanced modelling and innovations in water resources engineering: select proceedings of AMIWRE 2021. Springer Singapore, pp 419–428

Chandra S, Ahmed S, Auken E, Pedersen JB, Singh A, Verma SK (2016) 3D aquifer mapping employing airborne geophysics to meet India's water future. Lead Edge 35(9):770–774. https://doi.org/10.1190/tle35090770.1

Cosgrove WJ, Loucks DP (2015) Water management: current and future challenges and research directions. Water Resour Res 51(6):4823–4839

Cui Y, Chen X, Gao J, Yan B, Tang G, Hong Y (2018) Global water cycle and remote sensing big data: overview, challenge, and opportunities. Big Earth Data 2(3):282–297

Dabas J, Sarah S, Mondal NC, Ahmed S (2022) Geostatistical spatial projection of geophysical parameters for practical aquifer mapping. Sci Rep 12(1):4641

Dangar S, Asoka A, Mishra V (2021) Causes and implications of groundwater depletion in India: a review. J Hydrol 596:126103

Dogo EM, Salami AF, Nwulu NI, Aigbavboa CO (2019) Blockchain and internet of things-based technologies for intelligent water management system. In: Artificial intelligence in IoT, pp 129–150

Dolan F, Lamontagne J, Link R, Hejazi M, Reed P, Edmonds J (2021) Evaluating the economic impact of water scarcity in a changing world. Nat Commun 12(1):1915

Dong N, Guan W, Cao J, Zou Y, Yang M, Wei J et al (2023) A hybrid hydrologic modelling framework with data-driven and conceptual reservoir operation schemes for reservoir impact assessment and predictions. J Hydrol 619:129246

Doorn N (2021) Artificial intelligence in the water domain: opportunities for responsible use. Sci Total Environ 755:142561

Ewuzie U, Bolade OP, Egbedina AO (2022) Application of deep learning and machine learning methods in water quality modeling and prediction: a review. In: Current trends and advances in computer-aided intelligent environmental data engineering, pp 185–218

Fang B, Yu J, Chen Z, Osman AI, Farghali M, Ihara I et al (2023) Artificial intelligence for waste management in smart cities: a review. Environ Chem Lett 1–31

Ghobadi F, Kang D (2023) Application of machine learning in water resources management: a systematic literature review. Water 15(4):620

Gonzales-Inca C, Calle M, Croghan D, Torabi Haghighi A, Marttila H, Silander J, Alho P (2022) Geospatial artificial intelligence (GeoAI) in the integrated hydrological and fluvial systems modeling: review of current applications and trends. Water 14(14):2211

Haaf E, Giese M, Reimann T, Barthel R (2023) Data-driven estimation of groundwater level time-series at unmonitored sites using comparative regional analysis. Water Resour Res 59(7):e2022WR033470

Hauswirth SM, Bierkens MF, Beijk V, Wanders N (2021) The potential of data driven approaches for quantifying hydrological extremes. Adv Water Resour 155:104017

Hughes D, Birkinshaw S, Parkin G (2021) A method to include reservoir operations in catchment hydrological models using SHETRAN. Environ Model Softw 138:104980

Hunt DV, Shahab Z (2021) Sustainable water use practices: understanding and awareness of masters level students. Sustainability 13(19):10499

Huu Nguyen H, Dargusch P, Moss P, Tran DB (2016) A review of the drivers of 200 years of wetland degradation in the Mekong Delta of Vietnam. Reg Environ Change 16(8):2303–2315

Iyiola AO, Akinrinade AJ, Ajayi FO (2022) Effects of water pollution on biodiversity along the coastal regions. In: Biodiversity in Africa: potentials, threats and conservation. Springer Nature Singapore, Singapore, pp 345–367

Jha MK, Chowdhury A, Chowdary VM, Peiffer S (2007) Groundwater management and development by integrated remote sensing and geographic information systems: prospects and constraints. Water Resour Manage 21:427–467

Jiang D, Wang K (2019) The role of satellite-based remote sensing in improving simulated streamflow: a review. Water 11(8):1615

Joshi SK, Gupta S, Sinha R, Densmore AL, Rai SP, Shekhar S et al (2021) Strongly heterogeneous patterns of groundwater depletion in Northwestern India. J Hydrol 598:126492

Jury WA, Vaux H Jr (2005) The role of science in solving the world's emerging water problems. Proc Natl Acad Sci 102(44):15715–15720

Kamyab H, Khademi T, Chelliapan S, SaberiKamarposhti M, Rezania S, Yusuf M et al (2023) The latest innovative avenues for the utilization of artificial intelligence and big data analytics in water resource management. Results Eng 101566

Kenda K, Peternelj J, Mellios N, Kofinas D, Čerin M, Rožanec J (2020) Usage of statistical modeling techniques in surface and groundwater level prediction. J Water Supply Res Technol AQUA 69(3):248–265

Liu PW, Famiglietti JS, Purdy AJ, Adams KH, McEvoy AL, Reager JT et al (2022) Groundwater depletion in California's Central Valley accelerates during megadrought. Nat Commun 13(1):7825

Loucks DP, van Beek E (2017) Water resource systems modeling: its role in planning and management. Springer International Publishing, pp 51–72

Machiwal D, Cloutier V, Güler C, Kazakis N (2018) A review of GIS-integrated statistical techniques for groundwater quality evaluation and protection. Environ Earth Sci 77:1–30

Mallick J, Singh CK, Al-Wadi H, Ahmed M, Rahman A, Shashtri S, Mukherjee S (2015) Geospatial and geostatistical approach for groundwater potential zone delineation. Hydrol Process 29(3):395–418

Mallick J, Talukdar S, Pal S, Rahman A (2021) A novel classifier for improving wetland mapping by integrating image fusion techniques and ensemble machine learning classifiers. Eco Inform 65:101426

Mia MY, Haque ME, Islam ARMT, Jannat JN, Jion MMMF, Islam MS et al (2023) Analysis of self-organizing maps and explainable artificial intelligence to identify hydrochemical factors that drive drinking water quality in Haor region. Sci Total Environ 904:166927

Mishra BK, Kumar P, Saraswat C, Chakraborty S, Gautam A (2021) Water security in a changing environment: concept, challenges and solutions. Water 13(4):490

O'Grady J, Zhang D, O'Connor N, Regan F (2021) A comprehensive review of catchment water quality monitoring using a tiered framework of integrated sensing technologies. Sci Total Environ 765:142766

Pal S, Kundu S, Mahato S (2020) Groundwater potential zones for sustainable management plans in a river basin of India and Bangladesh. J Clean Prod 257:120311

Pandey A, Singh G, Chowdary VM, Behera MD, Prakash AJ, Singh VP (2022) Overview of geospatial technologies for land and water resources management. Springer International Publishing, pp 1–16

Pastor AV, Palazzo A, Havlik P, Biemans H, Wada Y, Obersteiner M et al (2019) The global nexus of food–trade–water sustaining environmental flows by 2050. Nat Sustain 2(6):499–507

Pavlides A, Varouchakis EA, Hristopulos DT (2023) Geostatistical analysis of groundwater levels in a mining area with three active mines. Hydrol J 31(6):1425–1441
Rahman A, Roy SS, Talukdar S, Shahfahad (2023) Remote sensing and artificial intelligence for urban environmental studies. In: Advancements in urban environmental studies: application of geospatial technology and artificial intelligence in urban studies. Springer International Publishing, Cham, pp 1–12
Rana R, Kalia A, Boora A, Alfaisal FM, Alharbi RS, Berwal P et al (2023) Artificial intelligence for surface water quality evaluation, monitoring and assessment. Water 15(22):3919
Rao NS, Gugulothu S, Das R (2022) Deciphering artificial groundwater recharge suitability zones in the agricultural area of a river basin in Andhra Pradesh, India using geospatial techniques and analytical hierarchical process method. CATENA 212:106085
Roy SS, Rahman A, Ahmed S, Ahmad IA (2022) Long-term trends of groundwater level variations in response to local level land use land cover changes in Mumbai, India. Groundw Sustain Dev 18:100797
Salehi M (2022) Global water shortage and potable water safety; today's concern and tomorrow's crisis. Environ Int 158:106936
Sarah S, Ahmed S, Boisson A, Violette S, de Marsily G (2014) Projected groundwater balance as a state indicator for addressing sustainability and management challenges of overexploited crystalline aquifers. J Hydrol 519:1409–1422. https://doi.org/10.1016/j.jhydrol.2014.09.016
Savelli E, Mazzoleni M, Di Baldassarre G, Cloke H, Rusca M (2023) Urban water crises driven by elites' unsustainable consumption. Nat Sustain 6:929–940
Shahfahad, Naikoo MW, Das T, Talukdar S, Asgher MS, Asif, Rahman A (2022) Prediction of land use changes at a metropolitan city using integrated cellular automata: past and future. Geol Ecol Landsc 1–19
Shahfahad, Talukdar S, Ghose B, Islam ARMT, Hasanuzzaman M, Ahmed IA et al (2023) Predicting long term regional drought pattern in Northeast India using advanced statistical technique and wavelet-machine learning approach. Model Earth Syst Environ 1–22
Shams MY, Elshewey AM, El-kenawy ESM, Ibrahim A, Talaat FM, Tarek Z (2023) Water quality prediction using machine learning models based on grid search method. Multimed Tools Appl 1–28
Sheffield J, Wood EF, Pan M, Beck H, Coccia G, Serrat-Capdevila A, Verbist K (2018) Satellite remote sensing for water resources management: potential for supporting sustainable development in data-poor regions. Water Resour Res 54(12):9724–9758
Shivanna KR (2022) Climate change and its impact on biodiversity and human welfare. Proc Indian Natl Sci Acad 88(2):160–171
Shu R, Cao X, Wu M (2021) Clarifying regional water scarcity in agriculture based on the theory of blue, green and grey water footprints. Water Resour Manage 35:1101–1118
Singh VP (2018) Hydrologic modeling: progress and future directions. Geosci Lett 5(1):1–18
Solecki W, Grimm N, Marcotullio P, Boone C, Bruns A, Lobo J et al (2019) Extreme events and climate adaptation-mitigation linkages: understanding low-carbon transitions in the era of global urbanization. Wiley Interdiscip Rev Clim Change 10(6):e616
Sun AY, Scanlon BR (2019) How can big data and machine learning benefit environment and water management: a survey of methods, applications, and future directions. Environ Res Lett 14(7):073001
Talukdar S, Singha P, Mahato S, Pal S, Liou YA, Rahman A (2020) Land-use land-cover classification by machine learning classifiers for satellite observations—a review. Remote Sens 12(7):1135
Talukdar S, Salam R, Samad A, Rihan M, Rahman A (2021) Improving wetland mapping techniques using the integration of image fusion techniques and artificial neural network (ANN). In: Mapping, monitoring, and modeling land and water resources. CRC Press, pp 149–164
Talukdar S, Ahmed S, Naikoo MW, Rahman A, Mallik S, Ningthoujam S et al (2023) Predicting lake water quality index with sensitivity-uncertainty analysis using deep learning algorithms. J Clean Prod 406:136885
Tao H, Hameed MM, Marhoon HA, Zounemat-Kermani M, Heddam S, Kim S et al (2022) Groundwater level prediction using machine learning models: a comprehensive review. Neurocomputing 489:271–308
Thanh NN, Thunyawatcharakul P, Ngu NH, Chotpantarat S (2022) Global review of groundwater potential models in the last decade: parameters, model techniques, and validation. J Hydrol 128501
Tung TM, Yaseen ZM (2020) A survey on river water quality modelling using artificial intelligence models: 2000–2020. J Hydrol 585:124670
Uddin MG, Nash S, Olbert AI (2021) A review of water quality index models and their use for assessing surface water quality. Ecol Ind 122:107218
Van Vliet MT, Jones ER, Flörke M, Franssen WH, Hanasaki N, Wada Y, Yearsley JR (2021) Global water scarcity including surface water quality and expansions of clean water technologies. Environ Res Lett 16(2):024020
Wada Y, Van Beek LP, Van Kempen CM, Reckman JW, Vasak S, Bierkens MF (2010) Global depletion of groundwater resources. Geophys Res Lett 37(20)
Whaley L, Cleaver F (2017) Can 'functionality' save the community management model of rural water supply? Water Resour Rural Dev 9:56–66
Xu Y, Liu X, Cao X, Huang C, Liu E, Qian S et al (2021) Artificial intelligence: a powerful paradigm for scientific research. Innovation 2(4)
Yang X, Jin W (2010) GIS-based spatial regression and prediction of water quality in river networks: a case study in Iowa. J Environ Manage 91(10):1943–1951

Yang S, Zhao W, Liu Y, Cherubini F, Fu B, Pereira P (2020) Prioritizing sustainable development goals and linking them to ecosystem services: a global expert's knowledge evaluation. Geogr Sustain 1(4):321–330

Yoon J, Klassert C, Selby P, Lachaut T, Knox S, Avisse N et al (2021) A coupled human–natural system analysis of freshwater security under climate and population change. Proc Natl Acad Sci 118(14):e2020431118

Zhu Z, Qiu S, Ye S (2022a) Remote sensing of land change: a multifaceted perspective. Remote Sens Environ 282:113266

Zhu M, Wang J, Yang X, Zhang Y, Zhang L, Ren H et al (2022b) A review of the application of machine learning in water quality evaluation. Eco-Environ Health 1(2):107–116

Part I
Innovations in Modeling and Mapping of Water Resources

Unveiling Baseflow Dynamics in Mountainous Catchments: Insights from Stable Isotopes and SWAT Modeling in the Upper Indus Basin

2

Sarah Sarah, Shah Waseem, and Shakeel Ahmed

Abstract

In mountainous catchments of the Upper Indus Basin (UIB), the role of baseflow in water balance remains insufficiently understood. Quantifying baseflow is pivotal for effective water resource management. This research elucidates baseflow dynamics and assesses spatio-temporal trends in baseflow fluxes within high-altitude glaciated and non-glaciated watersheds of the UIB. Leveraging stable water isotopes, the SWAT model, and digital filtering, our study unveils the critical role of baseflow in sustaining surface water resources. Notably, temporal variability in baseflow volume arises from seasonally dynamic aquifer recharge and discharge patterns. In the glaciated watershed (Liddar), baseflow substantially contributes to streamflow, reaching 71% and 55% during winter and autumn, respectively. Conversely, the minimum baseflow contributions of 25 and 28% occur in spring and summer. Likewise, within the non-glaciated watershed (Sandran), baseflow contribution peaks in winter (80%) and autumn (68%), while reaching a minimum in spring (39%) and summer (40%). Comparative analysis of tracer-based (δ18O, Cl−) and model-derived results reveals annual baseflow volumes of 54 and 39% for the glaciated watershed, and 57.68 and 33% for the non-glaciated watershed. This indicates an underestimation of baseflow by 15% and 24.68%, respectively, in model-based estimations compared to tracer-based outcomes. Furthermore, our study underscores the utility of stable water isotopes for understanding baseflow dynamics in ungauged high-altitude watersheds. The SWAT model demonstrates strong performance in non-glaciated watersheds, displaying superior model ranking when compared to glaciated watersheds. However, in contrast to tracer-based seasonal estimations, our findings reveal irregular and variable seasonal contributions of model-based baseflow volumes in these watersheds.

Keywords

Base flow · SWAT model · Mountainous catchments · Upper Indus Basin

Supplementary Information The online version contains supplementary material available at https://doi.org/10.1007/978-3-031-61121-6_2.

S. Sarah (✉) · S. Waseem
Deparment of Earth Sciences, University of Kashmir, Srinagar, Jammu and Kashmir 190006, India
e-mail: sarahqazi@uok.edu.in

S. Ahmed
CSIR-National Geophysical Research Institute, Hyderabad 500007, India

S. Ahmed
Centre for Disaster Risk Reduction, Islamic University of Science and Technology, Awantipora 192122, Jammu and Kashmir, India

S. Talukdar et al. (eds.), *Water Resource Management in Climate Change Scenario*, GIScience and Geo-environmental Modelling, https://doi.org/10.1007/978-3-031-61121-6_2

2.1 Introduction

Covering approximately 25% of Earth's land surface, mountainous terrain significantly contributes to 32% of surface runoff, serving as a primary water source for downstream populations (Chow et al. 2012). Baseflow, an integral component of the hydrological cycle, plays a crucial role in maintaining river systems during periods of low flow (Eckhardt 2008; Hall 1968; Nathan and McMahon 1990; Luo et al. 2012; Partington et al. 2012). Influenced by factors like hydrologic properties, aquifer size, soil distribution, recharge rates, topography, and climatic elements, baseflow's dynamics are complex and variable.

The heterogeneous nature of baseflow and the scarcity of data in high-altitude terrains, particularly in regions with significant seasonal variations, make its analysis challenging (Wittenberg and Sivapalan 1999). This complexity necessitates a thorough understanding for efficient water resource management.

The Himalayan Mountains, often called Asia's water towers (Immerzeel et al. 2010), are a crucial freshwater source, supporting over 1.2 billion people (Kaser et al. 2010; Ménégoz et al. 2013). While extensive research has explored the effects of snow and glacier melt on Himalayan watersheds and the Upper Indus Basin (UIB), these studies primarily focus on annual water budgets using high-resolution atmospheric-hydrology models. Predictions about river flow futures are varied, with some anticipating reductions due to climate change (Rees and Collins 2006; Briscoe and Qamar 2007; Akhtar et al. 2008; Immerzeel et al. 2009), and others suggesting decreases across the Himalaya–Karakoram–Hindukush regions from glacier retreat (Berthier et al. 2007; Eriksson et al. 2009; Bookhagen and Burbank 2010). However, research by Jeelani et al. (2012) indicates minimal glacier contribution to overall runoff in the UIB's Western Himalayan watersheds, projecting a shift from a glacier and snow-dominated regime to a nival-pluvial one.

Despite its importance, baseflow in the high-altitude UIB watersheds remains under-explored. Studies like those by Bastola et al. (2018) and Jeelani et al. (2017) report significant annual baseflow contributions but lack detailed seasonal analysis. The UIB, originating from the Himalayas, has over 60% of its river flow from snow and glacier melt (Immerzeel et al. 2010). The region's complex topography results in extreme climatic variability (Barros et al. 2004; Anders et al. 2006), complicating hydrological studies due to limited hydro-meteorological monitoring facilities in these high-altitude areas (Eeckman et al. 2017).

In this context, our study aims to estimate the contributions and dynamics of baseflow in such high-elevation watersheds of the UIB in the Western Himalayas (Kashmir region): the glaciated Liddar and the non-glaciated Sandran. Utilizing stable water isotopes (18O, 2H) and SWAT modeling combined with the baseflow digital filter program (BFlow), we hypothesize that baseflow is the dominant streamflow contributor during winter and autumn, following the cessation of snowmelt. This research, to our knowledge, is the first comprehensive analysis that combines stable water isotopes with distributed hydrological models in the UIB, aiming to enhance our understanding of baseflow estimates and dynamics in these high-altitude Western Himalayan watersheds.

2.2 Study Area

The study examines two distinct watersheds within the Upper Jhelum catchment of the Indus Basin. The Liddar watershed, positioned in the Kashmir valley, spans between 33° 40′ N–34° 12′ 35″ N in latitude and 75° 09′ E–75° 32′ E in longitude, covering 1243 km^2 (Fig. 2.1). It features a varied alpine terrain, with elevations ranging from 1450 to 5200 m above mean sea level (AMSL). The Liddar is characterized by its significant snow and glacier coverage, including the major Kolahoi and Sheshram glaciers, contributing to about 3% of the total area (Jeelani et al. 2012, 2016). In contrast, the Sandran watershed, also part of the upper Jhelum catchment, extends over a smaller area of 483.7 km^2.

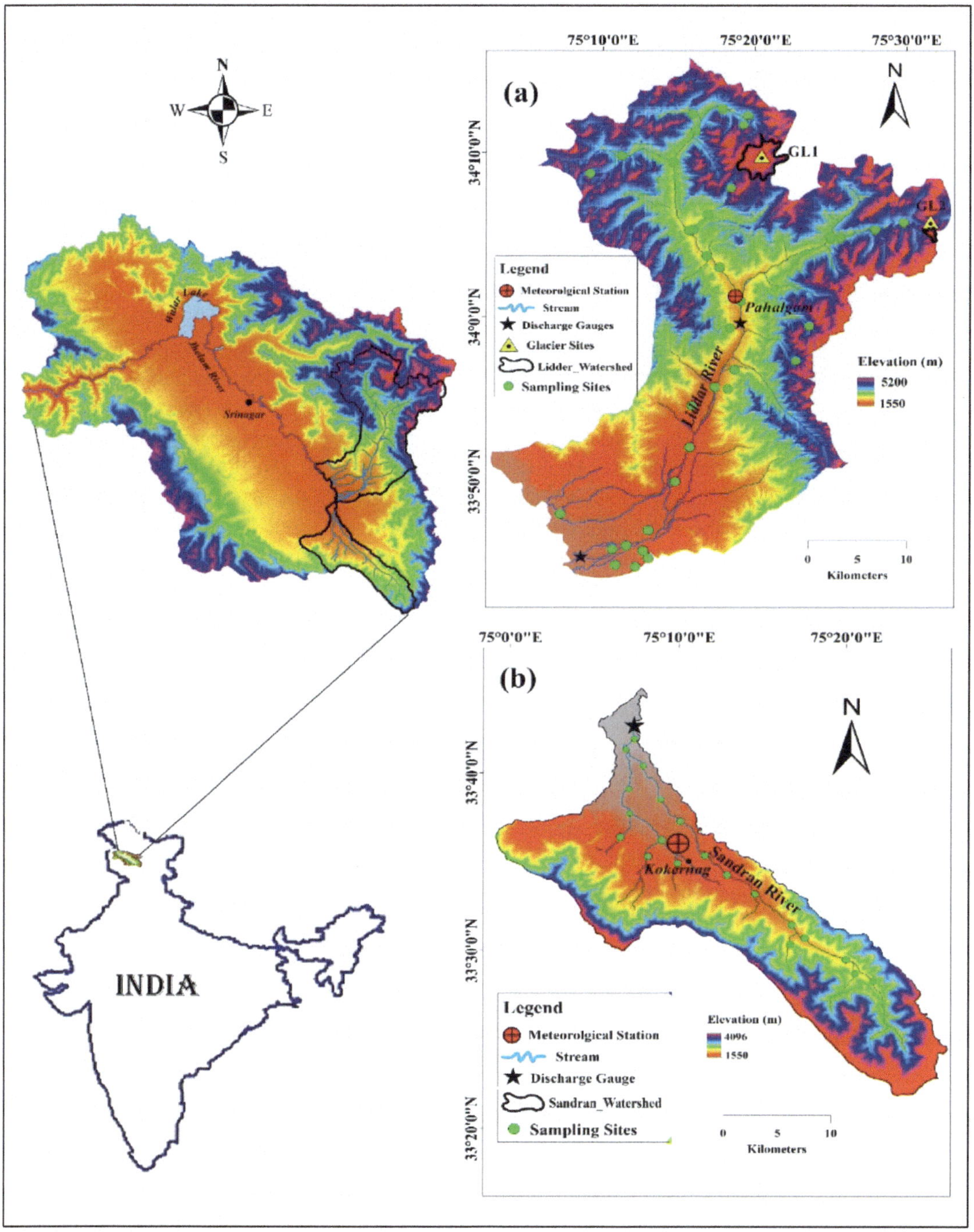

Fig. 2.1 Map showcasing the positions of the Liddar and Sandran watersheds: the perimeters of the Liddar and Sandran watersheds are delineated in black against the digital elevation model (DEM) of the Kashmir Valley, with individual representations in (**a**) and (**b**). In the Liddar watershed, GL1 and GL2 mark the extents of the Kolahai and Sheshram glaciers, respectively, outlined in black

It lies between 33° 20′ N–34° 15′ N and 74° 30′ E–75° 15′ E, with elevations varying from 1550 to 4096 m AMSL. Unlike the Liddar, the Sandran watershed lacks permanent glaciers and is distinguished by its elongated shape and diverse topography.

Both watersheds exhibit a complex geological composition, ranging from hard rock to alluvium, primarily consisting of Triassic limestone and Panjal traps. The prevalent rock formations are surrounded by Paleozoic sedimentary and volcanic rocks, and capped with Karewa sediments and alluvium (Middlemiss 1910; Wadia 1975). These formations contribute to the region's unique hydrological characteristics. The Liddar Valley, with its sub-humid temperate climate, receives an average annual precipitation of 1254 mm, recorded over 37 years at the Pahalgam weather station (Meher-Homji 1971a, b). Temperatures in this region fluctuate between a mean maximum of 19 °C in July and a mean minimum of – 1.7 °C in January (Sarah et al. 2021). The Sandran watershed also experiences a significant portion of its annual precipitation as snowfall, primarily in winter and spring (Meher 1971).

The upper Jhelum Basin houses four key aquifer types: Limestone, Volcanic, Alluvial, and Karewa. The Limestone and Volcanic aquifers are known for their fractured nature and high hydraulic conductivities, while the Alluvial and Karewa sedimentaries, comprising finer materials, exhibit lower hydraulic conductivities (Jeelani 2007, 2008). The presence of numerous springs in the region significantly contributes to the perennial nature of the streams.

2.3 Materials and Methods

2.3.1 Baseflow Modeling Using SWAT

We adopted the SWAT model for our study, a widely recognized tool for watershed-scale hydrological analysis. This involved integrating various datasets, including meteorological, soil texture, and land use land cover (LULC) data. We utilized topographic data to delineate watershed boundaries, define sub-basins, and create stream networks, along with determining elevation bands and slopes.

The LULC data for both watersheds were derived using a supervised classification approach. We identified eight distinct land use land cover classes, highlighting the dominance of forests and the prevalence of agricultural lands. The classification helped in understanding the impact of land use changes on the watershed's water balance (Schilling et al. 2008).

Soil maps at a 1:50,000 scale, based on existing surveys by the National Bureau of Soil Survey and Land Use Planning (NBSS&LUP) of the Indian Council of Agricultural Research (ICAR), were used to identify different soil types within the study area. These ranged from impermeable soils on rock outcrops to various well-drained loams and fine loams.

Meteorological datasets comprised daily precipitation measurements and simulated data for temperature, humidity, wind speed, and solar radiation, using a global weather generator within the SWAT model. Streamflow data from gauging stations in both watersheds were crucial for model validation.

2.3.2 Baseflow Separation Using Flow

We applied a digital filter approach for baseflow separation, a common technique in hydrological analysis (Lyne and Hollick 1979; Nathan and McMahon 1990; Arnold and Allen 1999; Szilagyi et al. 2003, 2004). This method involved processing streamflow data using a recursive filter and calculating the baseflow recession coefficient (αgw), a critical parameter indicating changes in groundwater storage (Brutsaert and Nieber 1977; Smedema and Rycroft 1983; Neitsch et al. 2011). The filter's iterative process allowed fine-tuning of baseflow separation to more accurately reflect watershed conditions.

The filter-based baseflow separation follows this equation:

$$q_t = \beta q_{t-1} + (1 + \beta)/2 \times (Q_t - Q_{t-1}), \quad (2.1)$$

$$b_t = Q_t - q_t. \quad (2.2)$$

In this context, q_t denotes the surface runoff (also known as quick flow) at a given time step t, Q_t symbolizes the aggregate streamflow, $\beta\beta$ is established as the filter parameter (fixed at 0.925), and b_t signifies the baseflow corresponding to time step t.

The selection of the filter parameter (β) and the calculation of baseflow (b_t) at each time step were crucial for achieving accurate baseflow estimates. The filter was applied in three passes—forward, backward, and forward again—to refine the baseflow component of total streamflow. This approach provided a range of αgw values under varying streamflow conditions, which were then integrated into the SWAT model's groundwater files to enhance the baseflow estimation (Figs. 2.2, 2.3, 2.4; Table 2.1).

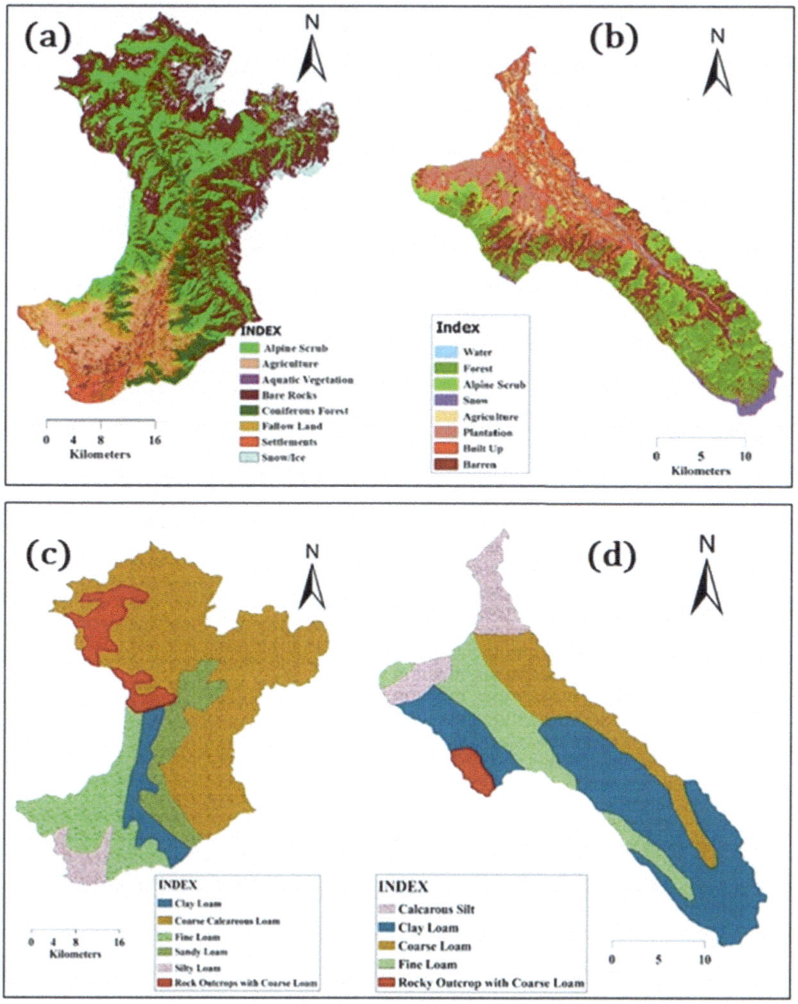

Fig. 2.2 Visual depictions of (**a**) and (**b**) show the land use and land cover (LULC) classes for the Liddar and Sandran watersheds, respectively, while (**c**) and (**d**) illustrate the soil classifications in these two watersheds

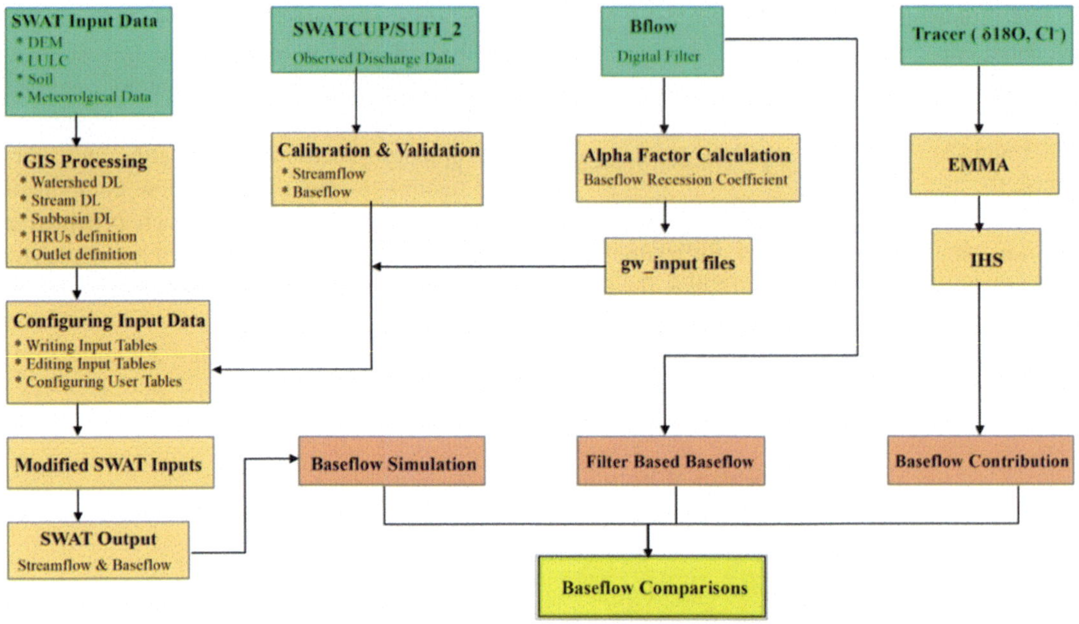

Fig. 2.3 Flowchart of the methodologies used for the estimation and comparisons of baseflow. *Note* DEM: digital elevation model; LULC: land use land cover; DL: delineation; HRUs: hydrological response units; gw: groundwater; EMMA: end member mixing analysis; IHS: isotope hydrograph separation

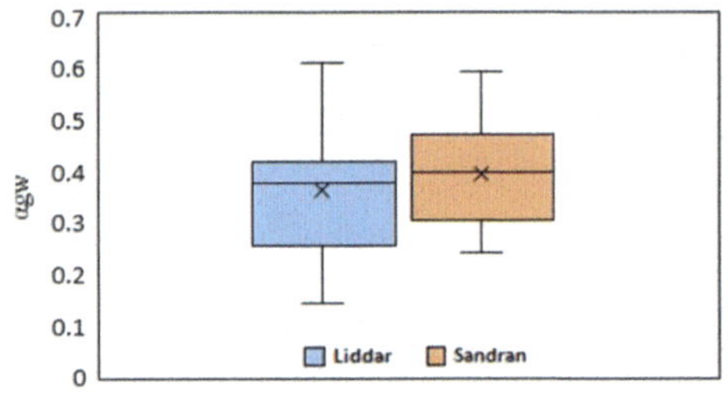

Fig. 2.4 Boxplots depicting the spread and average values of αgw for the Liddar and Sandran watersheds under various streamflow conditions

2.3.3 Calibration and Validation

To calibrate and validate our hydrological model, we implemented the aggregate parameter selection method, following the guidelines set by Yang et al. (2008). This technique assigns a combination of relative (rr), absolute (aa), and replace (vv) values to each model parameter. These values are defined by incorporating a specific term as a predefined parameter into a predetermined parameter, in line with Yang et al. (2008) and Zhang et al. (2014).

Table 2.1 Data types and description of the input datasets used for the SWAT model

Data type	Data source	Scale/resolution	Description
LU/LC	LANDSAT-8	30 m	Categorization of LU/LC classes
Soil types	NBSS&LUP	1:50,000	Soil classification
Slope	SRTM-DEM	30 m	Slope classes
Meteorological parameters	IMD Srinagar	Daily	Temperature, precipitation (relative humidity, solar radiation, wind speed is model-simulated)
River discharge data	I&FC Srinagar	Daily	Recorded streamflow

Our study involved running the model over an eight-year period, from 2009 to 2016, at both upstream and downstream gauging stations in the Liddar watershed. The initial two years, 2009 and 2010, were allocated as a preparatory phase for the model. Subsequent calibration and validation were conducted monthly from 2011 to 2013, with the validation phase extending to 2016. In the Sandran watershed, available in situ discharge data facilitated model calibration from 2014 to 2016 and validation from 2017 to 2018.

The calibration parameters were selected based on their impact on surface flow and baseflow within the watershed, a method supported by research from Abbaspour et al. (2007, 2009), Yang et al. (2008), and Zhang et al. (2011). Special attention was paid to parameters pertinent to mountainous catchment hydrology. The initial range of these parameters was established according to the watersheds' physical features, including their physiography, hydrometeorology, soil distribution, land use, elevation, and slope. Table 2.2 outlines the initial parameter ranges and their optimal values, while a detailed explanation of these parameters is provided in Supplementary Table S1.

Addressing the challenge of a single meteorological station in both watersheds, we divided each sub-basin into five elevation bands. This approach helped to localize the temperature lapse rate (TLAPS) and precipitation lapse rate (PLAPS) adjustments, critical factors due to their influence on temperature and precipitation variations (Zhang et al. 2008; Abbaspour et al. 2007). The SWAT model's capacity to simulate up to 10 elevation bands per sub-basin allowed us to accurately model spatial and temporal changes in snowpack and snowmelt (Fontaine et al. 2002). This division into elevation bands proved effective in capturing the orographic variations in precipitation and temperature, thus enhancing the streamflow simulations. Figure 2.5a, b illustrates

Table 2.2 Calibrated hydrological parameter values in the glaciated (Liddar) and non-glaciated (Sandran) watersheds (Sarah et al. 2021)

S. No.	Parameter	Initial range	Liddar (upstream) optimal values	Liddar (downstream) optimal values	Sandran optimal values
1	CN2	− 0.2, 0.2	− 0.17	− 0.13	− 0.09
2	ALFA_BF	0, 1	0.47	0.39	0.43
3	GW_DELAY	0, 500	108.75	117.3	99.65
4	GW_REVAP	0.02, 0.2	0.16	0.19	0.14
5	GWQMN	0, 1500	1031	1074	825
6	REVAPMN	0, 500	406.25	413	426
7	RCHRG_DP	0, 1	0.45	0.67	0.37
8	SURLAG	0.5, 10	1.23	2.07	0.23
9	SOL_AWC	0, 1	0.68	0.61	0.73
10	SOL_K	0, 500	18.75	21.43	21.45
11	SOL_BD	0.9, 2.5	1.2	1.83	0.99
12	TLAPS	− 20, 20	− 5.25	− 3.31	− 2.82
13	PLAPS	− 400, 400	75	62	21.11
14	SFTMP	− 6, 6	− 1.25	− 1.49	− 1.16
15	SMTMP	− 6, 6	3.75	3.04	4.17
16	SMFMX	0, 20	8.75	9.23	10.32
17	SMFMN	0, 20	1.25	1.09	3.14

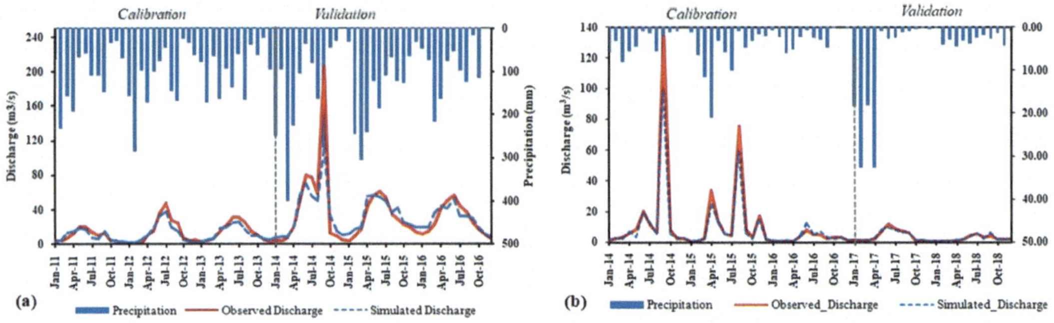

Fig. 2.5 Recorded and predicted monthly streamflow discharge in **a** glaciated Liddar watershed and **b** non-glaciated Sandran watershed (Sarah et al. 2021)

the graphical comparisons of observed versus simulated monthly streamflow during the calibration and validation periods.

Notably, the streamflow simulations in the non-glaciated watershed were characterized by linear and smooth peak flows, contrasting sharply with the uneven and sharp peaks observed in the glaciated watershed. This difference underscored the distinct runoff patterns between the two watershed types, highlighting the contrast in model performance in glaciated versus non-glaciated settings.

2.3.4 Statistical Model Performance Indices

The model's performance was assessed using widely recognized evaluation methods: Nash–Sutcliffe efficiency (NSE), percent bias (PBIAS), and the coefficient of determination (R^2). The NSE quantifies the statistical relationship between the model's simulated and observed values. Nash and Sutcliffe (1970) noted that NSE values range from $-\infty$ to 1, with higher values indicating better model performance.

PBIAS measures the tendency of the simulated values to be larger or smaller than the observed ones (Gupta et al. 1999). An optimal PBIAS value is 0.0, where lower values signify more accurate model simulation. Positive PBIAS values suggest model underestimation, while negative values indicate overestimation.

The coefficient of determination (R^2) evaluates the collinearity between simulated and observed values (Legates and McCabe 1999). R^2 values range from 0 to 1, with higher values signifying less error variance.

The criteria for evaluating model simulations on a monthly basis, as recommended by Moriasi et al. (2007), are presented in Table 2.3.

2.3.5 Baseflow Estimation with Tracers and Water Sampling and Analysis

To assess the model's accuracy, we utilized established statistical metrics: Nash–Sutcliffe efficiency (NSE), percent bias (PBIAS), and the coefficient of determination (R^2). NSE, which ranges from negative infinity to 1, measures the correlation between predicted and observed data, with higher values indicating better model accuracy (Nash and Sutcliffe 1970). PBIAS evaluates the average tendency of the simulated data to be higher or lower than the observed data, where a value of 0.0 reflects perfect accuracy (Gupta et al. 1999). R^2, which varies from 0 to 1, assesses the degree of alignment between the model's predictions and actual observations, with values closer to 1 indicating less variance (Legates and McCabe 1999). Monthly model evaluations were conducted following guidelines from Moriasi et al. (2007), with details outlined in Table 2.3.

Table 2.3 Criteria for statistical assessment of monthly model performance

	Very good	Good	Satisfactory	Unsatisfactory
NSE	0.75 < NSE ≤ 1.00	0.65 < NSE ≤ 0.75	0.50 < NSE ≤ 0.65	NSE ≤ 0.50
PBIAS	PBIAS < 10	10 ≤ PBIAS < 15	15 ≤ PBIAS < 25	PBIAS ≥ 25
RSR	0.00 ≤ RSR ≤ 0.50	0.50 < RSR ≤ 0.60	0.60 < RSR ≤ 0.70	RSR > 0.70

Source Moriasi et al. (2007)

For baseflow estimation, we conducted extensive water sampling in the Liddar and Sandran watersheds from 2017 to 2018, covering diverse sources: stream water (90 samples), rain (20 samples), snowmelt (40 samples), and groundwater (40 samples). Sampling occurred seasonally in March, June, September, and December to ensure a full representation of seasonal variations.

We collected stream water samples from various points along the watershed, from the headstream to the downstream outlet. Rainwater was gathered at multiple locations within the catchment during the study. Snow samples were taken from melting snowpacks at different sites, and in winter, snow cores were frozen, then thawed at room temperature for analysis. Groundwater samples were collected in 200 ml HDPE bottles during each field visit across the watershed.

For hydrochemical analysis, an additional 80 water samples were collected in 500 ml HDPE bottles throughout the Liddar watershed. These samples were analyzed and reported using the standard delta (δ) notation, expressed in parts per mil (‰) relative to the Vienna Standard Mean Ocean Water (VSMOW). The analytical uncertainties for these measurements are noted as ±0.2‰ for δ18O and ±0.5‰ for δ2H (Klaus and McDonnell 2013).

2.3.6 Isotopic Hydrograph Separation

For the separation of runoff components, we determined the isotopic (δ18O) and chemical (Cl−) makeup of the water. This process was essential to distinguish the unique signatures of water from various sources, including snowmelt, baseflow, and precipitation. We employed a three end-member mixing analysis, a technique outlined by Hooper et al. (1990) and Christophersen et al. (1990), which helped define the distinct isotopic and chemical characteristics of each water source.

Our approach assumed that these end members maintain consistent isotopic and chemical compositions and are notably different from one another, ensuring conservative mixing. We based our separation technique on the mass balance principle, involving both water and its isotopes. This process was executed by applying a set of linear mixing equations, as detailed by Hugenschmidt et al. (2014) and Engel et al. (2016).

$$Q_s = Q_i + Q_j + Q_k \tag{2.3}$$

$$Q_s = C_iX_i + C_jX_j + C_kX_k \tag{2.4}$$

$$Q_s = C_iY_i + C_jY_j + C_kY_k \tag{2.5}$$

$$\begin{aligned} C_i = \big(&X_sY_j - X_sY_k + X_jY_k - X_jY_s \\ &+ X_kY_s - X_kY_j | X_iY_j - X_iY_k \\ &+ X_jY_k - X_jY_i + X_kY_i - X_kY_j\big) \end{aligned}$$

$$\begin{aligned} C_j = (&X_sY_k - X_sY_i + X_iY_s - X_iY_k \\ &+ X_kY_i - X_kY_s | X_iY_j - X_iY_k \\ &+ X_jY_k - X_jY_i + X_jY_i - X_kY_j\big) \end{aligned}$$

$$\begin{aligned} C_k = \big(&X_sY_i - X_sY_j + X_iY_j - X_iY_s \\ &+ X_jY_s - X_jY_i | X_iY_j - X_iY_k \\ &+ X_jY_k - X_jY_i + X_kY - X_kY_j\big) \end{aligned}$$

In these equations, X represents the isotopic tracer δ18O, and Y signifies the chemical tracer Cl^-. Q_s is defined as the stream discharge, while

C_i, C_j, and C_k correspond to the proportions of rain, snowmelt, and baseflow, respectively. This methodology allowed us to effectively estimate the seasonal variations in the contributions of different water sources to the overall streamflow.

2.4 Results and Discussions

2.4.1 Investigating Baseflow Dynamics Using Isotopic Tracers

The streamflow in the Liddar and Sandran watersheds arises from a combination of baseflow, snowmelt or glacier melt, and rainfall. To analyze the relative contributions of these components, we utilized hydrograph separation based on tracers, specifically δ18O and Cl−, in line with Eqs. 2.3–2.5. Tracer selection was tailored to the unique characteristics of the study area (Barthold et al. 2011). Commonly, stable water isotopes δ18O and δD, along with chemical tracers like Ca, K, Na, Mg, Cl^-, $SO_4{}^{2}-$, alkalinity, and EC, are employed for accurate results (Elsenbeer et al. 1995; Burns et al. 2001; Soulsby et al. 2003; Katsuyama et al. 2009). We applied an end-member mixing analysis (EMMA) using δ18O and Cl^- to develop a three-component hydrograph separation suitable for both glaciated and non-glaciated watersheds. The three primary flow components identified were snowmelt, baseflow, and rain, each distinguished by unique δ18O and Cl− signals, leading to the creation of seasonal mixing plots (Figs. 2.6 and 2.7).

In the Liddar watershed, the δ18O levels in baseflow varied significantly, ranging from − 9.70 to − 6.29‰, with an average of − 8.20‰. Snowmelt showed less variation, ranging from − 9.52 to − 7.20‰. In contrast, the Sandran watershed's precipitation δ18O fluctuated more broadly, from − 2.02 to − 10.90‰, averaging − 7.02‰. This indicates lighter isotopic composition in winter precipitation, becoming heavier in summer.

In the Sandran watershed, baseflow's δ18O ranged from − 7.27 to − 8.52‰, averaging − 7.81‰, and displayed notable seasonal shifts. In the Liddar watershed during winter, baseflow,

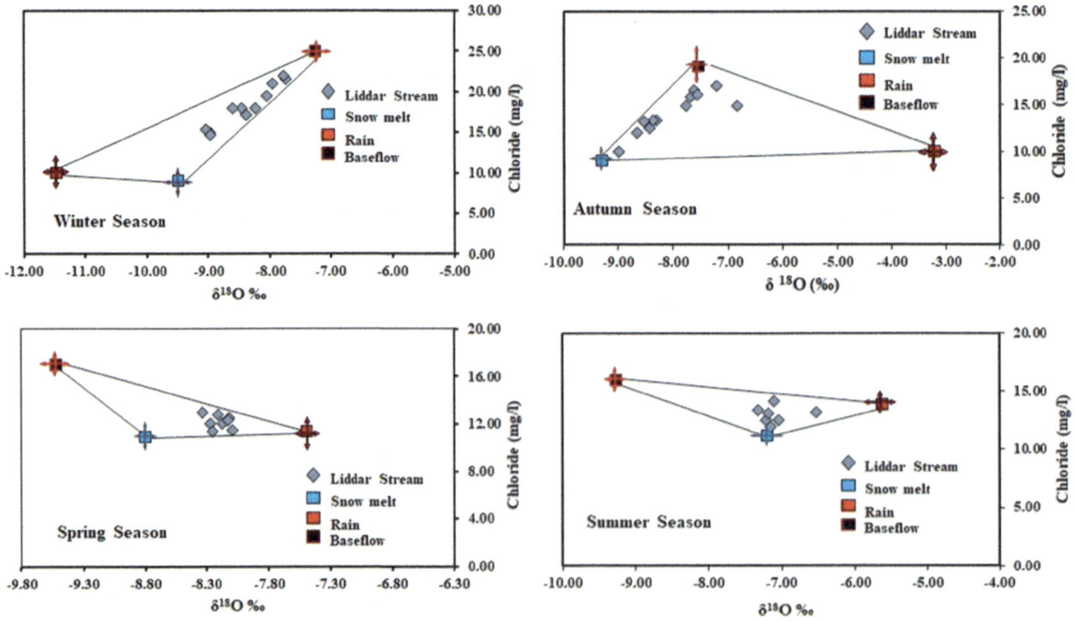

Fig. 2.6 Seasonal plots of end-member mixing depicting the average seasonal values of δ18O and Cl^- in stream water samples from the Liddar watershed

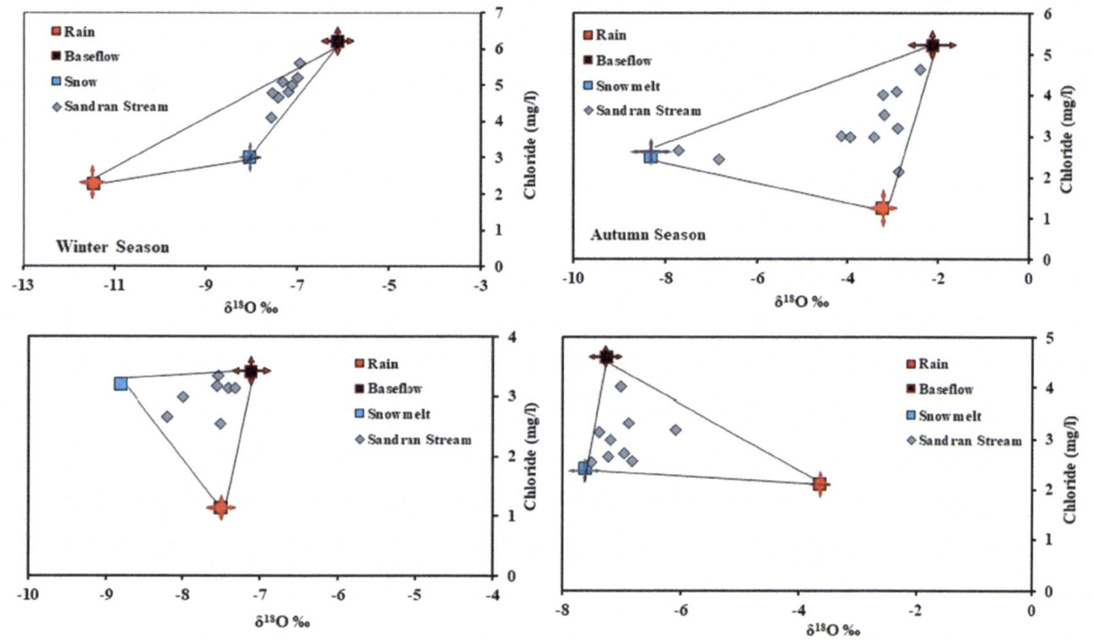

Fig. 2.7 Seasonal plots of end-member mixing depicting the average seasonal values of δ18O and Cl− in stream water samples from the Sandran watershed

indicated by groundwater contribution, is the primary component due to minimal snowmelt. An ‘altitude effect’ is observed, where δ18O in snowmelt is lighter at higher altitudes and heavier at lower elevations. As spring arrives, snowmelt commences at lower elevations with heavier isotopic values, which lighten with increasing elevation and time. In the glaciated regions of Liddar, snowmelt and glacier melt are combined in EMMA to minimize uncertainties in meltwater estimation.

Cl^- concentrations in both watersheds showed significant seasonal changes. Elevated chloride levels in baseflow suggest recharge from precipitation with high chloride concentrations, likely from winter precipitation influenced by western disturbances from the Mediterranean Sea. Glacier contributions to chloride levels are also significant. Conversely, the non-glaciated Sandran watershed showed lower chloride levels in snow and precipitation but higher in groundwater, hinting at a mix of recharge sources and the effects of evaporation and water–rock interactions.

Hydrograph separation using δ18O and Cl− tracers revealed that in the Liddar watershed, streamflow predominantly comprises baseflow, snowmelt, and rain (Fig. 2.8a). Baseflow is the dominant component in winter and autumn, decreasing in spring and summer. Annually, it contributes up to 54% of the streamflow. The Sandran watershed shows more consistent seasonal contributions (Fig. 2.8b), with baseflow being significant in winter and autumn and reducing in warmer months, contributing an estimated 57.68% annually. However, due to its smaller size and lower discharge volumes, the Sandran watershed’s overall baseflow magnitude is less than that of the Liddar watershed.

2.4.2 Baseflow Simulation and Estimates Through Distributed Hydrological Model (SWAT)

2.4.2.1 Glaciated (Liddar) Watershed

The Liddar watershed, spanning 1243 km^2 and characterized by its glaciated terrain, displays considerable spatial variations in aspects like topography, stream paths, soil types, land

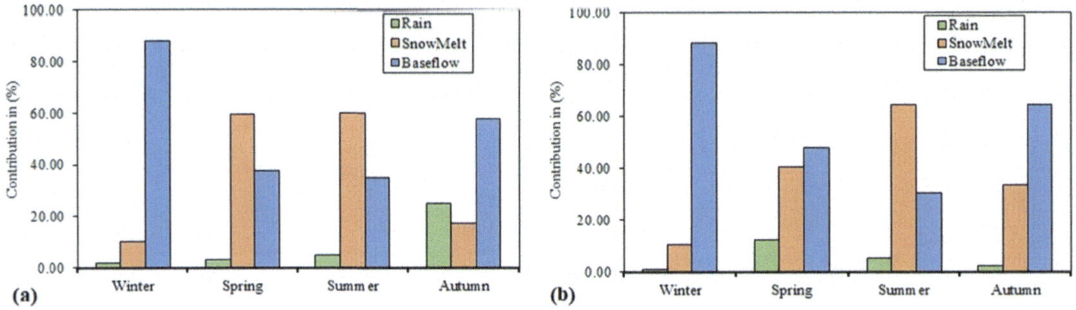

Fig. 2.8 Various sources of streamflow components and their contributions **a** Liddar and **b** Sandran watershed

cover, geological features, as well as elevation, slope, and changes due to orography. To effectively gauge the impact of these diverse factors on baseflow estimates, we strategically placed two gauging stations: one upstream and another downstream.

In the downstream location, baseflow analysis using the filter method showed an earlier initiation, quicker peak, and a faster transition to low flow phases compared to the baseflow predicted by the model, as illustrated in Fig. 2.9a, b. In contrast, the upstream station's filter-based baseflow displayed a rapid peak flow response and a quicker decline, differing from the downstream patterns. These variations are associated with differences in elevation, slope, hydraulic gradient, and the size of the drainage area, all influencing baseflow levels in various parts of the watershed (Tague and Grant 2004; Bloomfield et al. 2009; Price 2011).

The average annual baseflow volume in this glaciated watershed, calculated via both the model and filter methods, was found to be 39% and 46%, respectively, as detailed in Table 2.4. Our model's efficacy is evident in its robust simulation capabilities, reflected in R^2 values of 0.89 and 0.83 during calibration, and 0.86 and 0.75 in the validation phase. The percent bias (PBIAS) values recorded were − 11.80% in calibration and − 2.50% in validation, indicating the model's accuracy in simulating baseflow dynamics.

Our examination of the Liddar watershed revealed that baseflow contributions were higher

Table 2.4 Calculation of baseflow using various techniques in the glaciated (Liddar) and non-glaciated (Sandran) watersheds

Method	Liddar	Sandran
	Mean annual baseflow contribution (%)	
SWAT	39	33
Digital filter	46	42
HIS	54	57.68

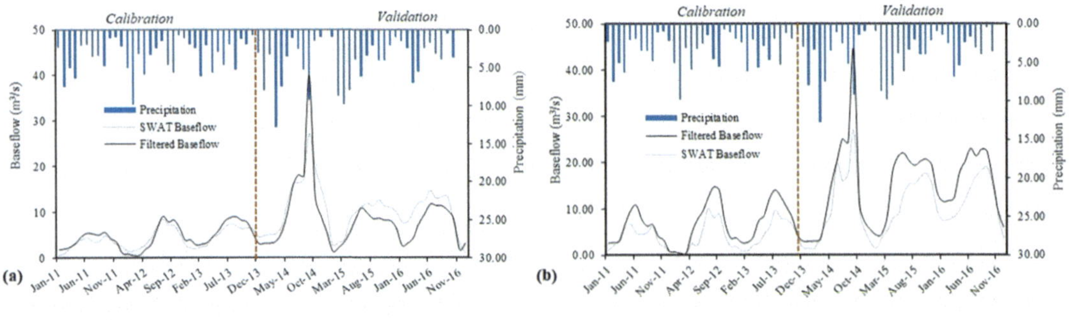

Fig. 2.9 Contrast of baseflow simulations produced through model and filter-based methods at **a** downstream and **b** upstream locations in the Liddar watershed

in areas of greater elevation. This is attributed to factors like steeper slopes, coarse soils, higher levels of precipitation (both rain and snow), coupled with lower air temperatures and reduced potential evapotranspiration (PET) at higher elevations (Rumsey et al. 2015). The significant precipitation in these elevated regions notably increases baseflow volumes (Santhi et al. 2008).

Within the mountainous sub-basins, much of the precipitation is stored in narrow valleys, alpine lakes, snowpacks, and basin fills, serving as key sources of baseflow by channeling water into subsurface flow paths (Queener and Stubblefield 2016; Segura et al. 2019). Furthermore, the coarse soils found at higher elevations, characterized by high infiltration rates, enhance water permeability compared to the finer soils at lower elevations (Price 2011).

On the contrary, at lower elevations, increased air temperatures and PET reduce the amount of precipitation available for infiltration, leading to decreased baseflow generation (Brutsaert 2005; Price 2011). Such variability, inherent to high-altitude Himalayan watersheds, results from diverse channel gradients, aquifer types, varying lengths of flow paths, and differing extents of snow cover. These characteristics contribute to a range of baseflow fluxes, influencing the overall water contribution from upstream to downstream areas.

2.4.2.2 Non-glaciated (Sandran) Watershed

In the Sandran watershed, comparisons between the filter method and the SWAT model for baseflow estimation revealed consistently higher values from the filter method (as shown in Fig. 2.10). The watershed's non-glaciated characteristics facilitate substantial baseflow contributions under varied streamflow conditions. On average, the filter method estimated annual baseflow contributions at 42%, while the SWAT model estimated them at 33%. The filter approach showed an immediate response to runoff at the beginning of the flow's rising phase, contrasting with the SWAT model's slower response (Luo et al. 2012).

Performance of the baseflow simulations in the Sandran watershed was notably better, as indicated by the R^2 and NSE values of 0.99 and 0.85 during the calibration phase, and 0.91 and 0.79 in the validation phase, respectively (refer to Fig. 2.11). The percent bias (PBIAS) recorded at -5.40% during calibration and

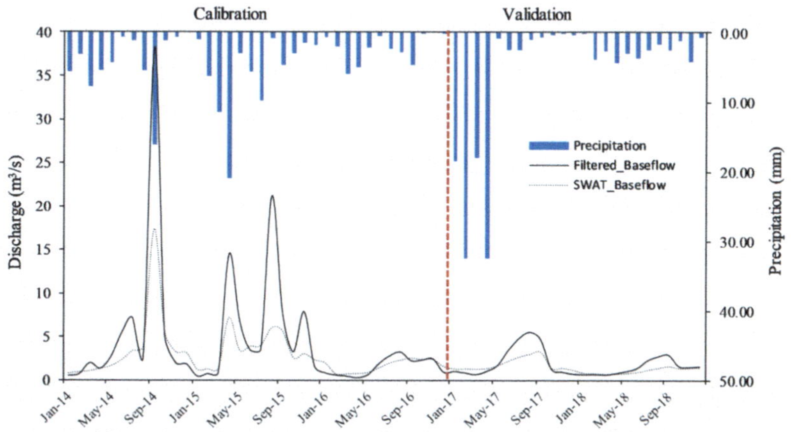

Fig. 2.10 Comparison of baseflow processes in the Sandran watershed as derived from model-based and filter-based methodologies

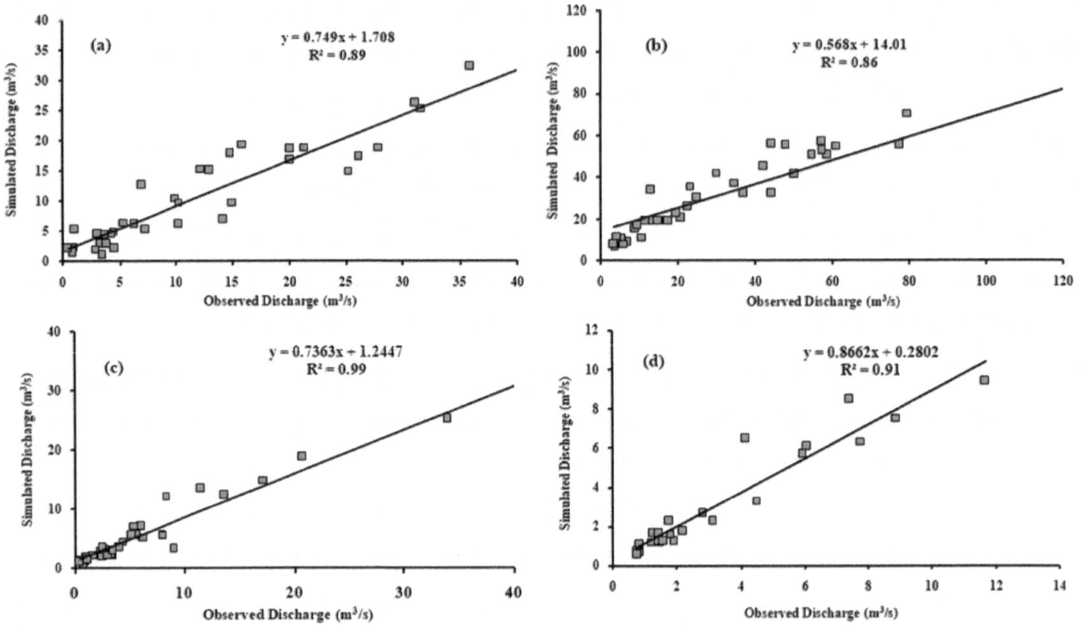

Fig. 2.11 **a**, **b** Comparison of observed and predicted discharge during the calibration and validation phases in the Liddar watershed is shown in (**a**) and (**b**). **c**, **d** Depict the alignment between actual and modeled discharge during these same stages in the Sandran watershed

− 1.50% during validation points to more accurate discharge simulations in this non-glaciated watershed, despite a minor tendency toward overestimation (average annual fraction of 1.36%).

The implementation of elevation bands in the model, which considered orographic changes in temperature and precipitation (TLAPS and PLAPS), markedly enhanced both streamflow and baseflow simulations. This enhancement was evident in the improved R^2 and NSE values, which rose from 0.92 to 0.99 and from 0.81 to 0.85 during calibration, and from 0.84 to 0.91 and from 0.73 to 0.79 during validation.

However, baseflow estimates from the SWAT model showed a significant divergence of 9% from the filter method results and were 24.68% less than those determined using the isotopic tracer method (57.68%). These variances highlight the difficulties the SWAT model faces in accurately representing the complex hydrogeologic characteristics of diverse environments like the Sandran watershed.

2.4.3 Temporal Variability of Baseflow and Its Role in Sustaining Surface Water Resources

Figure 2.12a displays the variation in baseflow contributions across different seasons over a six-year period in the glaciated Liddar watershed. The data reveals that baseflow reaches its peak during winter (71%) and autumn (55%), while its contributions are minimal in spring (25%) and summer (28%). This trend is in stark contrast to the runoff generated from precipitation, snowmelt, and glacier melt, which predominantly occurs in the spring and summer. According to Jeelani et al. (2016), snow and glacier melt account for up to 66% of streamflow in these seasons, surpassing baseflow during the spring and summer freshet. However, it is crucial to differentiate between the volumetric amount of baseflow and its relative share in the total streamflow. Notably, the proportion of baseflow in winter far exceeds its actual volume

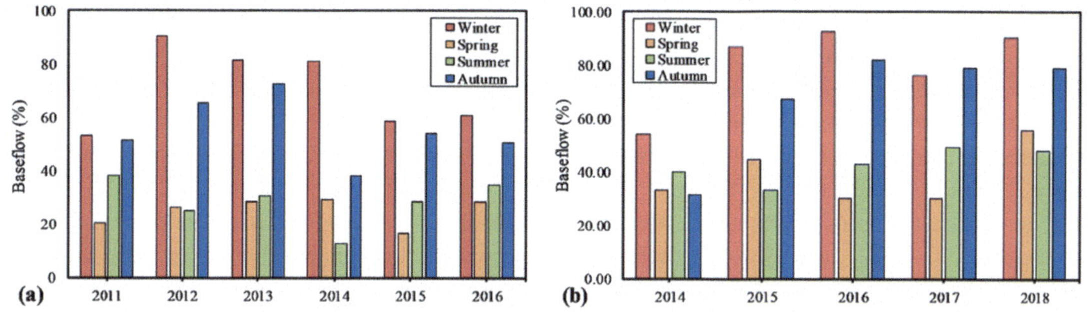

Fig. 2.12 Seasonal fluctuations in baseflow contributions within **a** glaciated and **b** non-glaciated Liddar and Sandran watersheds

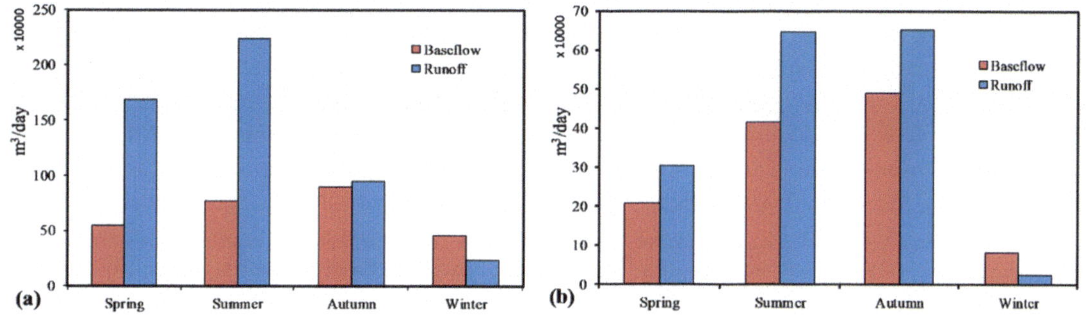

Fig. 2.13 Seasonal contribution of baseflow by volume in **a** glaciated and **b** non-glaciated watersheds

in spring and summer, as depicted in Fig. 2.13. Seasonal shifts in baseflow are significantly influenced by patterns of precipitation. During spring and summer, rainfall contributes to both immediate runoff and groundwater recharge, while in autumn and winter, the release of shallow groundwater enhances baseflow, resulting in a more balanced hydrograph (Fig. 2.9), as winter precipitation does not intensify the hydrograph.

In the Sandran watershed, as shown in Fig. 2.12b, the highest baseflow contributions are observed in winter (80%) and autumn (68%), with the lowest in spring (39%) and summer (40%). This distribution can be attributed to the aquifer's changing recharge and discharge behaviors, with slow and stable baseflow processes in late autumn and spring indicating rapid infiltration of rain, snowmelt, and glacier melt into subsurface storage, which is then gradually released during winter (Luo et al. 2012).

In the Liddar watershed, SWAT, and isotopic hydrograph separation (IHS) methods show the most significant seasonal baseflow variation in winter (±17%), followed by spring and summer (±10%), and the least in autumn (±3%). On the other hand, the non-glaciated Sandran watershed exhibits the largest variation in summer (±12%), then spring (±8.74%), winter (±8.34%), and the least in autumn (±3.52%). The more pronounced seasonal variations in the SWAT model's estimates are due to the annual calibration of streamflow and baseflow, which might not accurately capture the true low and high flow hydrological processes in watersheds with diverse environmental settings.

2.4.4 Sensitivity Analysis

Conducting a sensitivity analysis is essential to identify key parameters in a model that significantly affects outputs, aiming to reduce variability by fine-tuning these inputs. In our research, we pinpointed seventeen parameters as highly responsive based on their statistical significance (illustrated in Fig. 2.14). In both glaciated and non-glaciated areas like the Liddar and Sandran watersheds, parameters related to surface runoff (CN2) and soil characteristics (SOL_K, SOL_BD, SOL_AWC) were found to be particularly responsive. The CN2 parameter, reflecting surface runoff, is especially pertinent in mountainous areas with steep terrain and elevated sub-basins that substantially impact streamflow.

Parameters related to soil infiltration play a pivotal role in the recharge of aquifers, while those connected to groundwater (including ALPHA_BF, RCHRG_DP, GW_REVAP, REVAPMN, GW_DELAY, GWQMN) are crucial in determining the extent of baseflow contributing to the overall streamflow. These parameters are influenced by soil texture and moisture levels. The baseflow recession coefficient (ALPHA_BF), a vital marker of groundwater recharge, indicates that lower values correspond to reduced recharge and a consequent decline in streamflow (Neitsch et al. 2011).

In both the Liddar and Sandran watersheds, groundwater-related parameters exhibited similar levels of sensitivity. This observation mirrors the analogous nature of the aquifer systems in these regions, which include Limestone, Volcanic hard rock, and Alluvial aquifers (Jeelani 2007, 2008). However, elevation-based parameters (TLAPS, PLAPS) showed higher sensitivity in the Liddar watershed, a trend that was less pronounced in the Sandran watershed due to its more uniform topography, smaller catchment size, and lesser variation in elevation. Despite this, these parameters are crucial for capturing the orographic variations in temperature and precipitation.

Snow parameters, namely SMTMP, SFTMP, SMFMX, and SMFMN were observed to be highly sensitive in the glaciated Liddar watershed. This sensitivity was also notable in the snow-fed Sandran watershed, indicative of its environmental attributes. 'SMFMX' and 'SMFMN', representing maximum and minimum snowmelt factors, are critical in determining the dynamics of snowmelt, with any alterations in these values significantly impacting melt rates (Pradhanang et al. 2011). Similarly, 'SMTMP' and 'SFTMP', which relate to snowmelt and snowfall temperatures, contribute to parameter uncertainty as changes in these threshold temperatures can affect snowmelt and snowfall processes.

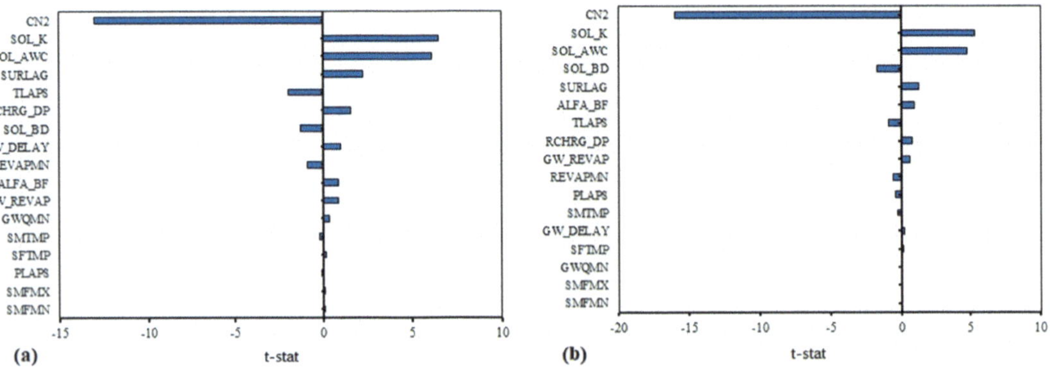

Fig. 2.14 **a** Hydrological sensitive parameters of glaciated (Liddar) and **b** non-glaciated (Sandran) watershed

2.5 New Insights Generated by the Study in the Context of High-Altitude Himalayan Watersheds

Understanding the role of baseflow in sustaining the perennial nature of surface water streams in the Western Himalayan watersheds of the UIB is crucial for developing effective water resource management strategies. Our findings indicate that in the Jhelum tributary of the upper Indus River Basin, baseflow primarily replenishes the water during winter and autumn. Though its contribution diminishes in spring and summer, it remains important, as snowmelt becomes the predominant source of discharge in these seasons.

A recent study by Rashid et al. (2020) highlights the concerning decrease in stream flows due to the loss of glacier area in the glaciated Liddar watershed. This reduction in stream flows, coupled with comparatively lower baseflow during critical periods, poses a potential risk of water shortages downstream. Our study contributes valuable methodological insights and identifies limitations in estimating baseflow contributions in both ungauged and high-altitude glaciated and non-glaciated watersheds of the Upper Indus Basin. These insights are vital for water policymakers, offering a comprehensive understanding of baseflow's role in water management planning for these fragile Western Himalayan watersheds. This knowledge is also foundational for assessing the impacts of climate change on mountainous water balances. However, further research is needed to delve deeper into the mechanisms and processes of baseflow generation, as well as to project baseflow contributions under evolving climatic conditions.

2.6 Summary and Conclusion

The comprehensive study conducted in the Liddar and Sandran watersheds within the Upper Indus Basin has yielded valuable insights into the dynamics of baseflow in high-altitude, mountainous environments. One of the key findings of this research is the considerable variability in baseflow contributions. It was observed that, especially in the glaciated Liddar watershed and the non-glaciated Sandran watershed, baseflow plays a crucial role in maintaining streamflow, with peak contributions during the winter and autumn months. This underlines the importance of baseflow in sustaining river systems during periods when precipitation is low. Another significant outcome of the study is the identification of seasonal variations in the sources of streamflow. In the Liddar watershed, snow and glacier melt predominantly contributes to streamflow during the spring and summer seasons. In contrast, the Sandran watershed exhibits significant baseflow contributions throughout the year, with the highest observed in winter and autumn. This seasonal differentiation in streamflow sources highlights the complex interplay between climatic conditions and hydrological processes in these regions. The study also emphasizes the impact of physiographic and climatic factors on baseflow and streamflow dynamics. Factors, such as topography, soil properties, and climate were found to significantly influence baseflow volumes, particularly in elevated regions. For instance, steeper slopes, coarse soils, higher levels of precipitation, and reduced potential evapotranspiration in these areas contributed to increased baseflow volumes.

Methodologically, the research employed the SWAT model, filter methods, and isotopic hydrograph separation to gain a comprehensive understanding of the hydrological processes in these watersheds. A sensitivity analysis of various parameters enhanced the models' accuracy and reliability, providing a more nuanced understanding of the watershed dynamics. However, the study also uncovered challenges in accurately modeling hydrological processes, particularly in diverse and complex environments like the Sandran watershed. The discrepancies observed between model-based estimates and filter-based measurements underscore the need for more refined modeling techniques in such terrains. Finally, the implications of this study for water resource management are significant, particularly in the context of global climate change. Understanding the seasonal and annual variability of baseflow is crucial for developing effective and sustainable water management strategies. This is especially important for regions that rely heavily on high-altitude watersheds for their water supply. In summary, this study makes a substantial contribution to our understanding of baseflow dynamics in high-altitude catchments. It offers critical insights for hydrologists, environmental scientists, and policymakers involved in the management and conservation of water resources, especially considering the impacts of changing climatic conditions on mountainous regions.

References

Abbaspour KC, Faramarzi M, Ghasemi SS, Yang H (2009) Assessing the impact of climate change on water resources in Iran. Water Resour Res 45(10)

Abbaspour KC, Yang J, Maximov I, Siber R, Bogner K, Mieleitner J, Zobrist J, Srinivasan R (2007) Modeling hydrology and water quality in the pre-alpine/alpine Thur watershed using SWAT. J Hydrol 333:413–430

Akhtar M, Ahmad N, Booij MJ (2008) The impact of climate change on the water resources of Hindukush–Karakorum–Himalaya region under different glacier coverage scenarios. J Hydrol 355:148–163

Anders AM, Roe GH, Hallet B, Montgomery DR, Finnegan NJ, Putkonen J (2006) Spatial patterns of precipitation and topography in the Himalaya. Spec Pap Geol Soc Am 398:39

Arnold JG, Allen PM (1999) Automated methods for estimating baseflow and groundwater recharge from streamflow records. J Am Water Resour Assoc 35:411–424

Barnes BS (1939) The structure of base flow recession curves. Trans AGU 20:721–725. https://doi.org/10.1029/TR020i004p00721

Barros AP, Kim G, Williams E, Nesbitt SW (2004) Probing orographic controls in the Himalayas during the monsoon using satellite imagery. Nat Hazards Earth Syst Sci 4(1):29–51

Barthold FK, Tyralla C, Schneider K, Vaché KB, Frede HG, Breuer L (2011) How many tracers do we need for end member mixing analysis (EMMA)? A sensitivity analysis. Water Resour Res 47(8)

Bastola S, Seong Y, Lee D, Youn I et al (2018) Contribution of baseflow to river streamflow: study on Nepal's Bagmati and Koshi Basins. Water Resour Hydrol Eng. https://doi.org/10.1007/s12205-018-0149-9

Berthier E, Arnaud Y, Kumar R, Ahmad S, Wagnon P, Chevallier P (2007) Remote sensing estimates of glacier mass balances in the Himachal Pradesh (Western Himalaya, India). Remote Sens Environ 108:327–338

Bloomfield JP, Allen DJ, Griffiths KJ (2009) Examining geological controls on baseflow index (BFI) using regression analysis: an illustration from the Thames Basin, UK. J Hydrol 373(2009):164–176

Bookhagen B, Burbank DW (2010) Toward a complete Himalayan hydrological budget: spatiotemporal distribution of snowmelt and rainfall and their impact on river discharge. J Geophys Res 115:F03019. https://doi.org/10.1029/2009JF001426

Briscoe J, Qamar U (2007) Pakistan's water economy running dry. Oxford University Press, Karachi

Brutsaert W (2005) Hydrology: an introduction. Cambridge University Press, New York

Brutsaert W, Nieber JL (1977) Regionalized drought flow hydrographs from a mature glaciated plateau. Water Resour Res 13:637–643

Burns DA, McDonnell JJ, Hooper RP, Peters NE, Freer JE, Kendall C, Beven K (2001) Quantifying contributions to storm runoff through end member mixing analysis and hydrologic measurements at the Panola Mountain Research Watershed (Georgia, USA). Hydrol Process 15(10):1903–1924

Chow FK, De Wekker SF, Snyder BJ (2012) Mountain weather research and forecasting: recent progress and current challenges. Springer Science & Business Media

Christophersen N, Neal C, Hooper RP, Vogt RD, Anderson S (1990) Modelling stream water chemistry as a mixture of soil water end-members: a step towards second-generation acidification models. J Hydrol 116:307–320

Eckhardt K (2008) A comparison of baseflow indices, which were calculated with seven different baseflow separation methods. J Hydrol 352:168–173

Eeckman J, Nepal S, Chevallier P, Camensuli G et al (2017) Assessing reliability of hydrological simulations through model intercomparison at the local scale in the Everest region. Hydrol Earth Syst Sci Discuss. https://doi.org/10.5194/hess-2017-401

Elsenbeer H, Lorieri D, Bonell M (1995) Mixing model approaches to estimate storm flow sources in an overland flow-dominated tropical rain forest catchment. Water Resour Res 31(9):2267–2278. https://doi.org/10.1029/95WR01651

Engel M, Penna D, Bertoldi G, Dell'Agnese A, Soulsby C, Comiti F (2016) Identifying runoff contributions during melt-induced runoff events in a glacierized Alpine basin. Hydrol Process 30:343–364

Eriksson M, Jianchu X, Shrestha AB, Vaidya RA, Nepal S, Sandstörm K (2009) The changing Himalayas: impact of climate change on water resources and livelihoods in the Greater Himalayas. ICIMOD, Kathmandu

Fontaine TA, Cruickshank TS, Arnold JG, Hotchkiss RH (2002) Development of a snowfall snowmelt routine

for mountainous terrain for the soil water assessment tool (SWAT). J Hydrol 262(1–4):209–223
Gupta HV, Sorooshian S, Yapo PO (1999) Status of automatic calibration for hydrologic models: comparison with multilevel expert calibration. J Hydrol Eng 4(2):135–143
Hall FR (1968) Baseflow recessions—a review. Water Resour Res 4(5):973–983
Hooper RP, Christophersen N, Peters J (1990) End member mixing analysis (EMMA): an analytical framework for the interpretation of stream water chemistry. J Hydrol 116:321–345
Hugenschmidt C, Ingwersen J, Sangchan W, Sukvanachaikul Y, Duffner A, Uhlenbrook S, Streck T (2014) A three-component hydrograph separation based on geochemical tracers in a tropical mountainous headwater basin in northern Thailand. Hydrol Earth Syst Sci 18:525–537
Immerzeel WW, Droogers P, de Jong SM, Bierkens MFP (2009) Large-scale monitoring of snow cover and runoff simulation in Himalayan river basins using remote sensing. Remote Sens Environ 113:40–49. https://doi.org/10.1016/j.rse.2008.08.010
Immerzeel WW, van Beek LP, Bierkens MF (2010) Climate change will affect the Asian water towers. Science 328:1382–1385
Javier SA, Francisco JA, Sitian L, Patricia JS (2020) Coupling SWAT model and CMB method for modeling of high-permeability bedrock basins receiving interbasin groundwater flow. Water 2020 12(3):657. https://doi.org/10.3390/w12030657
Jeelani G (2007) Hydrogeology of hard rock aquifer in Kashmir Valley: complexities and uncertainties. In: Ahmed S, Jayakumar R, Salih A (eds) Groundwater dynamics in hard rock aquifers: sustainable management and optimal monitoring network design. Capital, New Delhi, pp 243–248
Jeelani G (2008) Aquifer response to regional climate variability in a part of Kashmir Himalayas in India. Hydrogeol J 16:1625–1633
Jeelani G, Faddema J, Van der Veen C, Leigh S (2012) Role of snow and glacier melt in controlling river hydrology in Liddar watershed (western Himalaya). Water Resour Res 48(1):16
Jeelani G, Shah RA, Jacob N, Deshpande RD (2016) Estimation of snow and glacier melt contribution to Liddar stream in a mountainous basin, western Himalaya: an isotopic approach. J Isot Environ Health Stud 53:18–35
Jeelani G, Shah RA, Deshpande RD, Fryar AE, Perrin J, Mukherjee A (2017) Distinguishing and estimating recharge to karst springs in snow and glacier dominated mountainous basins of the western Himalaya, India. J Hydrol 550(2017):239–252
Kaser G, Großhauser M, Marzeion B (2010) Contribution potential of glaciers to water availability in different climate regimes. Proc Natl Acad Sci USA 107:20223–20227. https://doi.org/10.1073/pnas.1008162107
Katie P (2011) Effects of watershed topography, soils, land use, and climate on baseflow hydrology in humid regions: a review. Prog Phys Geogr 35(4):465–492
Katsuyama M, Kabeya N, Ohte N (2009) Elucidation of the relationship between geographic and time sources of stream water using a tracer approach in a headwater catchment. Water Resour Res 45(13):W06414. https://doi.org/10.1029/2008wr007458
Klaus J, McDonnell JJ (2013) Hydrograph separation using stable isotopes: review and evaluation. J Hydrol 505(2013):47–64
Legates DR, McCabe GJ Jr (1999) Evaluating the use of “goodness-of-fit” measures in hydrologic and hydroclimatic model validation. Water Resour Res 35(1):233–241
Luo Y, Arnold J, Allen P, Chen X (2012) Baseflow simulation using SWAT model in an inland river basin in Tianshan Mountains, Northwest China. Hydrol Earth Syst Sci 16:1259–1267
Lyne VD, Hollick M (1979) Stochastic time-variable rainfall-runoff modelling. In: Hydrology and water resources symposium. Engineering Institute of Technology, Australia, Perth, Australia, pp 89–92
Meher-Homji VM (1971a) On the Mediterranean climatic regime of West Pakistan. Theor Appl Climatol 19(3):277–286. ISSN 0177-798X
Meher-Homji VM (1971b) The climate of Srinagar and its variability. Geogr Rev India 33(1):1–14
Ménégoz M, Gallée H, Jacobi HW (2013) Precipitation and snow cover in the Himalaya: from reanalysis to regional climate simulations. Hydrol Earth Syst Sci 17(10):3921–3936
Middlemiss CS (1910) Revision of Silurian-Trias sequence of Kashmir. Rec Geol Surv India 40(3):206–260
Moriasi DN, Arnold JG, Van Liew MW, Bingner RL, Harmel RD, Veith TL (2007) Model evaluation guidelines for systematic quantification of accuracy in watershed simulations. Trans Am Soc Agric Biol Eng 50(3):885–900
Nash JE, Sutcliffe JC (1970) River flow forecasting through conceptual models. Part I—a discussion of principles. J Hydrol 10:282–290
Nathan RJ, McMahon TA (1990) Evaluation of automated techniques for baseflow and recession analysis. Water Resour Publ 26(7):1465–1473
Neitsch SL, Arnold JG, Kiniry JR, Williams JR (2011) Soil and water assessment tool theoretical documentation version 2009. Grassland, Soil and Water Research Laboratory, Agricultural Research Service and Blackland Research Center, Texas Agricultural Experiment Station, College Station, TX
Partington D, Brunner P, Simmons CT, Werner AD, Therrien R, Maier HR, Dandy GC (2012) Evaluation of outputs from automated baseflow separation methods against simulated baseflow from a physically based, surface water-groundwater flow model. J Hydrol 458–459:28–39

Pradhanang SM, Anandhi A, Mukundan R, Mark SZ, Donald CP, Eliot MS, Adao M, Allan F (2011) Application of SWAT model to assess snowpack development and streamflow in the Cannonsville watershed, New York, USA. Hydrol Process 25(21):3268–3277. https://doi.org/10.1002/hyp.8171

Price K (2011) Effects of watershed topography, soils, land use, and climate on baseflow hydrology in humid regions: a review. Prog Phys Geogr 35(4):465–492

Queener N, Stubblefield AP (2016) Spatial and temporal variability in baseflow in the Mattole river headwaters, California, USA. Hydrol Earth Syst Sci. https://doi.org/10.5194/hess-2016-300

Rashid I, Majeed U, Aneaus S, Pelto M (2020) Linking the recent glacier retreat and depleting streamflow patterns with land system changes in Kashmir Himalaya, India. Water 2020 12(4):1168. https://doi.org/10.3390/w12041168

Rees HG, Collins DN (2006) Regional differences in response of flow in glacier fed Himalayan rivers to climate warming. Hydrol Process 20:2157–2169

Rumsey CA, Miller MP, Susong DD, Tillman FD, Anning DW (2015) Regional scale estimates of baseflow and factors influencing baseflow in the Upper Colorado River Basin. J Hydrol Reg Stud 4(2015):91–107

Santhi C, Allen PM, Muttiah RS, Arnold JG, Tuppad P (2008) Regional estimation of base flow for the conterminous United States by hydrologic landscape regions. J Hydrol 351(1–2):139–153

Sarah S, Shah W, Ahmed S (2021) Modeling and comparing streamflow simulations in two different montane watersheds of western Himalayas. In: Groundwater for sustainable development, vol 15, p 100689, ISSN 2352-801X. https://doi.org/10.1016/j.gsd.2021.100689

Schilling KE, Jha MK, Zhang YK, Gassman PW, Wolter CF (2008) Impact of land use and land cover change on the water balance of a large agricultural watershed: historical effects and future directions. Water Resour Res 44:W00a09. https://doi.org/10.1029/2007wr006644

Segura C, Noone D, Warren D, Jones J, Tenny J, Ganio L (2019) Climate, landforms, and geology affect baseflow sources in a mountain catchment. Water Resour Res 55:5238–5254. https://doi.org/10.1029/2018WR023551

Smedema LK, Rycroft DW (1983) Land drainage—planning and design of agricultural drainage systems. Cornell University Press, Ithaca, NY, p 376

Soulsby C, Rodgers P, Smart R, Dawson J, Dunn S (2003) A tracer-based assessment of hydrological pathways at different spatial scales in a mesoscale Scottish catchment. Hydrol Process 17(4):759–777. https://doi.org/10.1002/hyp.1163

Tague C, Grant GE (2004) A geological framework for interpreting the low-flow regimes of Cascade streams, Willamette River Basin, Oregon. Water Resour Res 40:W0403

Wadia DN (1975) Geology of India. Tata McGraw Hill Co., New Delhi, 344 pp

Wittenberg H, Sivapalan M (1999) Watershed balance estimation using streamflow recession analysis and baseflow separation. J Hydrol 219(1–2):20–33

Yang J, Abbaspour KC, Reichert P, Yang H (2008) Comparing uncertainty analysis techniques for a SWAT application to Chaohe basin in China. J Hydrol 358(1–2):1–23

Zhang XS, Srinivasan R, Debele B, Hao FH (2008) Runoff simulation of the headwaters of the Yellow River using the SWAT model with three snowmelt algorithms. J Am Water Resour Assoc 44(1):48–61. https://doi.org/10.1111/j.1752-1688.2007.00137

Zhang X, Srinivasan R, Arnold J, Izaurralde RC, Bosch D (2011) Simultaneous calibration of surface flow and baseflow simulations: a revisit of the SWAT model calibration framework. Hydrol Process 25(14):2313–2320

Zhang P, Liu R, Bao Y, Wang J, Yu W, Shen Z (2014) Uncertainty of SWAT model at different DEM resolutions in a large mountainous watershed. Water Res 53:132–144

Wetland Habitat Health Condition Modeling Using Ensemble Machine Learning Algorithms

3

Swapan Talukdar, Rumki Khatun, and Swades Pal

Abstract

The construction of the Komardanga Dam on the Dhepa River at Birganj, Bangladesh, has precipitated significant ecological alterations in the downstream floodplain wetlands of the Punarbhaba River (Malda and Chapai Nawabganj of India-Bangladesh), marking a critical research problem. The substantial reduction in wetland area post-dam construction underscores the urgency to assess and model the health conditions of these habitats systematically. The major objective of this study was to leverage ensemble machine learning models to evaluate wetland habitat health, aiming to generate a robust classification system that delineates varying health conditions across the wetland ecosystem. Methodologically, this research incorporated a suite of hydrological, water quality, and land composition indicators to inform the modeling process. Parameters such as water presence frequency (WPF), range of variability (RVA), and change rate of water presence were meticulously prepared, alongside indicators like agricultural presence frequency, Chlorophyll-a (Chl-a) concentration, and Temperature Condition Index (TCI) estimations derived from land surface temperature (LST) data. These indicators were integrated into ensemble machine learning algorithms, including artificial neural network (ANN), support vector machine (SVM), random forest (RF), rotation forest, and bagging, to create a detailed model of wetland habitat health. The models' accuracy was rigorously validated using empirical and binormal ROC curves. Quantitative results revealed a stark decline in wetland areas, from 255.84 km^2 in 1993 to 72.2 km^2 by 2019. Indicators such as NDWI and WPF indicated healthier conditions in central wetland areas, in contrast to stressed peripheral zones. According to the random forest model, only 13.75% of the wetland area is classified as 'very good', while a substantial 54.8% is deemed 'poor' or 'very poor'. Comparatively, the ANN model identifies 13.31% as 'very good' and a combined 45.7% as 'poor' or 'very poor'. The ensemble machine learning models identified the most degraded regions, with the rotation forest algorithm displaying superior performance with a binormal ROC AUC of 0.936.

S. Talukdar (✉)
Department of Geography, Asutosh College, University of Calcutta, Kolkata 700026, India
e-mail: swapan.talukdar@asutoshcollege.in

R. Khatun
Department of Geography, Kazi Nazrul University, Asansol, Burdwan, India

S. Pal
Department of Geography, University of Gour Banga, Malda, West Bengal 732101, India

S. Talukdar et al. (eds.), *Water Resource Management in Climate Change Scenario*, GIScience and Geo-environmental Modelling, https://doi.org/10.1007/978-3-031-61121-6_3

The SVM and bagging models also showcased high accuracy, with binormal AUCs of 0.93 and 0.92, respectively, while the ANN and RF models demonstrated competent performance with binormal AUCs of 0.912 and 0.873. This study's integrated approach, combining detailed parameter analysis with sophisticated machine learning algorithms, offers a comprehensive tool for environmental monitoring and underscores the need for strategic wetland management to mitigate the adverse effects of hydraulic structures on these critical ecosystems.

Keywords

Wetland habitat health · Ensemble machine learning · Spatiotemporal dynamics · ROC curve analysis · Dam ecological impact

3.1 Introduction

A wetland, sometimes called a land-water ecotone, makes up only 6% of the earth's land area (Mitsch et al. 2009; Mitsch and Gosselink 2015), but it is home to endangered species and maintains 40% of ecosystem biodiversity. Furthermore, wetland habitats offer unparalleled quantities of products and services for mankind rather than other ecosystems (Malekmohammadi and Jahanishakib 2017). It offers a range of provisioning services, including woods, food, grasses, fibers, and medicinal plants; habitat services, including genetic diversity and nursery services; regulating services, including pollination, climate regulation, flood control, nutrient recycling, soil erosion control, storm moderation, and biological controls; habitat services, including genetic diversity and nursery services; and cultural services, including mental clarity, relaxation, spiritual peace, and cognitive development (Yang et al. 2021; Mitsch et al. 2013). Wetlands are becoming less diverse in terms of location and quality, even though they have great socio-ecological significance (Talukdar and Pal 2020). A number of goods and services are produced less frequently as a result of habitat loss and degradation in wetlands. Wetlands are the ecosystems that have been dehumanized the most recently, as Finlayson et al. (2019) correctly pointed out. Davidson et al. (2018) reported that the rate of extinction of wetlands is several times higher than that of other ecologies. An enormous amount of study has been conducted since then to identify the primary causes of wetland degradation. Expanding agriculture, urbanization, climate change, wetland fragmentation, the effects of damming, and reduction in groundwater levels are the main factors contributing to the conversion of wetlands (Talukdar and Pal 2020; Pal and Talukdar 2018; Pal et al. 2022). As an example, the construction of the Farakka Dam has led to significant alterations in the wetlands within the Padma River basin, resulting in a decline in water levels (Islam et al. 2022). Insufficient river flow characteristics may have adverse effects on the ecology of rivers (Kingsford 2000). Poisonous pesticide spills from urban trash, industrial effluent, and agricultural land degrade water quality, endangering ecosystem health and biodiversity. According to Lefebvre and Devreker (2020), a condition known as eutrophication occurs when native species are prevented from growing and alien plant growth is encouraged. This is a common issue in floodplain wetlands that are managed for agriculture (Khatun et al. 2021). Infrastructure development, such as urbanization, damming, and the installation of hydropower plans, causes bigger wetland areas to become fragmented, increasing the number of patches and edges and negatively affecting ecosystem services and habitats (Nielsen et al. 2020; Zhang et al. 2016). Wetland vulnerability is increased when there is direct contact between the wetland and the anthropogenic landscape (Pal and Paul 2020; Cui et al. 2015). Damming water flow has become essential for managing water supply and maintaining uniformity in the look of wetland areas. In wetlands, this regulation frequently results in ecohydrological drought because it is so strict (Pal and Sarda 2021). The Kaligandaki River in Nepal, where a hydrometric station was constructed across the river, is one example of how damming has also changed the flow of the environment (Suwal et al. 2020). Hydro-ecological droughts like these have affected many of the Atreyee River's wetlands, forcing flow management by damming (Saha et al. 2022).

Mapping wetland habitat vulnerability involves producing several precondition criteria for vulnerabilities delineating the driver-pressure-state-impact-response (DPSIR) framework through an effective methodology (Malekmohammadi and Jahanishakib 2017; Wantzen et al. 2019). A spatial database of this type is essential for planning based on priorities. Saha and Pal (2019) use limited surface and hydrological measurements to study the vulnerability of wetland ecosystems. The majority of cases involve the application of various statistical and data-driven methods, such as analytic hierarchy process (AHP), fuzzy logic (FL), logistic regression (LR), weights-of-evidence (WOE), and frequency ratio (FR) (Pal and Talukdar 2018; Talukdar and Pal 2020; Pal and Paul 2020). Nonlinear machine learning approaches include artificial neural networks (ANN), radial function bases (RBF), support vector machines (SVM), decision trees (DT), random forests (RF), k-nearest neighbor (KNN), and classification and regression trees (CART) (Islam et al. 2021; Merghadi et al. 2020; Huang et al. 2020; Mosavi et al. 2020; Talukdar et al. 2022; Shahfahad et al. 2022) have been extensively utilized in a variety of industries, including assessing susceptibility to forest fires, landslides, floods, and soil erosion. The results of these models are remarkable precision. Hybrid ensemble prediction is an innovative technique that has emerged recently for creating feature-driven new models. These models are constructed by integrating distinct machine learning models (Talukdar et al. 2022). Furthermore, a number of studies have demonstrated that ensemble prediction outperforms individual machine learning models (Saha et al. 2021; Alqadhi et al. 2022; Rihan et al. 2023). To the best of our knowledge, no comparable research has been conducted utilizing such advanced models to examine the vulnerability of wetland habitat. The effectiveness of these models depends critically on the development of trustworthy data layers and the selection of appropriate vulnerability conditioning criteria (Ahmed et al. 2022).

Current research in wetland ecosystem dynamics often overlooks the comprehensive integration of machine learning (ML) techniques in evaluating wetland habitat health. While numerous studies have focused on spatial and temporal changes in wetlands, there is a significant gap in utilizing advanced ML models to assess the multifaceted aspects of wetland health. This includes the incorporation of various ecological, hydrological, and anthropogenic factors into a cohesive model. Additionally, there is a lack of studies that systematically compare different ML models, such as artificial neural networks (ANN), support vector machine (SVM), random forest (RF), and others, in the context of their efficacy and accuracy in modeling wetland ecosystems. Furthermore, research often does not adequately address the impact of specific environmental interventions, such as the construction of dams, on wetland health using these advanced modeling techniques.

The primary objective of this study is to bridge these research gaps by developing a comprehensive wetland habitat health model using various ML techniques. This model aims to integrate a wide range of variables—including hydrological data, land use changes, water quality indices, and anthropogenic factors—to provide a holistic assessment of wetland health. The research intends to systematically apply and compare different ML models like ANN, SVM, RF, and others, to evaluate their effectiveness in predicting wetland habitat conditions accurately. The novelty of this study lies in its multidisciplinary approach, combining environmental science with advanced data analytics. This research is one of the first to compare multiple ML models in the context of wetland health, providing valuable insights into the strengths and limitations of each technique. Furthermore, it uniquely focuses on the implications of environmental interventions, like dam construction, within these models, offering a nuanced understanding of their impacts on wetland ecosystems. The outcomes of this research are expected to make a substantial contribution to the field of environmental management and conservation, offering a novel perspective and advanced tools for policymakers and researchers in wetland ecosystem assessment and preservation.

3.2 Materials and Methods

3.2.1 Study Area

The Punarbhaba River Basin, extending over the Barind plains and straddling the borders of India and Bangladesh, encompasses an area of approximately 5265.93 km^2 (Fig. 3.1). The river itself stretches about 160 km in length, with varying widths from 3 to 8 km. It originates in the lowlands of Bangladesh and traverses through the Gangarampur block, located in the Dakshin Dinajpur District of West Bengal, India. The river's course includes segments in West Bengal, flowing through districts like Panchagarh, Sadar Upazila, and Pirganj in Thakurgaon, as well as Bochaganj and Biral in the Dinajpur District of Bangladesh, before passing through Kushmandi and Bansihari blocks in Dakshin Dinajpur District, West Bengal. It eventually re-enters Bangladesh, joining with the Punarbhaba River in the Naogaon District at Rohanpur.

Historically, the Punarbhaba was a hydrologically vibrant river, partly fed by the Teesta River. However, neotectonic movements led to a significant eastward tilt in the Teesta River Basin, redirecting its flow directly into Bangladesh and thereby severing the Punarbhaba's connection. This geological shift profoundly altered the river's hydrological dynamics and the surrounding riparian environment, transforming it from an ice and rain-fed perennial flow into a rain-fed seasonal river. Moreover, the river underwent further changes in 1992 due to the construction of a dam. These modifications have had a substantial impact on the riparian ecosystem, particularly in the post-modification era, where dam construction and extensive water extraction in this densely populated area have been prevalent. Such activities have drastically affected the periodically inundated riparian wetlands, especially those adjacent to the primary river course. Additionally, the transboundary nature of the Punarbhaba River between Bangladesh and India adds complexity, making the basin a subject of significant interest for environmental and geopolitical studies.

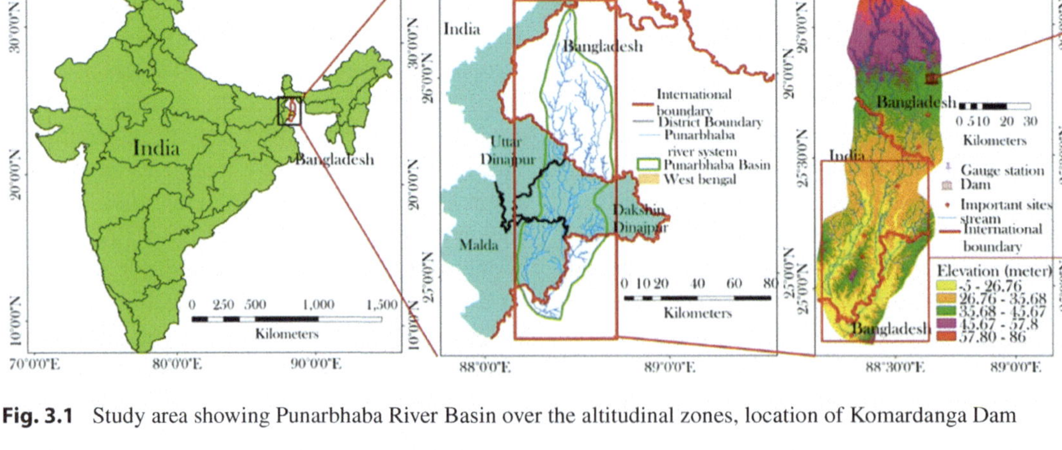

Fig. 3.1 Study area showing Punarbhaba River Basin over the altitudinal zones, location of Komardanga Dam

3.2.2 Materials

For this study, the United States Geological Survey (USGS) Landsat series imagery was used as an efficient and cost-effective source for analyzing time-series data. These imageries are crucial for mapping land features and tracking dynamic environmental changes. The study incorporated Landsat images acquired by various sensors, including the Thematic Mapper (TM) and Operational Land Imager (OLI) on the Landsat 8 satellite. These images, with a resolution of 30 m (path/line: 132/43), date from 1987 to 2019 and have been an integral part of creating detailed maps of wetlands in floodplains and identifying different land surface compositions, such as vegetated areas and urban developments, over a period of more than three decades.

In addition, the Landsat data was calibrated and used to calculate various environmental indicators. These indicators include Turbidity Index, total phosphorus content, Chlorophyll-a concentration, and temperature state index. The specific methods used for this calibration and subsequent analyses are described in detail in the Methods section of the study. This approach utilizes the high resolution and temporal span of Landsat imagery to gain a comprehensive understanding of environmental changes and trends in the study area.

3.2.3 Methods for Spatiotemporal Wetland Mapping

Spatiotemporal wetland mapping involves the systematic use of remote sensing techniques and spectral indices to identify and monitor aquatic bodies over time. Among the various spectral indices available, the Normalized Difference Water Index (NDWI) has been identified as particularly effective, especially in regions like the Barind area of Bangladesh and India (Das and Pal 2017). Other notable indices include the Modified Normalized Difference Water Index (MNDWI), Water Ratio Index (WRI), Automated Water Extraction Index (AWEI), and the Re-modified Normalized Difference Water Index (RmNDWI) (Debanshi and Pal 2020). Each of these indices has specific applications and advantages depending on the characteristics of the study area and the objectives of the research.

For the purpose of this study, the focus is on the application of NDWI, which is formulated as given in Eq. 3.1.

$$NDWI = bgreen + bNIR/bgreen - bNIR. \tag{3.1}$$

In this equation, *b*green represents the reflectance in the green band and *b*NIR is the near-infrared band reflectance. The NDWI values range from − 1 to + 1. A positive NDWI value indicates the presence of water pixels while a negative value suggests non-water pixels. The magnitude of the positive value is proportional to the depth of the water, with larger positive values indicating deeper water bodies.

The application of NDWI in spatiotemporal wetland mapping involves processing satellite imagery over the desired time frame. This process starts with the collection of time-series satellite images, followed by the computation of NDWI for each image. The resulting NDWI maps are then analyzed to detect changes in water bodies over time, such as changes in the size, shape, or location of wetlands. This method allows for the systematic tracking of wetland dynamics, providing valuable data for ecological studies, water resource management, and environmental monitoring.

3.2.4 Methods for Preparing Indicators for Wetland Habitat Health Modeling

3.2.4.1 Water Presence Frequency (WPF)

The water presence frequency (WPF) is calculated to quantify the temporal extent of water coverage in a pixel over time (Debanshi and Pal 2020). This is done by analyzing NDWI maps derived from Landsat imagery spanning from

1987 to 2019. Every NDWI map is converted into a binary representation, where wetland pixels are marked as '1' and non-wetland pixels as '0'. The annual WPF for a pixel is calculated by summing up the occurrences of water presence over the years and expressing it as a percentage. The is WPF calculated using Eq. 3.2

$$\mathrm{WPF}_p = \frac{\sum_{j=1}^{N} X_j}{N} \times 100, \tag{3.2}$$

where WPF_p is the water presence frequency for pixel p, X_j is the presence of water in the jth pixel, and N represents the number of years.

3.2.4.2 Range of Variability (RVA)

RVA is used to assess the ecological flow status of the wetland (Talukdar and Pal 2019). It involves calculating the standard deviation of NDWI values to establish the upper and lower threshold limits for water depth in each pixel. This method helps in understanding the variability in water depth and its ecological implications. The mean one standard deviation (1SD) is computed for each pixel, providing insights into the eco-hydrologically relevant state.

3.2.4.3 Change Rate of Water Presence

To assess the rate of change in wetland areas, remote sensing-based change detection methods are employed. This involves comparing the wetland area in earlier years with that in more recent years and identifying any transitions or conversions. The typical wetland map is utilized for developing change detection maps for different phases (pre-dam and post-dam periods).

3.2.4.4 Agricultural Land and Proximity to Agricultural Land

Agricultural land is identified using NDVI values extracted from satellite imagery. Areas with NDVI values ranging from 0.2 to 0.4 are classified as agricultural land. The Agricultural Presence Frequency (APF) is then calculated similarly to WPF by converting each NDVI map into a binary format (Saha and Pal 2019). The APF for a pixel is determined by using Eq. 3.3.

$$\mathrm{APF}_{sj} = \frac{\sum_{i=1}^{Nx} \mathrm{CF}_{pij}}{Nx} \times 100, \tag{3.3}$$

where APF_{sj} is the frequency of agricultural presence for pixel j, CF_{pij} is the frequency of agricultural land for the ith pixel in picture j, and Nx is the total number of images.

3.2.4.5 Built-Up Area and Proximity to Built-Up Area

The extraction of built-up land from satellite images is carried out using the Normalized Difference Built Index (NDBI). This index effectively distinguishes built-up areas based on their unique spectral characteristics. The NDBI is calculated using Eq. 3.4

$$\mathrm{NDBI} = \frac{b\mathrm{MIR} - b\mathrm{NIR}}{b\mathrm{MIR} + b\mathrm{NIR}}, \tag{3.4}$$

where bMIR and bNIR are the middle-infrared and near-infrared band reflectances, respectively. Proximity analysis is then conducted to determine the spatial relationship between built-up areas and wetlands, indicating potential impacts on the wetland ecosystem.

3.2.4.6 Temperature Condition Index (TCI)

The Temperature Condition Index (TCI) is used to assess the thermal characteristics of the wetland, which are crucial for understanding various ecological processes. The first step involves calculating the land surface temperature (LST) from the thermal band of Landsat imagery using the radiative transfer mode (RTM). TCI is then derived relying on a dimension index to standardize the pixel information. The TCI is calculated using Eq. 3.5.

$$\mathrm{TCI}_i = \frac{\mathrm{LST}_i - \mathrm{LST}_{\min}}{\mathrm{LST}_{\max} - \mathrm{LST}_{\min}} \times 100, \tag{3.5}$$

where LST_i is the land surface temperature for pixel i, $\mathrm{LST}_{\max}$ is the maximum LST value, and $\mathrm{LST}_{\min}$ is the minimum LST value across the dataset.

3.2.4.7 Hydro-Period

Hydro-period refers to the length of time that water is present within a wetland throughout the year and is indicative of the wetland's ecological character (permanent, seasonal, or ephemeral). To calculate the hydro-period, NDWI images from each month of the year are converted into binary maps where '1' indicates a water body and '0' a non-water body. The sum of these binary maps gives a value ranging from 0 (no water presence) to 12 (perennial water body), providing a measure of the hydro-period for each pixel.

3.2.4.8 Eco-Hydro Deficit and Surplus

The eco-hydro deficit and surplus are assessed using the Flow Duration Curve (FDC) method, with NDWI values substituting for flow values. This approach helps in understanding the differences in water presence before and after the dam construction. The eco-hydro surplus is determined by the difference between the median FDC of the regulated (post-dam) and unregulated (pre-dam) periods (Talukdar and Pal 2019). Conversely, an eco-hydro deficit is identified when the regulated median FDC is lower than the unregulated median FDC. This calculation provides insight into the potential quality of habitat in different wetland areas.

3.2.4.9 Wetland Fragmentation

Wetland fragmentation is analyzed to understand how large habitat areas are divided into smaller, isolated patches. This is carried out using the Landscape Fragmentation Tool, which identifies different types of fragmented habitats (e.g., patch, edge, perforated, small core, medium core, and large core). This tool is applied to recent NDWI images to assess the state of fragmentation in both the pre- and post-dam periods. Larger core areas are considered ecologically viable, whereas smaller patches or fragmented habitats are seen as ecologically vulnerable. This analysis is crucial for understanding the impact of urbanization and infrastructure development on wetland ecosystems.

3.2.4.10 Total Phosphorous (TP), Chlorophyll-a, and Secchi Disk Depth (SDD)

To determine the spatial distribution of TP, Chlorophyll-a, and SDD, in-situ data collection is carried out at 30 sample locations within the study area. Water samples are analyzed in the laboratory to obtain these parameters. The geographical coordinates of the sampling sites are recorded using a GPS device. To map these parameters at a spatial scale, the in-situ data is cross-verified with reflectance values from Landsat 8 OLI SR images. This process involves assessing the absorption spectra of Chlorophyll-a, typically showing attenuation from shorter to longer wavelengths with increased absorption in the near-infrared region. Pearson's correlation coefficient is used to establish the relationship between in-situ Chlorophyll-a concentration and satellite reflectance values. The most correlated bands and band ratios are then integrated using various regression functions (exponential, linear, power, logarithmic) to develop the best-fit models for predicting Chlorophyll-a, SDD, and TP (Talukdar et al. 2020).

$$\text{InSDD} = 2.925 - 0.629\text{In}(\text{TM4}) - 0.515\text{In}(\text{TM2})$$

$$\text{InChla} = 1.432 + 1.195\text{In}(\text{TM4}) + 1.239\text{In}(\text{TM2})$$

$$\text{InSS} = 0.675 + 1.4116\text{In}(\text{TM4}) + 0.01387\text{In}(\text{TM2})$$

3.2.4.11 Turbidity Index (TI)

Turbidity, indicating water clarity and the presence of suspended particles, is assessed using remote sensing data. The Normalized Difference Turbidity Index (NDTI) is formulated for calculate the turbidity of terrestrial water bodies using the red (b_{red}) and green (b_{green}) spectral bands. The formula for NDTI is given below (Eq. 3.6).

$$\text{NDTI} = \frac{b_{\text{red}} - b_{\text{green}}}{b_{\text{red}} + b_{\text{green}}}. \quad (3.6)$$

The index values are then categorized into low, moderate, and high turbidity levels, which are visualized and analyzed within a GIS environment to understand the spatial distribution of water turbidity.

3.2.4.12 Proximity to Road and River

The study also includes an analysis of the proximity of wetlands to road and river networks, which are mapped using tools like Google Earth. Buffer analysis is performed to determine the spatial relationship between wetlands and these features. Proximity to rivers is particularly important as it influences the hydrological connectivity and nutrient flow to the wetlands. The study focuses on significant streams that have the capacity to feed wetlands. In contrast, proximity to roads is analyzed to understand potential threats to wetlands due to urbanization and infrastructure development, as roads often lead to habitat fragmentation and increased human activity in these sensitive areas.

3.2.5 Methods for Modeling Wetland Habitat Health

For modeling the health condition of wetland habitats, several advanced machine learning algorithms can be employed for binary classification (healthy vs. unhealthy). These include artificial neural networks (ANN), support vector machines (SVM), random forest (RF), rotation forest, and bagging (codes are given in Appendix). Each algorithm offers a unique approach to classification based on the characteristics of the data.

3.2.5.1 Artificial Neural Networks (ANNs)

Artificial neural networks (ANNs) are computational models inspired by the human brain's structure and functioning. In the context of wetland habitat health modeling, ANN can be used to identify patterns and relationships in complex environmental datasets. The model consists of layers of interconnected nodes (neurons) where each connection represents a weight. Input data, such as hydrological, land composition, and water quality parameters, are fed into the network. These inputs are processed through hidden layers where the neurons apply activation functions to the weighted inputs and pass the output to the next layer or the output layer. The output layer then provides the classification results, indicating the health status of the wetland (healthy or not healthy). The network's weights are adjusted through a process called backpropagation, using a training dataset to minimize the difference between the predicted output and the actual output.

3.2.5.2 Support Vector Machine (SVM)

Support vector machine (SVM) is a supervised machine learning algorithm widely used for classification tasks. In the assessment of wetland habitat well-being, support vector machines (SVMs) operate by identifying a hyperplane within a multi-dimensional space to effectively separate data points into two distinct categories (healthy or not healthy). Key features influencing wetland health, such as water quality indices and land use patterns, are used as input vectors. SVM performs classification by finding the optimal hyperplane that maximizes the margin between the two classes. The decision function is defined by a subset of training samples known as support vectors. SVM is particularly effective in high-dimensional spaces and is capable of handling nonlinear relationships through the use of kernel functions.

3.2.5.3 Random Forest (RF)

Random forest (RF) is an ensemble learning method that operates by constructing multiple decision trees during training and outputting the mode of the classes (classification) of individual trees. In wetland habitat health modeling, RF takes various environmental parameters as input and creates numerous decision trees. Each tree gives a classification result, and the final output is determined by majority voting. RF is known for its high accuracy, ability to handle large datasets with higher dimensionality, and its robustness to overfitting. It also provides insights into the importance of different features contributing to the health of the wetland ecosystem.

3.2.5.4 Rotation Forest

Rotation forest is an ensemble classifier that enhances diversity among the classifiers (decision trees) through feature extraction and rotation. The training dataset is split into subsets, and principal component analysis (PCA) is applied to each subset to rotate the feature space. Individual decision trees are then trained on these rotated subsets. In the context of wetland habitat health modeling, this approach allows the model to encompass diverse aspects and interactions of environmental factors affecting wetland health. The final classification is obtained by aggregating the predictions from all the decision trees, thus ensuring robustness and reducing the likelihood of overfitting.

3.2.5.5 Bagging

Bagging, or bootstrap aggregating, is an ensemble machine learning algorithm used to improve the stability and accuracy of machine learning algorithms. It works by creating multiple versions of a predictor (such as a decision tree) on random subsets of the original training set. For wetland habitat health modeling, bagging would involve creating multiple decision trees on randomly selected subsets of environmental data. Each tree votes and the final classification (healthy or not healthy) are determined by majority voting. Bagging is effective in reducing variance and avoiding overfitting, making it suitable for complex datasets like those used in wetland habitat health assessment.

3.2.6 Accuracy Assessment of Wetland Habitat Health Model

The empirical ROC curve is a widely used tool for assessing the effectiveness of binary classification models, such as those assessing wetland habitat health. It plots the True Positive Rate (TPR) against the False Positive Rate (FPR) at various threshold settings. In the context of wetland health modeling, the TPR (sensitivity) is the proportion of correctly identified healthy wetlands, while the FPR (1 − specificity) is the proportion of unhealthy wetlands incorrectly classified as healthy. To construct the empirical ROC curve, the model's predictions are compared against the actual data. For each possible cutoff point, a TPR and FPR value is calculated and plotted, creating a curve. The area under this curve (AUC) provides a measure of the model's accuracy. An AUC of 1 indicates perfect classification, while an AUC of 0.5 suggests performance no better than random chance. This method allows for the assessment of the model's discriminative ability across various classification thresholds.

The binormal ROC curve is an extension of the traditional ROC curve and is particularly useful when dealing with continuous or ordinal diagnostic test results. It assumes that the distributions of test results for both positive and negative cases are normally distributed. In the case of wetland habitat health assessment, this approach involves transforming the empirical scores of the model into a standardized z-score, which then follows a normal distribution. The binormal ROC curve analysis involves estimating the parameters of the normal distributions for both healthy and unhealthy wetlands. Like the empirical ROC curve, the TPRs and FPRs are calculated and plotted to form the ROC curve. However, the binormal ROC curve smoothens the empirical ROC curve, especially useful when dealing with small sample sizes or when the empirical ROC curve is too jagged. The AUC in the binormal ROC curve also serves as an accuracy measure, providing a more refined analysis of the model's performance.

Using both empirical and binormal ROC curves for accuracy assessment allows for a comprehensive evaluation of the wetland habitat health model. These methods offer insights into the model's effectiveness at different thresholds and under various assumptions, ensuring a robust validation of its predictive capability.

3.3 Results and Discussion

3.3.1 Spatiotemporal Wetland Dynamics

The spatiotemporal analysis of wetland dynamics in the Punarbhaba River region, particularly following the construction of the Komardanga Dam in 1992 on the Dhepa River, a tributary and upstream source of the Punarbhaba River, reveals significant changes in wetland area. This dam, being a key supplier of water to the Punarbhaba River and its downstream floodplain wetland, has had a profound impact on the wetland ecosystem. Wetland mapping for the years 1993, 1998, 2003, 2008, 2013, and 2019, as depicted in Fig. 3.2, indicates a substantial reduction in wetland area over this period. In 1993, shortly after the dam's construction, the total wetland area was recorded at 255.84 km^2. However, this area experienced a steady decline over the subsequent years. By 1998, the area had reduced to 166.36 km^2, and the downward trend continued, reaching 146.52 km^2 in 2003, 99.16 km^2 in 2008, 83.95 km^2 in 2013, and finally, a significant decrease to only 72.2 km^2 by 2019.

This reduction in the wetland area can be largely attributed to the effects of damming. The dam's construction has not only impacted the wetland area but also its depth and overall ecological health. The peripheral regions of the wetland have been particularly affected, showing a marked decrease. The dramatic reduction in both the extent and fringe areas of the wetland indicates a fragile current state of this ecosystem. The quantitative analysis clearly demonstrates the substantial impact that damming and subsequent hydrological alterations can have on wetland ecosystems. The Komardanga Dam, by altering the natural flow of the Dhepa River, has significantly reduced the water supply to the downstream Punarbhaba River and its floodplain wetland, leading to drastic changes in the wetland's spatial extent. This analysis underscores the importance of considering ecological impacts in the planning and management of hydraulic structures, especially in ecologically sensitive areas like floodplain wetlands.

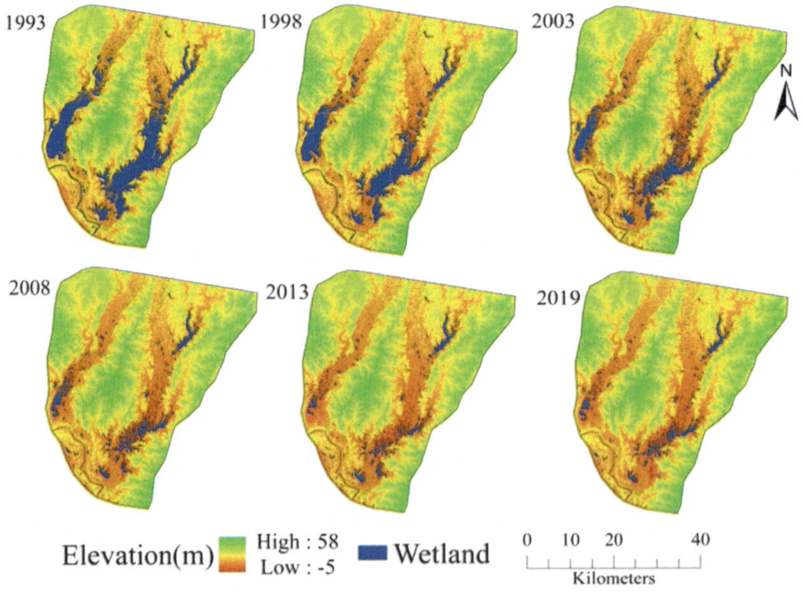

Fig. 3.2 Spatiotemporal dynamics of wetland dynamics from 1993 to 2019

3.3.2 Assessment of Wetland Habitat Health Indicators

Figure 3.3 presents a multifaceted view of wetland habitat health through the lens of fourteen distinct indicators. These indicators collectively reflect the varying conditions of the wetland habitat, illustrating a gradient from healthier central areas to more stressed peripheral zones, where proximity to human activities exerts noticeable pressure. The hydrological integrity of the wetland, as indicated by the Normalized Difference Water Index (NDWI) and water presence frequency (WPF), is crucial for sustaining the wetland's ecological functions. In Fig. 3.3, central areas show a higher NDWI (A) and consistent WPF (D), indicating stable aquatic environments vital for wildlife and plant communities. Conversely, at the edges where the distance from the river increases (N), the NDWI and WPF values decrease, suggesting reduced water availability, which can lead to a loss of habitat for aquatic and moisture-dependent species. The hydro-period (C) and change rate of water depth (E) further confirm this trend, as the reduced number of months with water and higher rates of depth change at the periphery reflect a more volatile water regime, likely

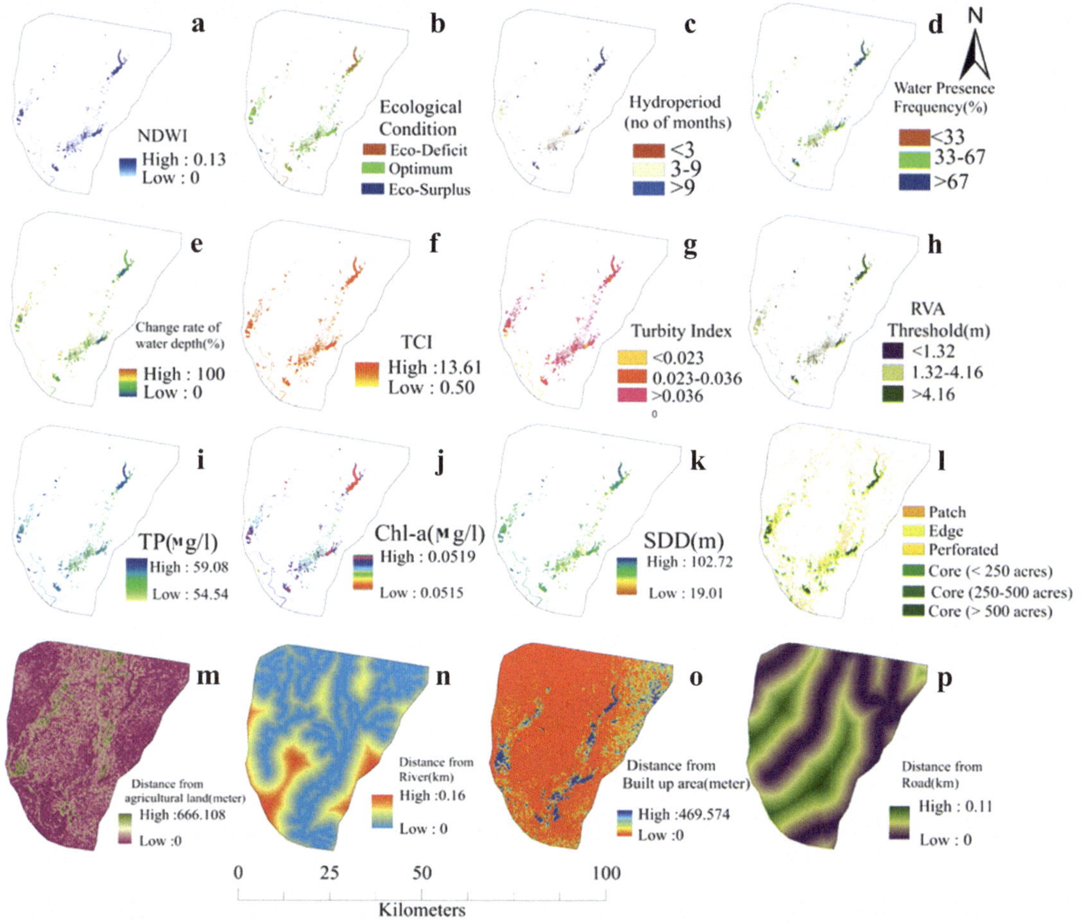

Fig. 3.3 Parameters for wetland habitat vulnerability assessment; **a** NDWI, **b** eco-deficit and surplus, **c** hydroperiod, **d** water presence frequency, **e** change rate of water depth, **f** temperature condition index, **g** Turbidity Index, **h** range of variability approach, **i** total phosphorus, **j** Chlorophyll-a, **k** Secchi disk depth, **l** wetland fragmentation, **m** proximity to agricultural land, **n** proximity to river, **o** proximity to built-up, **p** proximity to road

influenced by damming and water withdrawal for human use. The proximity to agricultural (M) and built-up areas (O) plays a significant role in wetland health. Agricultural runoff can introduce nutrients and pesticides, as evidenced by the elevated total phosphorus (TP) levels (I) and Chlorophyll-a concentrations (J) near these zones. This eutrophication process can lead to harmful algal blooms and oxygen depletion, negatively impacting aquatic life. The proximity to built-up areas is closely associated with higher temperatures (F), as seen in the Temperature Condition Index (TCI), potentially altering the wetland's microclimate and affecting species that are sensitive to temperature changes. The fragmentation of the wetland habitat (L) is especially pronounced near roads (P) and built-up areas, leading to smaller, isolated patches that can compromise the resilience and ecological connectivity of wetland species. Turbidity (G) and Secchi disk depth (SDD) (K) are indicative of water clarity and sedimentation levels. High turbidity at the wetland's edges can obstruct sunlight penetration, affecting photosynthesis in aquatic plants and visibility for predators and prey, disrupting the food web. Low SDD values near areas of intense human activity suggest increased sedimentation, likely from construction and surface runoff, further impacting water quality and habitat suitability.

Each indicator in Fig. 3.3, from hydrological fluctuations to land use changes, interweaves to paint a picture of a wetland system under stress at its boundaries. The edge effects are evident, where water quality deteriorates, habitat fragmentation is more acute, and the ecological conditions swing toward a deficit. These findings underscore the need for integrated wetland management that considers the riparian buffer zones and the impact of adjacent land use practices.

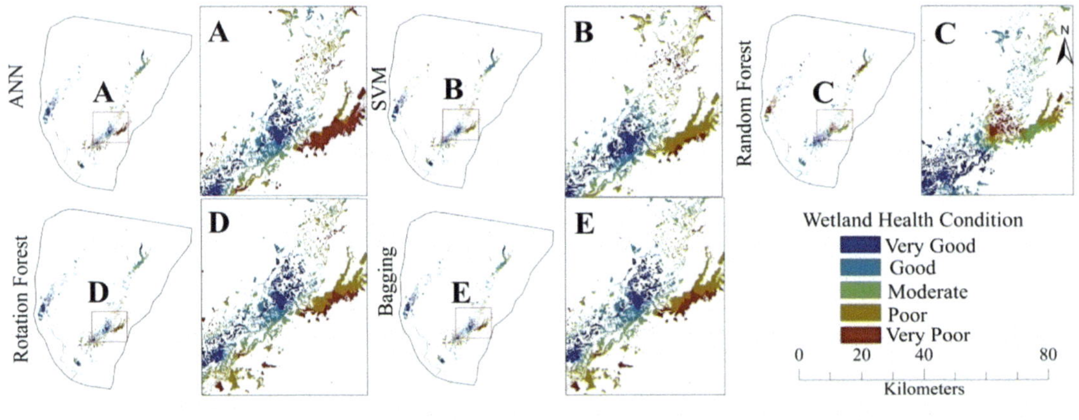

Fig. 3.4 Wetland habitat health condition modeling using ensemble machine learning algorithms

Table 3.1 Computation of area coverage under different health zones

Wetland health condition	ANN		SVM		Random forest		Rotation forest		Bagging	
	Area in km²	Area in %	Area in km²	Area in %	Area in km²	Area in %	Area in km²	Area in %	Area in km²	Area in %
Very good	9.61	13.31	13.11	18.12	9.93	13.75	11.6	16.06	6.8	9.41
Good	13.3	18.42	7.76	10.75	10.36	14.35	18.9	26.17	22.32	30.92
Moderate	15.97	22.11	15.92	22.02	12.99	17.99	23.22	32.16	10.51	14.57
Poor	20.95	29.02	19.98	27.68	16.42	22.74	6.64	9.20	22.39	31.01
Very poor	12.05	16.68	15.36	21.28	23.15	32.06	11.26	15.56	10.43	14.44

Protecting the core wetland areas while mitigating the influences at the periphery could be key to maintaining the health and functionality of these vital ecosystems.

3.3.3 Wetland Habitat Health Modeling Using ML Models

Figure 3.4 provides and the data in Table 3.1 offer a quantitative view of the wetland habitat health condition as assessed by various ensemble machine learning algorithms, such as ANN, SVM, RF, rotation forest, and bagging. These models have been trained on sixteen factors to classify the wetland health into five categories: very good, good, moderate, poor, and very poor. From the analysis of Fig. 3.4, we can infer that the different models, despite varying in the spatial extent of the classified health categories, generally agree on the regions that are characterized by poor habitat quality. Particularly along the rivers Punarbhaba and Tangon, pockets of healthy riparian wetland are identified, signifying a vital corridor for the local ecosystem. However, these areas do not uniformly fall under the 'excellent' health category, reflecting the nuanced impact of hydrological changes post-dam construction. These changes have led to a reduction in the size and continuity of high-quality wetland areas, as evidenced by an increase in fragmented patches and edge lengths.

Table 3.1 quantifies the area coverage under different health zones for each model. According to the random forest model, only 13.75% of the wetland area is classified as 'very good', while a substantial 54.8% is deemed 'poor' or 'very poor'. Comparatively, the ANN model identifies 13.31% as 'very good' and a combined 45.7% as 'poor' or 'very poor'. The SVM model appears slightly more optimistic with 18.12% as 'very good', but still 48.96% as 'poor' or 'very poor'. The rotation forest model presents a different distribution with the largest area, 32.16%, classified as 'moderate', and a relatively lower combined percentage of 'poor' and 'very poor' zones at 24.76%. Lastly, the bagging model shows the highest percentage of 'good' wetland at 30.92%, yet still has a significant 45.45% as 'poor' or 'very poor'.

This quantitative analysis reveals the challenges faced by the wetland ecosystem, particularly highlighting the areas that require urgent conservation efforts. The models collectively underscore the vulnerable conditions at the periphery of larger wetlands, in small patchy wetlands, and along the upper reaches of the river. They also emphasize the disparity in health conditions within the wetland, necessitating targeted actions to address the area's most at risk. The ensemble machine learning approach, by integrating multiple modeling techniques, offers a robust tool for environmental monitoring and decision-making.

3.3.4 Accuracy Assessment of the Models

Figure 3.5 displays ROC curves (empirical and binormal curves) for five different models, such as (a) ANN, (b) SVM, (c) RF, (d) rotation forest, and (e) bagging used in assessing the accuracy of wetland habitat health classification.

The empirical ROC curve has an AUC of 0.874, while the binomial ROC curve has an AUC of 0.912 (Fig. 3.5a). The higher AUC of the binomial curve suggests that when the data is modeled assuming a binomial distribution, the model's performance improves. The 90% confidence intervals (CI) for both empirical and binomial ROC curves are fairly tight, which indicates precision in the model's performance estimates. This model shows an empirical AUC of 0.88 and a binomial AUC of 0.93, which are both high, reflecting a strong predictive performance (Fig. 3.5b). The narrow confidence intervals, especially for the binomial ROC curve, imply that the model consistently discriminates between the two classes with high accuracy. Here, the empirical ROC curve has an AUC of 0.861 and the binomial ROC has an AUC of 0.873 (Fig. 3.5c). Both AUCs are slightly lower than those in panels a and b but still indicate a good classification capability. The overlapping confidence intervals suggest that the empirical and binomial models have similar predictive performances for this model. This model possesses an empirical AUC of 0.883 and a binomial AUC of 0.936, indicating excellent classification performance, especially for the binomial ROC curve (Fig. 3.5d). The wide gap between the empirical and binomial AUCs suggests that the binomial assumption significantly enhances the model's discriminatory power. The empirical ROC curve has an AUC of 0.89 and the binomial ROC curve has an AUC of 0.92 (Fig. 3.5d). Both are indicative of high accuracy in the model's predictive ability. The confidence intervals are again quite narrow, reinforcing the reliability of the model's performance.

The ROC analysis for various ensemble machine learning algorithms reveals their efficacy in modeling wetland habitat health, with the rotation forest algorithm (d) emerging as the top performer with a binomial AUC of 0.936, followed by Bagging (e) with a binomial AUC of 0.92, indicating robust predictive power. SVM (b) also shows high accuracy with a binomial AUC of 0.93. ANN (a) and RF (c) demonstrate competent performance with binomial AUCs of 0.912 and 0.873, respectively. The tight confidence intervals across all models underscore the reliability of these methods in

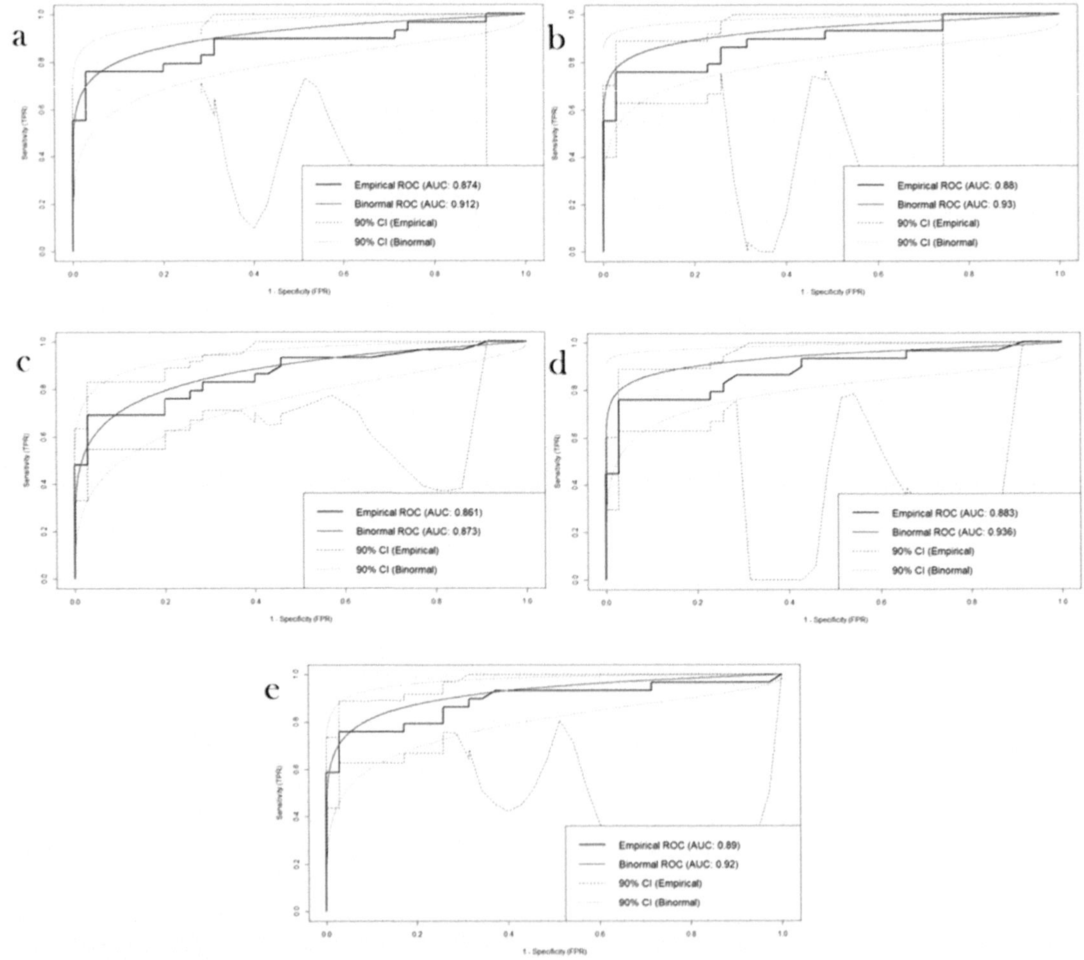

Fig. 3.5 Validation of wetland health condition models using empirical and binormal ROC curve for

accurately classifying the health of wetland habitats, with the binomial ROC generally providing a more optimistic assessment compared to the empirical ROC, suggesting that the binomial assumption may offer a slight edge in classification accuracy.

3.4 Conclusion

This study provided a comprehensive analysis of the ecological changes in the floodplains of the Punarbhaba River, focusing on the consequences of the construction of the Komardanga Dam. Using machine learning modeling, the condition of the wetlands was classified into five categories from 'very good' to 'very poor' based on a range of hydrological parameters, soil characteristics, and water quality. The spatiotemporal analysis highlighted a worrying trend of habitat degradation, particularly in the fringes of the wetland. The significance of this work lies in its integrative approach that combines field data with advanced machine learning algorithms to understand and predict the state of wetland ecosystems. The use of ensemble models such as ANN, SVM, RF, rotation forest, and bagging enabled a robust classification system underpinned by empirical and binormal ROC curve validation, with the rotation forest model showing the highest accuracy. This methodology is a valuable tool for environmental monitoring and management, providing insight into the impact of human intervention on natural habitats. However, the study is not without limitations. The use of remote sensing data can introduce errors due to resolution and sensor limitations, and machine learning models are inherently subject to uncertainty in their predictions. In addition, model performance can vary depending on environmental conditions and datasets. Future research could extend this work by incorporating more diverse data sources, including higher-resolution imagery and real-time surveillance data, to improve modeling accuracy. Further studies could also investigate the integration of socio-economic factors to gain a more holistic understanding of wetland health. Longitudinal studies could examine the long-term impacts of damming and other anthropogenic activities, providing valuable information for policy decisions and conservation efforts. There is also a need to develop adaptive management strategies that can respond to the modeling results and aim to mitigate the negative impacts on wetland ecosystems and conserve their biodiversity.

Appendix: Code for Developing Wetland Health Habitat Model

###Data source

```
from google.colab import drive
drive.mount('/content/drive')
%cd /content/drive/MyDrive/Swades Sir Experiment/
```

###Data import

```
import pandas as pd
df = pd.read_csv('Hydrology.csv')
print(df.columns)


# Assuming your dataset is stored in a DataFrame named df
missing_values = df.isnull().sum()
print(missing_values)
!pip install pyswarm
```

Import libraries

```
import numpy as np
from sklearn.model_selection import train_test_split
from sklearn.preprocessing import StandardScaler
from sklearn.metrics import accuracy_score
from keras.models import Sequential
from keras.layers import Dense, Dropout
from pyswarm import pso
import matplotlib.pyplot as plt
from mpl_toolkits.mplot3d import Axes3D
```

Creation of target and predictor dataset

```
X = df.drop('FID', axis=1)
y = df['FID']
X_train, X_test, y_train, y_test = train_test_split(X, y,
test_size=0.2, random_state=42)
print("X_train shape:", X_train.shape)
print("X_test shape:", X_test.shape)
print("y_train shape:", y_train.shape)
print("y_test shape:", y_test.shape)
```

##############Development of ANN model##############################

```
# Define the model function
def create_model(hidden_units, dropout_rate, activation, optimizer):
    model = Sequential()
    model.add(Dense(hidden_units[0], input_dim=X_train.shape[1],
activation=activation))
```

```
    model.add(Dropout(dropout_rate))
    for units in hidden_units[1:]:
        model.add(Dense(units, activation=activation))
        model.add(Dropout(dropout_rate))
    model.add(Dense(1, activation='sigmoid'))
    model.compile(loss='binary_crossentropy', optimizer=optimizer,
metrics=['accuracy'])
    return model

import matplotlib.pyplot as plt

# Lists to store the values
fitness_values = []
hidden_units_values = []
dropout_rate_values = []

# Cost function for PSO
def cost_func(params):
    hidden_units = int(params[0])
    dropout_rate = params[1]
    model = create_model([hidden_units], dropout_rate, 'relu', 'adam')
    model.fit(X_train, y_train, epochs=10, batch_size=32, verbose=0)
    predictions_proba = model.predict(X_test)
    predictions = (predictions_proba > 0.5).astype('int32')
    fitness = -accuracy_score(y_test, predictions)
    fitness_values.append(fitness)
    hidden_units_values.append(hidden_units)
    dropout_rate_values.append(dropout_rate)
    return fitness

# Constraints for PSO
lb = [5, 0.1] # Lower bounds
ub = [100, 0.5] # Upper bounds
# Use PSO to optimize the hyperparameters
params, _ = pso(cost_func, lb, ub, swarmsize=12, maxiter=12) # You may
need to adjust these parameters
# Create a 3D plot for the fitness value during PSO training
fig = plt.figure(figsize=(12, 10))
ax = fig.add_subplot(111, projection='3d')

# Scatter plot for all iterations
ax.scatter(hidden_units_values, dropout_rate_values, fitness_values,
c='b', marker='o')

# Highlight the best parameters
ax.scatter([best_hidden_units], [best_dropout_rate],
[min(fitness_values)], c='r', marker='*', s=100)
```

```
# Set labels and title
ax.set_xlabel('Hidden Units', labelpad=10)
ax.set_ylabel('Dropout Rate', labelpad=10)
ax.set_zlabel('Fitness Value (Negative Accuracy)', labelpad=10)
ax.set_title('3D Plot of Fitness Value During PSO Training')
plt.savefig('3DANN.png')
# Show the plot
plt.show()
# Extract the best hyperparameters
best_hidden_units, best_dropout_rate = int(params[0]), params[1]
# Print the best hyperparameters and accuracy outside of the plot
print('Best Hyperparameters:')
print('Hidden Units:', best_hidden_units)
print('Dropout Rate:', best_dropout_rate)
print('Final Accuracy:', accuracy)

# Train the best model
best_model = create_model(hidden_units=[best_hidden_units],
dropout_rate=best_dropout_rate,
                              activation='relu', optimizer='adam')
history = best_model.fit(X_train, y_train, batch_size=32, epochs=20)
# Evaluate the model
predictions_proba = best_model.predict(X_test)
predictions = (predictions_proba > 0.5).astype('int32') # Apply
threshold
accuracy = accuracy_score(y_test, predictions)
print('Final Accuracy:', accuracy)
```

```
###############################RF####################################################

from sklearn.ensemble import RandomForestClassifier
from sklearn.tree import plot_tree
from sklearn.impute import SimpleImputer
# Impute missing values with the mean of the column
imputer = SimpleImputer(strategy='mean')
X_train_imputed = imputer.fit_transform(X_train)
X_test_imputed = imputer.transform(X_test)
# Lists to store the values
fitness_values = []

# Cost function for PSO
# Update the cost function to use the imputed data
def cost_func(params):
    n_estimators, max_depth, min_samples_split = map(int, params)
    model = RandomForestClassifier(n_estimators=n_estimators,
max_depth=max_depth, min_samples_split=min_samples_split,
random_state=42)
    model.fit(X_train_imputed, y_train)
```

```
    predictions = model.predict(X_test_imputed)
    fitness = -accuracy_score(y_test, predictions)
    fitness_values.append(fitness)
    return fitness

# Constraints for PSO
lb = [10, 2, 2] # Lower bounds
ub = [200, 20, 10] # Upper bounds
# Use PSO to optimize the hyperparameters
params, _ = pso(cost_func, lb, ub, swarmsize=12, maxiter=12)
# Extract the best hyperparameters
best_params = [int(p) for p in params]
best_n_estimators, best_max_depth, best_min_samples_split = best_params

# Print the best parameters
print('Best Parameters:')
print('Number of Estimators:', best_n_estimators)
print('Max Depth:', best_max_depth)
print('Min Samples Split:', best_min_samples_split)
# Train the best model
best_model1 = RandomForestClassifier(n_estimators=best_n_estimators,
max_depth=best_max_depth, min_samples_split=best_min_samples_split,
random_state=42)
best_model1.fit(X_train, y_train)

# Evaluate the model
predictions = best_model.predict(X_test)
accuracy = accuracy_score(y_test, predictions)
print('Final Accuracy:', accuracy)
from matplotlib import pyplot as plt

# Plot one tree from the Random Forest
plt.figure(figsize=(20, 10))
plot_tree(best_model1.estimators_[0], filled=True,
feature_names=df.drop('FID', axis=1).columns, class_names=['Class 0',
'Class 1'], rounded=True, fontsize=5) # Increase fontsize as needed
plt.title('Random Forest - Tree Visualization', fontsize=16) # Increase
title fontsize as needed
# Save the plot to a file
plt.savefig('RF_tree.png')
plt.show()

#####################SVM##########################################################

from pyswarm import pso
from sklearn.svm import SVC
```

```
# Lists to store the values
fitness_values = []
C_values = []
gamma_values = []

# Cost function for PSO
def cost_func(params):
    C, gamma = params
    model = SVC(C=C, gamma=gamma, kernel='rbf', random_state=42)
    model.fit(X_train_imputed, y_train)
    predictions = model.predict(X_test_imputed)
    fitness = -accuracy_score(y_test, predictions)
    fitness_values.append(fitness)
    C_values.append(C)
    gamma_values.append(gamma)
    return fitness

# Constraints for PSO
lb = [0.1, 0.001] # Lower bounds
ub = [10, 1] # Upper bounds

# Use PSO to optimize the hyperparameters
params, _ = pso(cost_func, lb, ub, swarmsize=10, maxiter=10)
# Extract the best hyperparameters
best_C, best_gamma = params

# Print the best parameters
print('Best Parameters:')
print('C:', best_C)
print('Gamma:', best_gamma)
# Train the best model
best_model = SVC(C=best_C, gamma=best_gamma, kernel='rbf',
random_state=42)
best_model.fit(X_train_imputed, y_train)
# Evaluate the model
predictions = best_model.predict(X_test_imputed)
accuracy = accuracy_score(y_test, predictions)
print('Final Accuracy:', accuracy)
# Plot the fitness values
plt.plot(fitness_values)
plt.xlabel('Iteration')
plt.ylabel('Fitness Value (Negative Accuracy)')
plt.title('Fitness Value During PSO Training')
# Save the plot to a file
plt.savefig('SVM_PSO1.png')
plt.show()
```

################Bagging##

```
from sklearn.ensemble import BaggingClassifier
from sklearn.tree import DecisionTreeClassifier

# Lists to store the values
fitness_values = []
n_estimators_values = []
max_samples_values = []  # Add this line
def cost_func(params):
    n_estimators, max_samples, max_features_fraction = params
    max_features = max(int(max_features_fraction *
X_train_imputed.shape[1]), 1) # Ensure max_features is at least 1
    base_estimator = DecisionTreeClassifier()
    model = BaggingClassifier(base_estimator=base_estimator,
n_estimators=int(n_estimators), max_samples=max_samples,
max_features=max_features, random_state=42)
    model.fit(X_train_imputed, y_train)
    predictions = model.predict(X_test_imputed)
    fitness = -accuracy_score(y_test, predictions)
    fitness_values.append(fitness)
    n_estimators_values.append(n_estimators)  # Add this line
    max_samples_values.append(max_samples)  # Add this line
    return fitness

# Constraints for PSO
lb = [5, 0.1, 0.01] # Lower bounds
ub = [100, 1, 0.99] # Upper bounds for max_features_fraction
# Use PSO to optimize the hyperparameters
params, _ = pso(cost_func, lb, ub, swarmsize=15, maxiter=15)
# Extract the best hyperparameters after PSO
best_n_estimators, best_max_samples, best_max_features_fraction =
params
best_n_estimators = int(best_n_estimators)  # Convert to integer
best_max_features = max(int(best_max_features_fraction *
X_train_imputed.shape[1]), 1) # Compute max_features for best model

# Print the best hyperparameters
print('Best Hyperparameters:')
print('Number of Estimators:', best_n_estimators)
print('Max Samples:', best_max_samples)
print('Max Features Fraction:', best_max_features_fraction)
print('Max Features:', best_max_features)
# Train the best model
base_estimator = DecisionTreeClassifier()
best_model = BaggingClassifier(base_estimator=base_estimator,
n_estimators=best_n_estimators, max_samples=best_max_samples,
max_features=best_max_features, random_state=42)
best_model.fit(X_train, y_train)
# Evaluate the model

predictions = best_model.predict(X_test_imputed)
accuracy = accuracy_score(y_test, predictions)
print('Final Accuracy:', accuracy)
```

References

Ahmed IA, Talukdar S, Shahfahad, Parvez A, Rihan M, Baig MRI, Rahman A (2022) Flood susceptibility modeling in the urban watershed of Guwahati using improved metaheuristic-based ensemble machine learning algorithms. Geocarto Int 37(26):12238–12266

Alqadhi S, Mallick J, Talukdar S, Bindajam AA, Saha TK, Ahmed M, Khan RA (2022) Combining logistic regression-based hybrid optimized machine learning algorithms with sensitivity analysis to achieve robust landslide susceptibility mapping. Geocarto Int 37(25):9518–9543

Cui L, Ge Z, Yuan L, Zhang L (2015) Vulnerability assessment of the coastal wetlands in the Yangtze Estuary, China to sea-level rise. Estuar Coast Shelf Sci 156:42–51

Das RT, Pal S (2017) Exploring geospatial changes of wetland in different hydrological paradigms using water presence frequency approach in Barind Tract of West Bengal. Spat Inf Res 25:467–479

Davidson NC, Fluet-Chouinard E, Finlayson CM (2018) Global extent and distribution of wetlands: trends and issues. Mar Freshw Res 69(4):620–627

Debanshi S, Pal S (2020) Effects of water richness and seasonality on atmospheric methane emission from the wetlands of deltaic environment. Ecol Ind 118:106767

Finlayson CM, Davies GT, Moomaw WR, Chmura GL, Natali SM, Perry JE et al (2019) The second warning to humanity—providing a context for wetland management and policy. Wetlands 39:1–5

Huang F, Cao Z, Guo J, Jiang SH, Li S, Guo Z (2020) Comparisons of heuristic, general statistical and machine learning models for landslide susceptibility prediction and mapping. CATENA 191:104580

Islam ARMT, Talukdar S, Mahato S, Kundu S, Eibek KU, Pham QB et al (2021) Flood susceptibility modelling using advanced ensemble machine learning models. Geosci Front 12(3):101075

Islam ARMT, Talukdar S, Akhter S, Eibek KU, Rahman MM, Pal S et al (2022) Assessing the impact of the Farakka Barrage on hydrological alteration in the Padma River with future insight. Sustainability 14(9):5233

Khatun R, Talukdar S, Pal S, Kundu S (2021) Measuring dam induced alteration in water richness and eco-hydrological deficit in flood plain wetland. J Environ Manage 285:112157

Kingsford RT (2000) Ecological impacts of dams, water diversions and river management on floodplain wetlands in Australia. Austral Ecol 25(2):109–127

Lefebvre A, Devreker D (2020) First comprehensive quantitative multi-parameter assessment of the eutrophication status from coastal to marine French waters in the English channel, the Celtic Sea, the Bay of Biscay, and the Mediterranean Sea. J Mar Sci Eng 8(8):561

Malekmohammadi B, Jahanishakib F (2017) Vulnerability assessment of wetland landscape ecosystem services using driver-pressure-state-impact-response (DPSIR) model. Ecol Ind 82:293–303

Merghadi A, Yunus AP, Dou J, Whiteley J, ThaiPham B, Bui DT et al (2020) Machine learning methods for landslide susceptibility studies: a comparative overview of algorithm performance. Earth Sci Rev 207:103225

Mitsch WJ, Gosselink JG (2015) Wetlands. Wiley

Mitsch WJ, Gosselink JG, Zhang L, Anderson CJ (2009) Wetland ecosystems. Wiley

Mitsch WJ, Bernal B, Nahlik AM, Mander Ü, Zhang L, Anderson CJ et al (2013) Wetlands, carbon, and climate change. Landscape Ecol 28:583–597

Mosavi A, Sajedi-Hosseini F, Choubin B, Taromideh F, Rahi G, Dineva AA (2020) Susceptibility mapping of soil water erosion using machine learning models. Water 12(7):1995

Nielsen DL, Merrin LE, Pollino CA, Karim F, Stratford D, O'Sullivan J (2020) Climate change and dam development: effects on wetland connectivity and ecological habitat in tropical wetlands. Ecohydrology 13(6):e2228

Pal S, Paul S (2020) Assessing wetland habitat vulnerability in moribund Ganges delta using bivariate models and machine learning algorithms. Ecol Ind 119:106866

Pal S, Sarda R (2021) Measuring the degree of hydrological variability of riparian wetland using hydrological attributes integration (HAI), histogram comparison approach (HCA) and range of variability approach (RVA). Ecol Ind 120:106966

Pal S, Talukdar S (2018) Drivers of vulnerability to wetlands in Punarbhaba river basin of India-Bangladesh. Ecol Ind 93:612–626

Pal S, Sarkar R, Saha TK (2022) Exploring the forms of wetland modifications and investigating the causes in lower Atreyee river floodplain area. Eco Inform 67:101494

Rihan M, Bindajam AA, Talukdar S, Naikoo MW, Mallick J, Rahman A (2023) Forest fire susceptibility mapping with sensitivity and uncertainty analysis using machine learning and deep learning algorithms. Adv Space Res 72(2):426–443

Saha TK, Pal S (2019) Exploring physical wetland vulnerability of Atreyee river basin in India and Bangladesh using logistic regression and fuzzy logic approaches. Ecol Ind 98:251–265

Saha TK, Pal S, Talukdar S, Debanshi S, Khatun R, Singha P, Mandal I (2021) How far spatial resolution affects the ensemble machine learning based flood susceptibility prediction in data sparse region. J Environ Manage 297:113344

Saha TK, Pal S, Sarda R (2022) Impact of river flow modification on wetland hydrological and morphological characters. Environ Sci Pollut Res 29(50):75769–75789

Shahfahad, Talukdar S, Das T, Naikoo MW, Rihan M, Rahman A (2022) Forest fire susceptibility mapping by integrating remote sensing and machine learning algorithms. In: Advances in remote sensing for forest monitoring, pp 179–195

Suwal N, Kuriqi A, Huang X, Delgado J, Młyński D, Walega A (2020) Environmental flows assessment in Nepal: the case of Kaligandaki River. Sustainability 12(21):8766

Talukdar S, Pal S (2019) Effects of damming on the hydrological regime of Punarbhaba river basin wetlands. Ecol Eng 135:61–74

Talukdar S, Pal S (2020) Wetland habitat vulnerability of lower Punarbhaba river basin of the uplifted Barind region of Indo-Bangladesh. Geocarto Int 35(8):857–886

Talukdar S, Pal S, Chakraborty A, Mahato S (2020) Damming effects on trophic and habitat state of riparian wetlands and their spatial relationship. Ecol Ind 118:106757

Talukdar S, Roy SK, Sarkar SK, Mahato S, Pal S, Rahman A et al (2022) Application of hybrid machine learning algorithms for flood susceptibility modeling. In: Spatial modelling of flood risk and flood hazards: societal implications. Springer International Publishing, Cham, pp 105–118

Wantzen KM, Alves CBM, Badiane SD, Bala R, Blettler M, Callisto M et al (2019) Urban stream and wetland restoration in the Global South—a DPSIR analysis. Sustainability 11(18):4975

Yang X, Liu S, Jia C, Liu Y, Yu C (2021) Vulnerability assessment and management planning for the ecological environment in urban wetlands. J Environ Manage 298:113540

Zhang X, Dong Z, Gupta H, Wu G, Li D (2016) Impact of the Three Gorges Dam on the hydrology and ecology of the Yangtze River. Water 8(12):590

Impact of Inundation Inconsistency on Temperature in Seasonally Inundated Floodplain Wetland

4

Rumki Khatun, Somen Das, and Swades Pal

Abstract

Hydrological consistency analysis in wetland area was widely explored but how does it control water surface temperature was not explored yet, but this analysis is ecologically very crucial. The present study aimed to analyze how inundation consistency gradient in seasonally inundated wetland could control thermal character of the wetland tract taking lower catchment of Punarbhaba River Basin as an example. For inundation consistency analysis, water presence frequency (WPF) and for water surface temperature analysis, land surface temperature (LST) method was applied. For linking them linear regression and ordinary least square (OLS) regression were employed. The entire study period (1990–2022) was subdivided into three phases and analysis was done based on these three phases. Result clearly revealed that over the phases wetland area was substantially reduced (60%) and existing wetland area was witnessed increasing inundation inconsistency in wetland. In phase 1, 47.26% wetland area was characterized by high WPF but in phase 3, no area was detected under this class indicating growing inundation inconsistency. Maximum, minimum and mean water surface temperature was enhanced by 6, 17 and 4.63 °C from phases 1 to 3. Temperature was found quite high in the shallower part of the wetland outskirt. Linear regression between WPF and temperature revealed statistically significant control of WPF on reducing temperature in all the phases. OLS also exhibited same over 73% sample wetland sites. So, it can be inferred that growing inundation inconsistency may increase threat to the thermal ambience of the aquatic ecosystem and neighbouring human ecosystem.

Keywords

Inundation inconsistency gradient · Seasonal floodplain wetland · Punarbhaba river basin · Water surface temperature · Ordinary least square regression · Inundation transformation and temperature

R. Khatun · S. Das
Department of Geography, Kazi Nazrul University, Asansol, West Bengal, India

S. Pal (✉)
Department of Geography, University of Gour Banga, Malda, West Bengal, India
e-mail: swadeshpal82@gmail.com

4.1 Introduction

Linkage between blue space and temperature moderation is well-explored across the world (Völker et al. 2013; Song et al. 2022; Ampatzidis

S. Talukdar et al. (eds.), *Water Resource Management in Climate Change Scenario*, GIScience and Geo-environmental Modelling, https://doi.org/10.1007/978-3-031-61121-6_4

and Kershaw 2020). Hathway and Sharples (2012) reported that presence of river, water body within urban area supplies moisture to the atmosphere through evaporation leading to decrease of temperature during summer period. Hydrological types like seasonal and perennial wetland play crucial role towards temperature control (Lee et al. 2015; Junk et al. 2013). Perennial wetland reduces temperature during the summer and raises the temperature during the winter seasons at the neighbouring areas, particularly in the monsoon climate (Xue et al. 2019; Neubauer and Verhoeven 2019). Seasonal wetland only seasonally controls thermal environment (Schedlbauer et al. 2011). At broad scale, coastal area witnesses moderate climate due to presence of ocean at very proximity and exchange of heat through land and sea breezes (Davis and Fitzgerald 2009). Urban planners, designers therefore, treat such environmental component as crucial for mediating urban climate. Loss of blue space is a big issue not only in the urban and per-urban area but also in the rural environment (Let and Pal 2023). Several studies have noted steady loss of blue space over time primarily as a result of agriculture and urbanization encroachment that are major drivers of wetland loss (Pal et al. 2023; Bratley and Ghoneim 2018). Peng et al. (2021) reported built area encroachment and blue space reclamation. Although Let and Pal (2023) reported agriculture encroachment is the major driver of wetland loss in the economically less advance countries like India, Bangladesh, Thailand, etc., however, the former driver is rather active in the rural areas while later in the urban areas (Rojas et al. 2019; Mao et al. 2018; McCauley et al. 2013). Loss of wetland area means less chance of evaporation and controlling air movement (Mohamed et al. 2004). At the same time, wetland areas, those are reclaimed by other land uses which have greater potentiality of enhancing temperature. Fragmentation of wetland area means increasing edge area ratio leading to enhancing human landscape effects on the wetland (Jiang et al. 2014; Knutson et al. 1999). Such conditions can also increase thermal intensity within wetland scape.

Wetland water depth is another established driver which can control water temperature. Khatun and Pal (2021) worked on the hydrological effect dynamics on suitable fish habitat and pointed out that due to damming water availability within wetland was declined in the flood plain river basin of Barind tract of India and Bangladesh. It causes decreasing water depth. Such new hydrological paradigm is caused for increasing thermal condition in summer season. They identified due to decreasing water depth, thermal hostility was inflated adverting the fishing environment over larger part of the wetland. Kundu et al. (2021) also reported the same in this region.

From the above studies it is very evident that presence or absence of wetland, areal shrinkage, shallowing water depth, conversion of larger wetland into smaller wetland patches found some major drivers which can control thermal environment. Since monsoon climatic region is characterized by seasonal dry and wet spell, most of the non-ice melting rivers and flood plain wetlands concerned are seasonally inundated (Junk et al. 2013). Generally during June–September (monsoon months) and October–November (post-monsoon months), wetlands retain water (Dash et al. 2007). Rest periods of a year wetlands and rivers become dry except in some cases where ground water support is very strong. Moreover, due to highly arrhythmic rainfall pattern during monsoon season, water presence and water availability within seasonally inundated wetland is very irregular. Many studies carried out by Talukdar and Pal (2019) proved this irregular behaviour of water appearance within flood plain wetlands of monsoon climate. Is this irregular seasonal water appearance within seasonally inundated flood plain wetland matter on thermal environment? Is a wetland patch with consistent water appearance register relatively lesser temperature than the wetland patches that record highly inconsistent water presence? The current study attempted to investigate the influence of water presence frequency in a seasonally inundated wetland on temperature. No such direct studies are available in this issue but this is crucially important for receiving insight into the linkage between hydrothermal environment controlled by the water presence frequency/consistency and ecological ekistics.

4.2 Study Area

Punarbhaba River Basin of Barind tract in India and Bangladesh possesses a wider tract of seasonally inundated floodplain wetland covering about 85.99 km^2 as of 2022. Seasonal dynamics of wetland area is very prominent majorly driven by the seasonal rainfall fluctuation. The total annual rainfall in this region is 1500 mm, with approximately 80% falling during the monsoon season (June–September). Since the wetlands are mainly driven by irregular rainfall amount during monsoon months, areal dynamics at temporal scale is very prominent. The damming over Punarbhaba and Tangan River Basin in 1992 and 1989 are caused for major hydrological modification of this river basin leading to significant squeeze of wetland area (Pal and Talukdar 2018) (Fig. 4.1).

4.3 Materials and Methods

The Landsat 4–5 thematic mapper (TM) and 8 operational land imager (OLI) data (path/row: 139/43 and spatial resolution: 30 m) were retrieved from the USGS Earth Explorer website (https://earthexplorer.usgs.gov) from 1990 to 2022. The images from 1990 to 2022 were employed for wetland consistency mapping, and 25 field-based water depth data were used to confirm the results. For extracting the land surface temperature, the same images were used.

4.3.1 Wetland Mapping and Validation

There are numerous spectral indices produced from satellite imageries that can be used to identify water bodies. The Automated Water Extraction Index (AWEI) (Feyisa et al. 2014), Water Ratio Index (WRI) (Shen and Li 2010), RmNDWI (Debanshi and Pal 2020), and NDWI (McFeeters 1996) are few crucial techniques. The best method for locating water bodies in the Barind region of Bangladesh and India, according to Das and Pal (2018), is NDWI (Eq. 4.1). This information led to the current investigation's use of the NDWI technique to extract water bodies and water intensities.

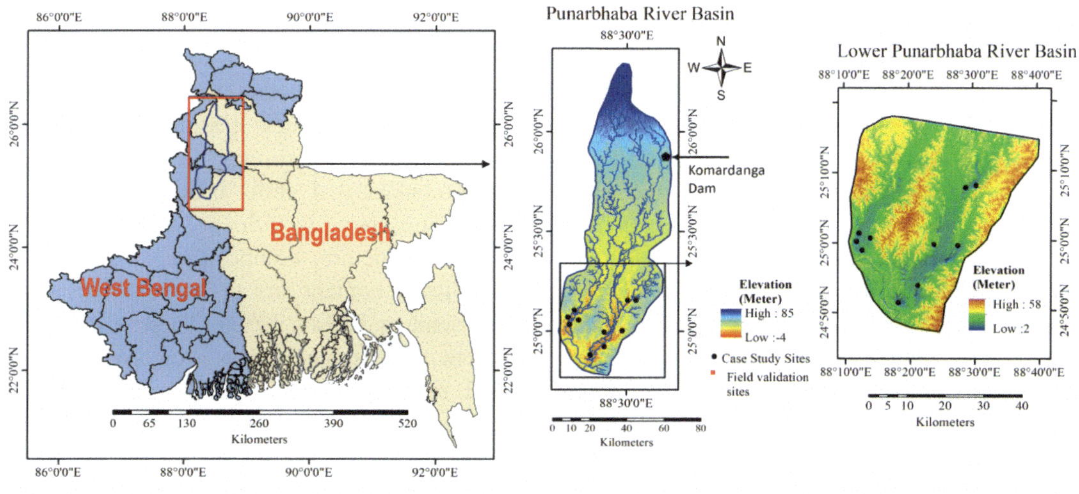

Fig. 4.1 Location of the study area

$$\text{NDWI} = \frac{b\text{green} - b\text{NIR}}{b\text{green} + b\text{NIR}} \quad (4.1)$$

The NDWI value varies between − 1 and 1. Water and non-water pixels are indicated by positive and negative values, respectively (Guo et al. 2017). Greater availability or water depth are indicated by higher positive values (Gao 1996). Since 1990–2022, water pixels were extracted using the NDWI approach from time series satellite images taken during the October and November (post-monsoon season). The consistency of water appearance map using water presence frequency (WPF) approach was created for three stages, 1990–1999 (Phase 1); 2000–2009 (Phase 2); and 2010–2022 (Phase 3), using the time series data of water bodies retrieved from satellite images. 105–231 samples from the respective years' Google Earth images were used, the time sequence NDWI images (2005–2022) were verified using kappa coefficient. 55 sites were chosen for the years before 2005 that contained wetlands in NDWI images. After that data about the water appearance during those times was gathered from relevant stakeholders and local residents. The time series wetland maps were validated by computing the kappa coefficient (K) (Eq. 4.2).

$$K = \frac{N\sum_{i-1}^{r} Xii - \sum_{i-1}^{r}(Xi + x + i)}{N^2 - \sum_{i-1}^{r}(Xi + x + i)} \quad (4.2)$$

4.3.2 Inundation Frequency Gradient of the Wetland

According to Borro et al. (2014), the frequency of water in a pixel over time is referred to as its water presence frequency. For all the years (1990 through 2022), Landsat images are used to construct NDWI maps, which show how frequently water is visible. Each map is turned into a binary map, with 1 designating wetland and 0 representing other types of pixels. Binary wetland maps from time series were included and transformed to percentages using Eq. (4.3). WPF value is between 0 and 100% indicating inconsistent to consistent water appearance. If water is present in a pixel in all the years throughout the chosen time period, the WPF value will be 100%. A wetland with a higher WPF rating indicates consistent water presence. The value 0 or near to 0 indicates inconsistent water appearance.

$$\text{WPF} = \frac{\sum_{n=1}^{i} Xj}{N} \quad (4.3)$$

4.3.3 Land Surface Temperature (LST) Derivation from Satellite Images and Validation

When an object's temperature rises above absolute zero (K), it emits thermal electromagnetic radiation. According to this idea, LST is calculated for various objects. Transforming digital numbers (DN) to spectral radiation (L), spectral radiation (L) to at-satellite brightness temperatures (TB), and LST from Kelvin to degrees Celsius is a simple procedure for calculating land surface temperature (LST). Numerous studies (Chatterjee et al. 2017; Asgarian et al. 2015; Weng and Fu 2014) provided detailed illustrations of this methodology.

4.3.3.1 Emissivity and Fractional Vegetation Cover (FVC) Derivation Method

Every substance on Earth is, in theory, composed of molecules that include atoms connected by molecular bonds. The atoms at the end of a bond vibrate when the infrared spectrum hits a molecule. In a nutshell, the molecule continually emits the same light spectrum. Every molecule vibrates at a unique frequency, and the surface determines the make-up of a natural contact that emits infrared light. Land surface emissivity is influenced by aspects of the soil's structure, composition, organic matter, moisture content and vegetation cover (Van De Griend and Owe 1993). There are several ways to forecast land surface emissivity using NDVI readings. In this portion, Sobrino and Raissouni (2000) first introduced the NDVI thresholding method, abbreviated NDVITHM. The method for deriving emissivity values from

NDVI in various scenarios was described. As an illustration, consider the following examples: (i) NDVI 0.2; in this case, the pixel is assumed to be bare soil, and the emissivity is calculated from the reflectivity values in the red region; (ii) NDVI>0.5; pixels with NDVI values higher than 0.5 are assumed to be fully vegetated; and (iii) 0.2 NDVI 0.5; the pixel is comprised of bare soil and vegetation. To determine emissivity, apply Eq. (4.4) as follows:

$$\varepsilon = \varepsilon_v P_v + \varepsilon_s (1 - P_v) + d\varepsilon \tag{4.4}$$

The Carlson and Ripley (1997) equation can be used to compute the vegetation proportion P_v, where v is the vegetation emissivity and s is the soil emissivity.

$$P_v = \left(\frac{NDVI - NDVI_{min}}{NDVI_{max} - NDVI_{min}} \right) \tag{4.5}$$

The range of land surface emissivity is 0.97–0.99, and the final formula for LSE is provided in Eq. (4.6).

$$\varepsilon = 0.004 * P_v + 0.986 \tag{4.6}$$

Based on the proportion of vertically projected vegetation area to total ground area, fractional vegetation cover (FVC) is a crucial quantitative indication of the status of vegetation coverage (Bian et al. 2017). Many environmental and climatic modelling applications, such as land surface models, hydrological models, soil erosion models, and weather-related models rely on FVC (Mallick et al. 2008; Zhao et al. 2014). FVC is typically extracted from remotely sensed data (Li et al. 2017; Jia et al. 2016). This number typically varies from 0 to 1, with 1 indicating greater greenery intensity and integrity and 0 indicating bare ground.

4.3.3.2 Digital Number (DN) to Spectral Radiance (*L*) Conversion

When an object's temperature exceeds absolute zero (*K*), thermal electromagnetic radiation starts to be released from it. The signals from the thermal sensors (ETM+) can be converted to at-sensor radiance using this approach. The spectral radiance (*L*) is calculated using Eq. (4.7) (Landsat Project Science Office 2002).

$$L\lambda = \text{“gain”} * \text{QCAL} + \text{“offset”} \tag{4.7}$$

where DN is the digital number associated with a particular pixel, "gain" is the slope of the radiance/DN conversion function, and bias is the intercept. Also stated is as follows:

$$L\lambda = L\min\lambda + \left[\{ (L\max\lambda - L\min\lambda) / (\text{QCAL}_{max} - \text{QCAL}_{min}) \} * \text{QCAL} \right] \tag{4.8}$$

where QCAL = pixel digital number (DN), QCALMIN = 0, and QCALMAX = 255. The LMIN and LMAX spectral radiances for band 6 are at digital digits 0 and 255, respectively. These measure, respectively, are 3.2 and 12.65 W m^{-2} sr. Equation (4.9) is produced by substituting the appropriate integers in Eq. (4.8).

$$L\lambda = (0.037059 * DN) + 3.2 \tag{4.9}$$

4.3.3.3 Spectral Radiance (*L*) to At-Satellite Brightness Temperatures (TB) Conversion

Based on the kind of land cover, the radiant temperatures have been adjusted for emissivity. According to Nichol (1994), vegetated regions have an average value of 0.95 and non-vegetated areas have an average value of 0.92. An emissivity adjusted surface temperature was determined after Artis and Carnahan (1982).

$$T_b = \frac{K_2}{\ln\left(\frac{K_1}{L\lambda} + 1\right)} \tag{4.10}$$

where L = spectral radiance in W m$-^2$ sr^{-1} m$-^1$, and TB = at-satellite brightness temperature (K). Two calibration parameters used before launch are K1 and K2 = K1 and K2. These compute to 1282.71 K and 666.09 W m^{-2} sr$-^1$ m, respectively, for the Landsat 7 ETM+ 6.2 band.

4.3.3.4 Land Surface Temperature (LST)

The collected temperature readings are set against a black body in the previous section. As a result, spectral emissivity modifications are required. These can be carried out based on the kind of land cover (Snyder et al. 1998) or by calculating the corresponding emissivity values for each pixel from the NDVI readings. The land surface temperatures (LST) have been calculated in accordance with Artis and Carnahan (1982).

$$\mathrm{LST} = T_{\mathrm{B}}/\left[1 + \{(\lambda * \mathrm{TB}/\rho) * \mathrm{In}\varepsilon\}\right] \quad (4.11)$$

where, LST = land surface temperature (LST) in Kelvin, λ = wavelength of emitted radiance in metres (for which the peak response and the average of the limiting wavelengths (λ = 11.5 μm) (Markham and Barker 1985) is used, $\rho = h * c/\sigma$ (1.438×10^{-2} m K), σ = Boltzmann constant (1.38×10^{8} m/s) $\times 10^{-23}$ J/K), h = Planck's constant ($6.626 \times 10{-}^{34}$ J s), and c = velocity of light (2.998) and ε = emissivity (ranges between 0.97 and 0.99). For calculating emissivity of land surface Eq. (4.11) is used. For simplicity, the above-derived LST unit is converted to degrees Celsius using the Kelvin–degree-Celsius relationship (0 °C = 273.15 K). Degree Celsius data is very standard for determining temperature intensity.

4.3.4 Compositing and Averaging of LST Images

The average of a specific time period is not represented by a single date's image state (Mondal and Pal 2018). In order to provide a single, accurate representation of a given season, it is therefore preferable to combine numerous dates both in a single year and over several years. The credibility of the image will be increased in addition to being taken into consideration (Mondal and Pal 2018). But there is a data scarcity in the study considered. For this, we composited the images separately for the year of 1990–1999 (phase 1), 2000–2009 (phase 2) and 2010–2022 (phase 3).

$$\mathrm{WLC} = \sum_{j=1}^{n} a_{ij} w_j \quad (4.12)$$

where a_{ij} is the jth attribute's ith rank and w_j is the attribute's weight.

4.3.5 Relation Between Inundation Frequency Gradient and LST

In order to establish a spatial link between WPF and LST, ordinary least square (OLS) regression was performed. An effective method for calculating the coefficients of a linear regression equation that illustrates the link between one or more independent variables and a dependent variable is the ordinary least squares (OLS) method. In this case, wetland pixels were converted into point data and used to access both the WPF and LST levels for information. WPF was chosen as an independent variable and LST as a dependent variable for OLS regression in order to stick with the OLS methodology. A typical regression residual was used to demonstrate how much influence WPF had over the land surface temperature. A low residual value (Eq. 4.13) denotes a considerable influence of the independent variable on the dependent variable.

$$\hat{\beta} = \left(X^{\mathrm{T}}X\right)^{-1}X^{\mathrm{T}}y \quad (4.13)$$

where Y is the vector containing the value of the response variable, X, T is the transposed matrix, X is the matrix regression variable, and $\hat{\beta}$ is the ordinary least square estimation.

4.4 Result

4.4.1 Changing Scenario of Wetland

NDWI-based wetland area delineation from Landsat satellite image in different years clarified that at the lower part of rivers Punarbhaba and Tangon, prominent flood plain wetland tracts are found. Flooding during monsoon months shows spread of water cover area and found almost dry except some small patches

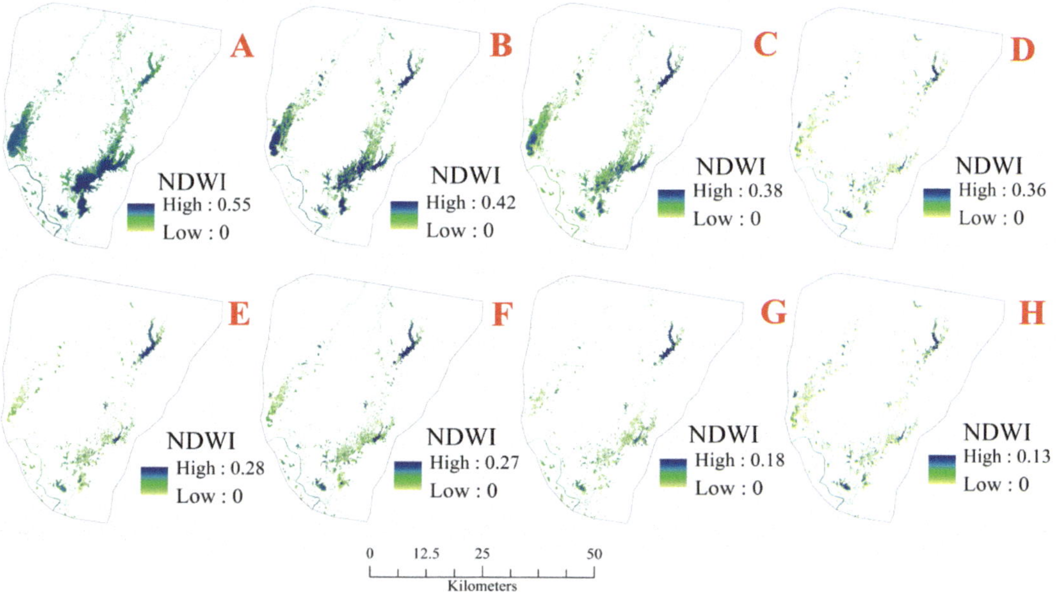

Fig. 4.2 NDWI maps of different years **a** 1990, **b** 1995, **c** 2000, **d** 2005, **e** 2010, **f** 2015, **g** 2020, **h** 2022

during pre-monsoon season due to rainlessness and excessive evaporation. Figure 4.2 shows the wetland area of post-monsoon season in 1990, 1995, 2000, 2005, 2010, 2015, 2020 and 2022. In 1990, total wetland area was 219.86 km^2 which was declined to 85.99 km^2 in 2022. Not only fringe area, but core wetland tract also witnessed remarkable wetland loss. Validation of wetland maps using Google Earth and field-based reference sites proved the validity of the wetland maps. The range of the kappa coefficient over all years is between 0.81 and 0.92, indicating very good to excellent agreement between the map and the ground truth.

4.4.2 Consistency in Inundation Frequency Gradient

Inundation frequency (IF or WPF) gradient at three consecutive phases represents very prominent change. IF at each phase was classified into five consistency classes like 0–20, 20–40, 40–60, 60–80 and 80–100% (Fig. 4.3). Higher IF signifies consistency in inundation in different years and lower IF contrarily signifies inconsistency in wetland inundation. In phase 1 (1990–1999), 54.18% wetland area witnessed high to very high (60–100%) consistency in IF and 24.31% area recorded low to very low consistency in IF. The condition was adversely changed in two successive phases. In phase 3 (2000–2022), no area was found with high IF and conversely, area under low IF consistency was enhanced to 67.78%. Over the recent phase not only the wetland area was squeezed due to water scarcity and land use transformation, existing wetland also witnessed growing inconsistency in IF. This change is against of ecological viability of the wetland.

4.4.3 Spatio-Temporal Temperature Changes

Landsat image-driven LST of different phases have been represented in Fig. 4.4. This map shows clear trend of mean, maximum and minimum temperature rise in the wetland area. For instance, in phase 1 LST was ranged from 4.19 to 20.43 °C with a mean of 19.46 °C and this range was enhanced from 21.86 to 26.92 °C

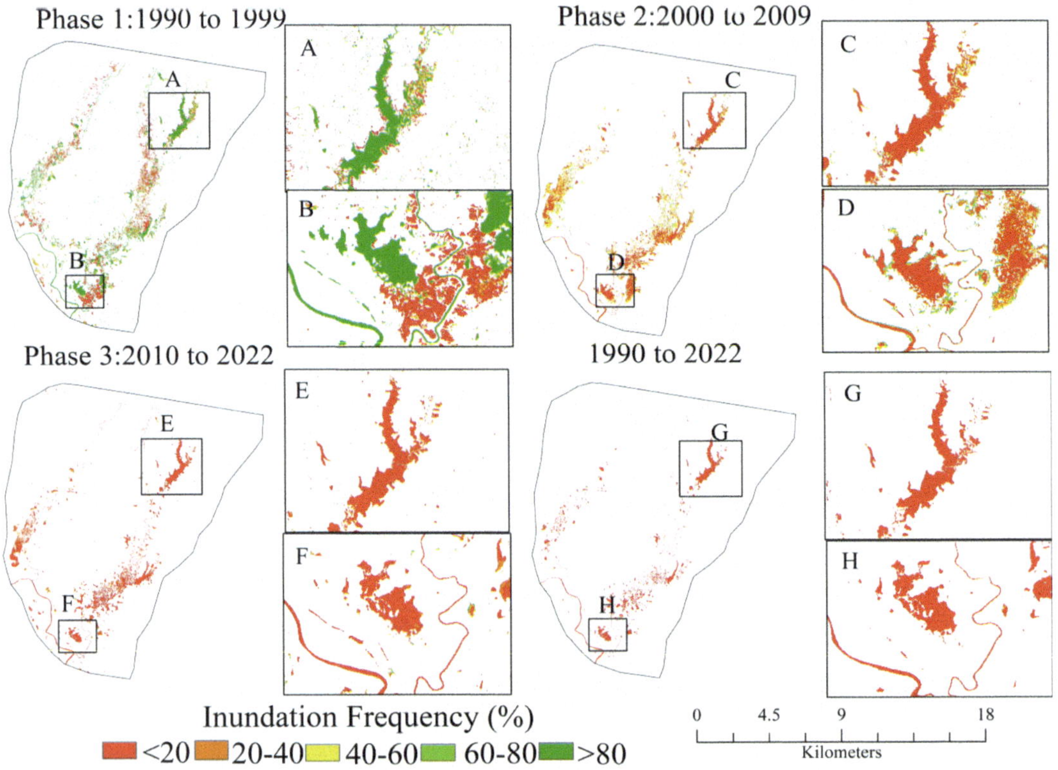

Fig. 4.3 Inundation frequency (IF) of wetland in different phases (1) phase 1 (1990–1999), (2) phase 2 (2000–2009) and (3) phase 3 (2010–2022) and (4) IF of entire period (1990–2022)

with a mean of 24.09 °C in phase 3. It shows considerably high increase of wetland water surface temperature both at upper and lower limits. The degree of change was found considerably high at its lower limit (17.67 °C) than upper limit (6.49 °C) since 1990–2022. Larger wetland and core wetland parts were registered relatively low temperature in phase 1 which was even found more in phase 3. Fringe wetlands parts recorded high temperature during phase 1 were mostly wiped out in successive phases.

Water surface temperature of all the individual years was not validated due to lack of historical spatial data but validation was done for the years of 2020, 2021 and 2022 based on laser infrared thermometer-driven LST collected from field. Between Landsat and field-driven LST data during a period of three years, the Pearson correlation coefficient ranged from 0.73 to 0.89, and this was significant at the 0.05 level of significance. This proves the acceptability of the LST maps.

4.4.4 Relation Between WPF and Temperature

The linear regression between WPF and LST in different phases with regression equation and coefficient of determination (R^2) has been presented in Fig. 4.5. In all the phases, statistically acceptable influence of IF on LST was detected. R^2 values were found as 0.527, 0.337 and 0.323, respectively, in phases 1, 2 and 3. It signifies that the influence was gradually declined to some extent. Table 4.1 represents correlation and regression status between WPF and LST at different WPF zones in order to explore the fact that is

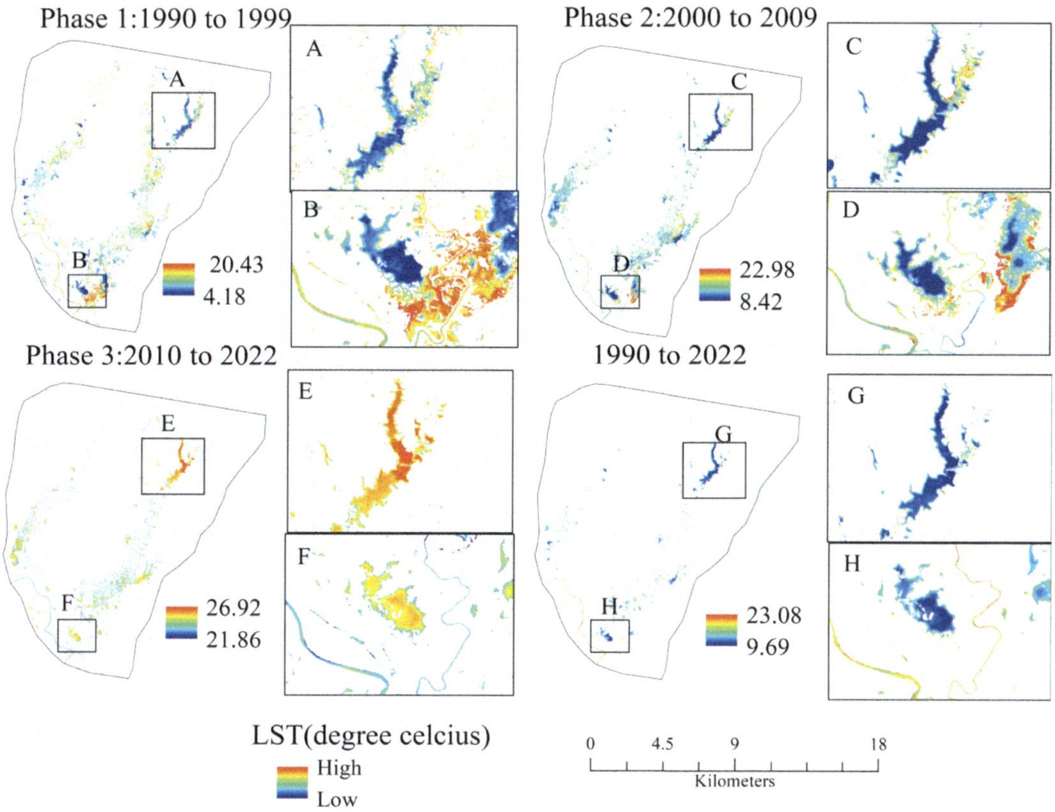

Fig. 4.4 LST of wetland area in different phases (1) phase 1 (1990–1999), (2) phase 2 (2000–2009) and (3) phase 3 (2010–2022) and (4) IF of entire period (1990–2022)

there any difference of control of WPF on LST on those WPF zones. In all the WPF zones negative relation was detected and the relations were statistically significant (0.01 level of significance). It does mean high water presence consistency can reduce temperature more in the concerned wetland than low consistent areas. Linear regression equation, coefficient of determination also revealed that WPF has strong control on temperature in all WPF zones (Table 4.1) irrespective of WPF zones. R^2 value in all the cases was >0.45 with a p value of <0.01. Phase-wise analysis of control did not show any specific trend. Computed explained variance in all the WPF zones over the phases ranges between 72 and 94% which also exhibits commendable control of consistency of inundation frequency on temperature (Table 4.1).

Ordinary least square (OLS) regression between WPF and temperature across the wetland area as displayed in Fig. 4.6 depicts that in most part of the wetland areas, a strong relation was found with very negligible standard residual. About 49%, 73% and 45% cases, respectively, in phases 1, 2 and 3 such strong control was found and it is irrespective of wetland core or periphery, upper or lower reaches.

4.4.5 Temperature in the Wetland Drained Part

In order to explore the effect of phase-wise WPF conversion and conversion of water body to other land uses on temperature, first conversion matrix was prepared and showed in Fig. 4.7. This figure depicts that conversion of water body to other land uses is more common in the wetland outskirts. Wetland interior or transition part in phase

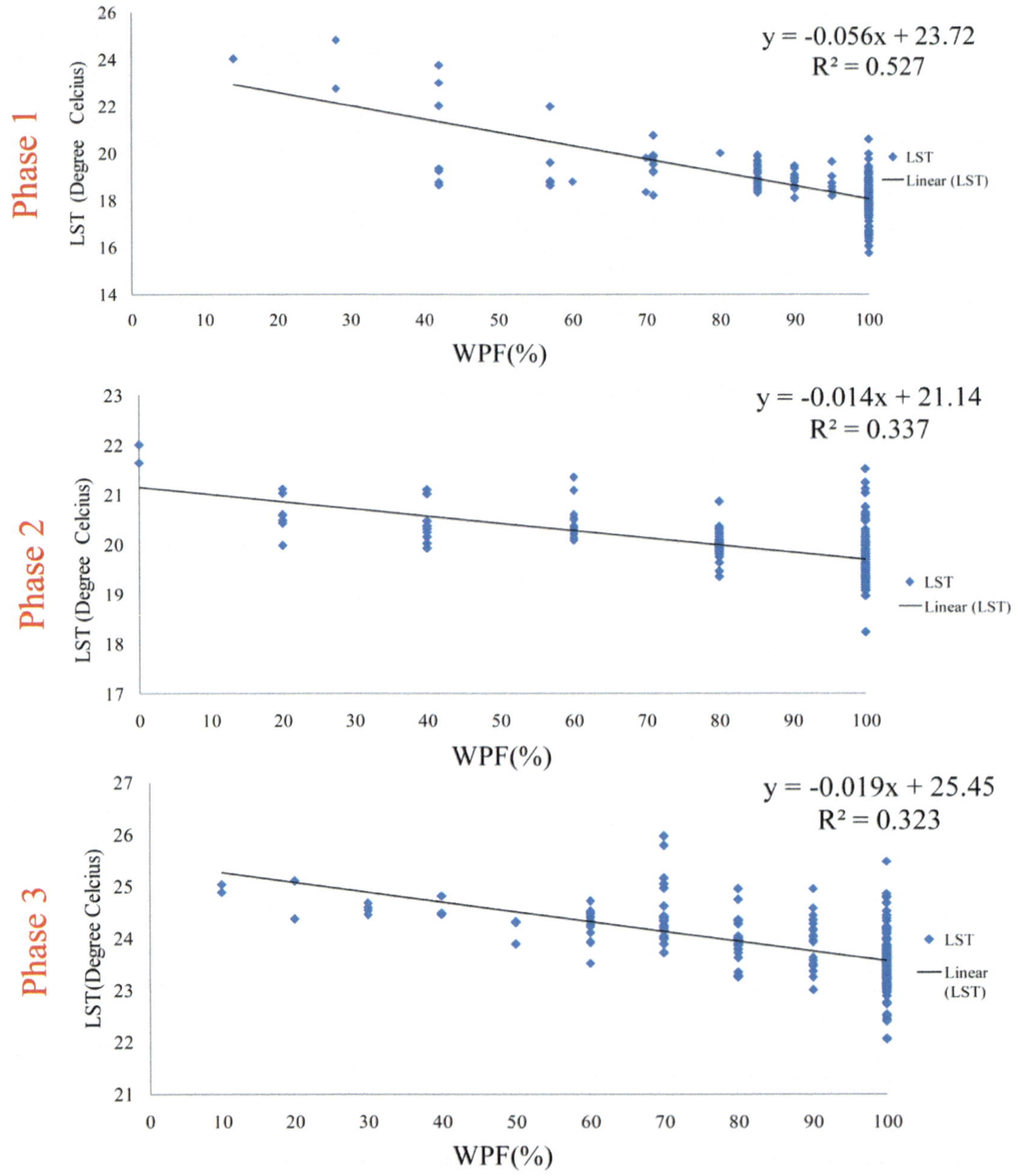

Fig. 4.5 Linear regression between IF (WPF) and LST in phase 1, phase 2 and phase 3

1 has become outskirt in successive phases. At the same way WPF transformation is very prominent in the wetland outskirt areas over the phases (Fig. 4.7). About 32.07% area was transformed from high WPF (60–100%) to <40% in between phases 1 to 2. This proportion was 46.24% in between phases 1 to 3. Temperature change and dispersion between phases 1–3 at different transformation units have been illustrated in Fig. 4.8. From these diagrams, it is very clear that degree of temperature change and dispersion was found very less in the unchanged units. Contrarily

Table 4.1 Pearson's correlation coefficient, significance level, variance and linear regression between WPF and LST at different WPF zones in successive three phases

	WPF class (%)	Correlation between WPF and LST	Significance level (confidence level = 95%)	Total variance explained (%)	Regression equation	*p* value	*r* square
Phase 1	0–20	– 0.454	0.001	72.689	$Y = -0.083x + 21.19$	1.33E−98	0.52
	20–40	– 0.519	0.001	75.926	$Y = -0.061x + 21.45$	6.06E−87	0.59
	40–60	– 0.539	0.001	76.938	$Y = -0.036x + 21.03$	1.49E−88	0.62
	60–80	– 0.722	0.001	86.104	$Y = -0.079x + 24.54$	2.91E−67	0.85
	80–100	– 0.574	0.001	78.705	$Y = -0.064x + 24.29$	3.12E−50	0.62
Phase 2	0–20	– 0.688	0.001	84.393	$Y = -0.082x + 22.08$	2E−124	0.73
	20–40	– 0.749	0.001	87.446	$Y = -0.054x + 22.7$	3.2E−115	0.83
	40–60	– 0.696	0.001	72.45	$Y = -0.026x + 21.9$	2.46E−95	0.54
	60–80	– 0.704	0.001	85.224	$Y = -0.048x + 23.58$	7.17E−84	0.67
	80–100	– 0.504	0.001	76.989	$Y = -0.044x + 23.71$	7.66E−58	0.59
Phase 3	0–20	– 0.835	0.001	91.737	$Y = -0.217x + 27.22$	8.9E−119	0.69
	20–40	– 0.886	0.001	94.315	$Y = -0.088x + 27.19$	3E−127	0.85
	40–60	– 0.643	0.001	82.138	$Y = -0.103x + 29.57$	5.22E−71	0.69
	60–80	– 0.65	0.001	82.478	$Y = -0.048x + 27.49$	5.32E−85	0.71
	80–100	– 0.46	0.001	72.98	$Y = -0.045x + 27.5$	3.89E−56	0.42

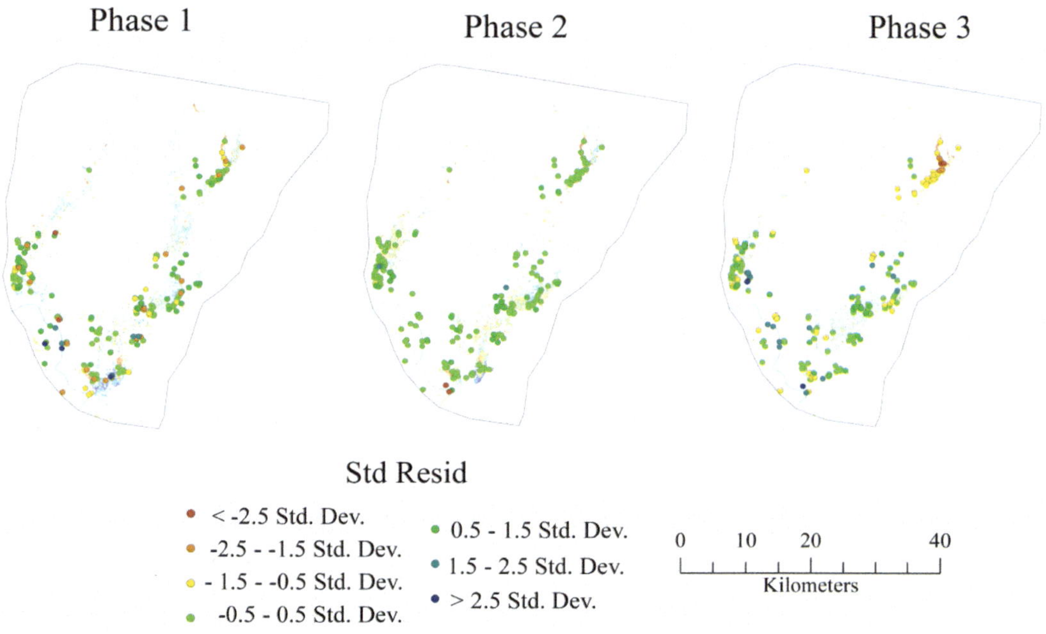

Fig. 4.6 OLS between WPF and LST in three consecutive phases

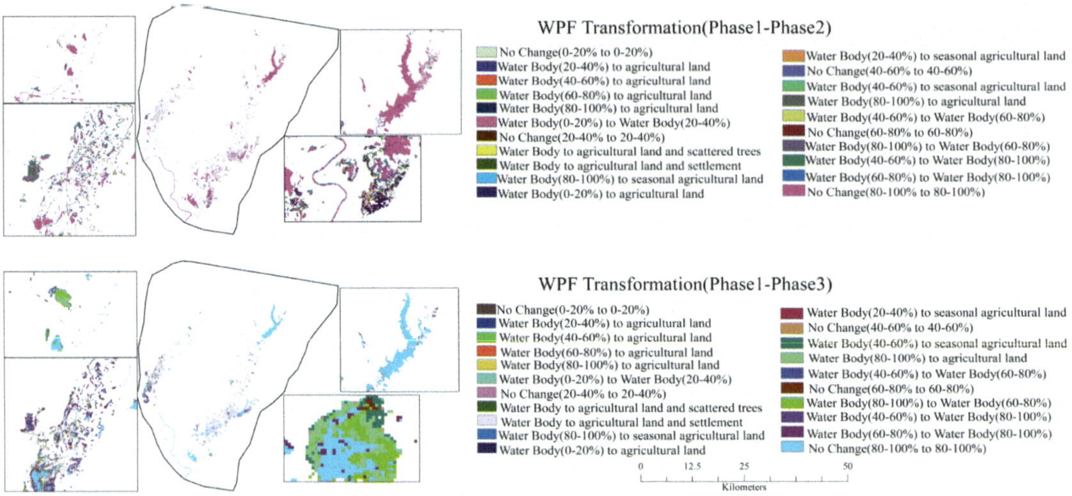

Fig. 4.7 Change matrix of WPF

higher temperature change with greater dispersion was found very high on those areas where water body of different WPF classes to agriculture land (4–5.5 °C). But this rate of temperature rise is relatively less in the case of higher frequency WPF to low frequency WPF (confined within 1.5–3 °C). When WPF with >80% consistency was transformed into 20%, temperature rise was found quite greater than when it was transformed from >80 to 60%. Such information clarified the fact that not only land use transformation, transformation in water presence consistency also plays vital role behind temperature dynamics over the phases.

Table 4.2 presents the temperature change between phases 1 and 2 and phases 2 and 3 in different land use and WPF transformation matrices in order to explain how time distance affects the temperature change in a transformed unit. For instance, a wetland patch transformed from perennial to seasonal water body between phases 1 and 2, temperature rise was confined to <2 °C but the same land recorded >5 °C rise of temperature. A patch of wetland transformed from perennial to seasonal wetland between phases 1 and 2 registered about 2 °C temperature, but when the same unit was turned into agriculture or built land in between phases 2 and 3, its rate of temperature rise was found >3 °C more than former transformation regime. When wetland with different consistency was transformed into agriculture or built land, no discriminated result was detected.

4.4.6 LST Change in WPF Conversion Matrices

Table 4.2 describes the LST change between phases 1 and 2 and phases 1 and 3 in different WPF consistency matrices. Degree of LST change between phases 1 and 3 was found considerably higher than LST change between phases 1 and 2. It was also accounted that degree of LST change was found comparatively low when WPF conversion occurred between two proximate classes than distant classes. For instance, LST change was recorded 1.77° when WPF was changed from 60–80 to 0–20% classes. It was 1.33° when WPF changes from 60–80 to 40–60%. In phases 2–3, when water body with different consistency classes converted to other land uses, degree of temperature was found considerably high.

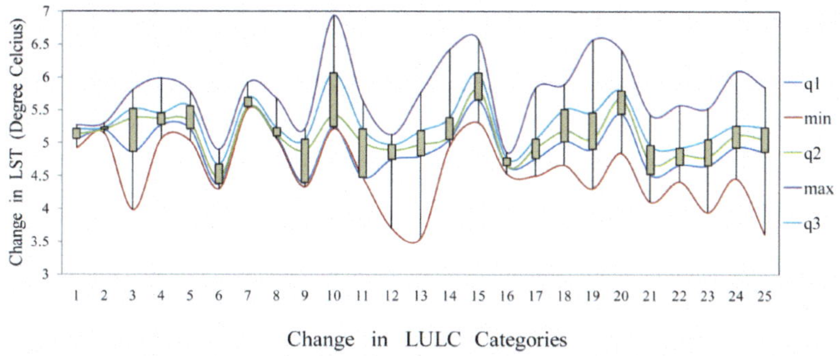

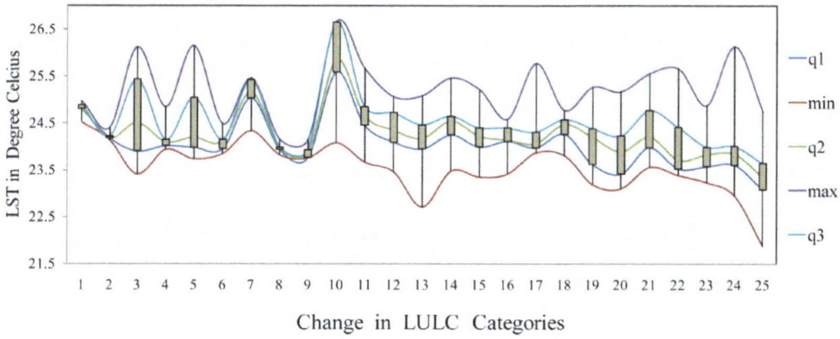

Fig. 4.8 Showing box plot; *1* no change (0–20%), *2* water body (20–40%) to agricultural land, *3* water body (40–60%) to agricultural land, *4* water body (60–80%) to agricultural land, *5* water body (80–100%) to agricultural land, *6* water body to water body (0–20 to 20–40%), *7* no change (20–40 to 20–40%), *8* water body to agricultural land and scattered trees, *9* water body to agricultural land and settlement, *10* water body (80–100%) to agricultural land, *11* water body (0–20%) to agricultural land, *12* water body (20–40%) to agricultural land, *13* no change (40–60 to 40–60%), *14* water body (60–80%) to agricultural land, *15* water body (80–100%) to agricultural land, *16* insignificant change, *17* insignificant change, *18* water body (40–60%) remains water body (60–80%), *19* no change (60–80 to 60–80%), *20* water body remains water body (80–100 to 60–80%), *21* insignificant change, *22* insignificant change, *23* water body (40–60%) remains water body (80–100%), *24* water body (60–80%) remains water body (80–100%), *25* no change (80–100 to 80–100%)

Table 4.2 Temperature change in different WPF conversion matrices

Classes	Change in LULC (phase 1–phase 2)	LST change (degree Celsius)	Change in LULC (phase 1–phase 3)	LST change (degree Celsius)
60–80 to 0–20	Perennial to seasonal	1.77	Water body to agricultural land	5.08
80–100 to 0–20	Perennial to seasonal	1.51	Water body to agricultural land	4.97
60–80 to 20–40	Perennial to seasonal	1.35	Water body to agricultural land and settlement	5.45
80–100 to 20–40	Perennial to seasonal	1.69	Water body to agricultural land	5.55
60–80 to 40–60	Water to water	1.33	Water body to agricultural land and settlement	5.38
80–100 to 40–60	Water to water	0.95	Water body to agricultural land	5.52

4.5 Discussion

Rapid wetland loss was detected in the present study area over the phases. Significant reduction of wetland area (60%) was detected both in two major rivers Punarbhaba and Tangon River. The trend of wetland loss across the world is very prominent as presented by the scholars like Das and Pal (2017), Talukdar and Pal (2017), Pal and Talukdar (2019). Pal and Saha (2018) reported 268.33 km^2 wetland loss in Atreyee area caused by mainly agriculture extension. Basu et al. (2021) reported 60% wetland loss caused by built area encroachment in Malda District. Some works like Singha and Pal (2023), Talukdar and Pal (2019) also reported hydrological medication caused by damming; water diversion through barrage was also dominant reason behind wetland squeeze. They reported this based on the case of this river basin. Han et al. (2018) and Xu et al. (2019) also condemned the same driver of wetland transformation in other regions. So the present result is very concomitant with the previous literatures.

Result further reported that not only wetland area was declined over the advances of time, consistency in inundation frequency in existing wetland area was also declined (Talukdar and Pal 2020; Saha et al. 2021a, b). Yu et al. (2018) also reported similar result in flood plain regions of Northeast China. Now the question why inundation frequency decreases in the seasonally inundated flood plain wetlands? Arrhythmic rainfall is the major cause behind this (Pal and Singha 2023). But there is no clear signature of change in irregularity of rainfall over the study period. Apart from rainfall, flood water supply from surrounding river, water supply to the wetland via tie channels are also very dependable water source of flood plain wetland. Reduction of discharge and water level in the cognitive streams is due to damming and other forms of water harvesting from rivers (Pal 2016). Pal and Talukdar (2019) reported significant discharge reduction following damming in Punarbhaba River and it was caused for shrinkages of active flood domain. Due to this, a wide part of the wetland left behind active flood domain and thereby deprived from receiving regular water supply. Moreover, the wetland areas that fall under active flood domain also suffer from water scarcity from river due to significant reduction of flood frequency (Saha et al. 2022). Pal and Talukdar (2019), Singha and Pal (2022) documented 39.72% reduction of flood frequency reduction in Punarbhaba and Tangon river basins, respectively. Water supply through tie channels was also reported interrupted due to blockage and reclamation in this region (Pal et al. 2019). All these caused decreasing consistency of inundation frequency in the present study area.

A bulk of researches were done regarding change of temperature in relation to land use transformation across the world including Das and Angadi (2020) and Majumdar and Biswas (2016). Most of the researchers explored temperature rise incidents both in urban and rural areas. The present study also reported the temperature rise incident within wetland areas which was not yet transformed into other land uses. The wetland transformed from high inundation consistency gradient to low inundation consistency gradient that was witnessed about 2–5 °C temperature rise in last 33 years. Low consistency was found in the shallower part of the wetland. In the shallower water body, sunrays penetrate more easily up to the wetland bed and heats up the water body in comparison with the deeper water body. Since the wetland tract was gradually shallowing over time and hasten the same after hydrological modification (Kundu et al. 2022), changes of thermal environment become very easier. Due to increasing shallowness and inconsistent water appearance in these areas, lower limit of temperature rise was recorded very high in the present study area. Inconsistent water appearance in a wetland pixel means, some years it retains water and some years it is devoid of water. Wetland devoid of water behaves like a land and therefore temperature is recorded higher on those years than the years that retain water. Therefore, average temperature in inconsistent inundation gradient was found higher. Since such area increases over the phases, mean water surface temperature was also found enhanced.

When water body transformed into agriculture land or built area, temperature rise was found rather high as reported by other scholars like Pal and Ziaul (2017) and Basu et al. (2021). Temperature change between phases 1 and 3 was quite greater and 1–2 when a water body converted into agriculture land. So, the present study revealed that time distance is related with magnitude of temperature rise. Immediate after transformation of water body into agriculture land, the transformed unit retains high moisture within soil pore space and support evaporation (Edwards et al. 2015). It causes relatively less temperature rise in this case. But over the time since soil moisture decreases in the transformed areas, temperature rise was found faster.

4.6 Conclusion

The present study has established a viable linkage between inundation consistency in seasonally inundated flood plain wetland and water surface temperature. Growing inundation consistency was condemned as a cause behind growing water surface temperature. Correlation coefficient and OLS analysis proved this linkage. This finding is quite unique and valuable for ecological management perspective in particular and human well-being in general. Previous studies showed linkage between water depth, water body size, fragmentation of wetland, with temperature but in the seasonally inundated wetland consistency of water presence is a big issue and how it could control the water surface temperature was explored by this study. Also, the study explained how thermal character of a wetland converted from one to another inundation consistency gradient changes with time distance. This approach of analysis is also innovative. Along with novelty of the work, there are some limitations of the work that need to be pointed out for future scope of further study. For analyzing inundation consistency and water surface temperature always images of same dates in different years were not found. For depicting post-monsoon season, images of both October and November months were taken. So, such time difference may intrude some arbitrariness in result. This analysis focused mainly post-monsoon season, but other seasons, particularly pre-monsoon season could be taken for analysis in order to show the seasonal variation in the control of inundation consistency on water surface temperature.

References

Ampatzidis P, Kershaw T (2020) A review of the impact of blue space on the urban microclimate. Sci Total Environ 730:139068

Artis DA, Carnahan WH (1982) Survey of emissivity variability in thermography of urban areas. Remote Sens Environ 12(4):313–329

Asgarian A, Amiri BJ, Sakieh Y (2015) Assessing the effect of green cover spatial patterns on urban land surface temperature using landscape metrics approach. Urban Ecosyst 18:209–222

Basu T, Das A, Pham QB, Al-Ansari N, Linh NTT, Lagerwall G (2021) Development of an integrated peri-urban wetland degradation assessment approach for the Chatra Wetland in eastern India. Sci Rep 11(1):1–22

Bian J, Li A, Zhang Z, Zhao W, Lei G, Yin G et al (2017) Monitoring fractional green vegetation cover dynamics over a seasonally inundated alpine wetland using dense time series HJ-1A/B constellation images and an adaptive endmember selection LSMM model. Remote Sens Environ 197:98–114

Borro M, Morandeira N, Salvia M, Minotti P, Perna P, Kandus P (2014) Mapping shallow lakes in a large South American floodplain: a frequency approach on multitemporal Landsat TM/ETM data. J Hydrol 512:39–52

Bratley K, Ghoneim E (2018) Modeling urban encroachment on the agricultural land of the eastern Nile Delta using remote sensing and a GIS-based Markov chain model. Land 7(4):114

Carlson TN, Ripley DA (1997) On the relation between NDVI, fractional vegetation cover, and leaf area index. Remote Sens Environ 62(3):241–252

Chatterjee RS, Singh N, Thapa S, Sharma D, Kumar D (2017) Retrieval of land surface temperature (LST) from landsat TM6 and TIRS data by single channel radiative transfer algorithm using satellite and ground-based inputs. Int J Appl Earth Obs Geoinf 58:264–277

Das S, Angadi DP (2020) Land use-land cover (LULC) transformation and its relation with land surface temperature changes: a case study of Barrackpore Subdivision, West Bengal, India. Remote Sens Appl Soc Environ 19:100322

Das RT, Pal S (2017) Exploring geospatial changes of wetland in different hydrological paradigms using water presence frequency approach in Barind tract of West Bengal. Spat Inf Res 25:467–479

Das RT, Pal S (2018) Investigation of the principal vectors of wetland loss in Barind tract of West Bengal. GeoJournal 83:1115–1131

Dash SK, Jenamani RK, Kalsi SR, Panda SK (2007) Some evidence of climate change in twentieth-century India. Clim Change 85(3–4):299–321

Davis RA Jr, FitzGerald DM (2009) Beaches and coasts. Wiley

Debanshi S, Pal S (2020) Wetland delineation simulation and prediction in deltaic landscape. Ecol Ind 108:105757

Edwards PJ, Williard KW, Schoonover JE (2015) Fundamentals of watershed hydrology. J Contemp Water Res Educ 154(1):3–20

Feyisa GL, Meilby H, Fensholt R, Proud SR (2014) Automated Water Extraction Index: a new technique for surface water mapping using Landsat imagery. Remote Sens Environ 140:23–35

Gao BC (1996) NDWI—a normalized difference water index for remote sensing of vegetation liquid water from space. Remote Sens Environ 58(3):257–266

Guo M, Li J, Sheng C, Xu J, Wu L (2017) A review of wetland remote sensing. Sensors 17(4):777

Han X, Feng L, Hu C, Chen X (2018) Wetland changes of China's largest freshwater lake and their linkage with the Three Gorges Dam. Remote Sens Environ 204:799–811

Hathway EA, Sharples S (2012) The interaction of rivers and urban form in mitigating the Urban Heat Island effect: a UK case study. Build Environ 58:14–22

Jia K, Liang S, Gu X, Baret F, Wei X, Wang X et al (2016) Fractional vegetation cover estimation algorithm for Chinese GF-1 wide field view data. Remote Sens Environ 177:184–191

Jiang P, Cheng L, Li M, Zhao R, Huang Q (2014) Analysis of landscape fragmentation processes and driving forces in wetlands in arid areas: a case study of the middle reaches of the Heihe River, China. Ecol Ind 46:240–252

Junk WJ, An S, Finlayson CM, Gopal B, Květ J, Mitchell SA et al (2013) Current state of knowledge regarding the world's wetlands and their future under global climate change: a synthesis. Aquat Sci 75:151–167

Khatun R, Pal S (2021) Effects of hydrological modification on fish habitability in riparian flood plain river basin. Ecol Inf 65:101398

Knutson MG, Sauer JR, Olsen DA, Mossman MJ, Hemesath LM, Lannoo MJ (1999) Effects of landscape composition and wetland fragmentation on frog and toad abundance and species richness in Iowa and Wisconsin, USA. Conserv Biol 13(6):1437–1446

Kundu S, Pal S, Talukdar S, Mandal I (2021) Impact of wetland fragmentation due to damming on the linkages between water richness and ecosystem services. Environ Sci Pollut Res 28(36):50266–50285

Kundu S, Pal S, Talukdar S, Mahato S, Singha P (2022) Integration of satellite image–derived temperature and water depth for assessing fish habitability in dam controlled flood plain wetland. Environ Sci Pollut Res 1–15

Lee SY, Ryan ME, Hamlet AF, Palen WJ, Lawler JJ, Halabisky M (2015) Projecting the hydrologic impacts of climate change on montane wetlands. PLoS ONE 10(9):e0136385

Let M, Pal S (2023) Socio-ecological well-being perspectives of wetland loss scenario: a review. J Environ Manage 326:116692

Li F, Lawrence DM, Bond-Lamberty B (2017) Impact of fire on global land surface air temperature and energy budget for the 20th century due to changes within ecosystems. Environ Res Lett 12(4):044014

Majumdar DD, Biswas A (2016) Quantifying land surface temperature change from LISA clusters: an alternative approach to identifying urban land use transformation. Landsc Urban Plan 153:51–65

Mallick J, Kant Y, Bharath BD (2008) Estimation of land surface temperature over Delhi using Landsat-7 ETM+. J Ind Geophys Union 12(3):131–140

Mao D, Wang Z, Wu J, Wu B, Zeng Y, Song K et al (2018) China's wetlands loss to urban expansion. Land Degrad Dev 29(8):2644–2657

Markham BL, Barker JL (1985) Spectral characterization of the landsat thematic mapper sensors. Int J Remote Sen 6(5):697–716

McCauley LA, Jenkins DG, Quintana-Ascencio PF (2013) Isolated wetland loss and degradation over two decades in an increasingly urbanized landscape. Wetlands 33:117–127

McFeeters SK (1996) The use of the Normalized Difference Water Index (NDWI) in the delineation of open water features. Int J Remote Sens 17(7):1425–1432

Mohamed YA, Bastiaanssen WGM, Savenije HHG (2004) Spatial variability of evaporation and moisture storage in the swamps of the upper Nile studied by remote sensing techniques. J Hydrol 289(1–4):145–164

Mondal D, Pal S (2018) Monitoring dual-season hydrological dynamics of seasonally flooded wetlands in the lower reach of Mayurakshi River, Eastern India. Geocarto Int 33(3):225–239

Neubauer SC, Verhoeven JT (2019) Wetland effects on global climate: mechanisms, impacts, and management recommendations. In: Wetlands: ecosystem services, restoration and wise use, pp 39–62

Nichol JE (1994) A GIS-based approach to microclimate monitoring in Singapore's high-rise housing estates. Photogramm Eng Remote Sens 60(10):1225–1232

Pal S (2016) Impact of Massanjore dam on hydro-geomorphological modification of Mayurakshi river, Eastern India. Environ Dev Sustain 18:921–944

Pal S, Saha TK (2018) Identifying dam-induced wetland changes using an inundation frequency approach: the case of the Atreyee River basin of Indo-Bangladesh. Ecohydrol Hydrobiol 18(1):66–81

Pal S, Singha P (2023) Linking river flow modification with wetland hydrological instability, habitat condition, and ecological responses. Environ Sci Pollut Res 30(5):11634–11660
Pal S, Talukdar S (2018) Application of frequency ratio and logistic regression models for assessing physical wetland vulnerability in Punarbhaba river basin of Indo-Bangladesh. Hum Ecol Risk Assess Int J 24(5):1291–1311
Pal S, Talukdar S (2019) Impact of missing flow on active inundation areas and transformation of parafluvial wetlands in Punarbhaba–Tangon river basin of Indo-Bangladesh. Geocarto Int 34(10):1055–1074
Pal S, Ziaul SK (2017) Detection of land use and land cover change and land surface temperature in English Bazar urban centre. Egypt J Remote Sens Space Sci 20(1):125–145
Pal S, Saha A, Das T (2019) Analysis of flow modifications and stress in the Tangon river basin of the Barind tract. Int J River Basin Manag 17(3):301–321
Pal S, Singha A, Mondal S, Debanshi S (2023) Ecological consequences of urban blue space transformation. Environ Sci Pollut Res 30(12):34115–34134
Peng J, Liu S, Lu W, Liu M, Feng S, Cong P (2021) Continuous change mapping to understand wetland quantity and quality evolution and driving forces: a case study in the Liao River Estuary from 1986 to 2018. Remote Sens 13(23):4900
Rojas C, Munizaga J, Rojas O, Martínez C, Pino J (2019) Urban development versus wetland loss in a coastal Latin American city: lessons for sustainable land use planning. Land Use Policy 80:47–56
Saha TK, Pal S, Sarkar R (2021a) Prediction of wetland area and depth using linear regression model and artificial neural network based cellular automata. Eco Inform 62:101272
Saha TK, Pal S, Talukdar S, Debanshi S, Khatun R, Singha P, Mandal I (2021b) How far spatial resolution affects the ensemble machine learning based flood susceptibility prediction in data sparse region. J Environ Manage 297:113344
Saha TK, Pal S, Sarda R (2022) Impact of river flow modification on wetland hydrological and morphological characters. Environ Sci Pollut Res 29(50):75769–75789
Schedlbauer JL, Oberbauer SF, Starr G, Jimenez KL (2011) Controls on sensible heat and latent energy fluxes from a short-hydroperiod Florida Everglades marsh. J Hydrol 411(3–4):331–341
Shen L, Li C (2010) Water body extraction from Landsat ETM+ imagery using adaboost algorithm. In: 2010 18th international conference on geoinformatics, June 2010. IEEE, pp 1–4
Singha P, Pal S (2022) Predicting wetland area and water depth in Barind plain of India. Environ Sci Pollut Res 29(47):70933–70949
Singha P, Pal S (2023) Wetland transformation and its impact on the livelihood of the fishing community in a flood plain river basin of India. Sci Total Environ 858:159547
Snyder WC, Wan Z, Zhang Y, Feng YZ (1998) Classification-based emissivity for land surface temperature measurement from space. Int J Remote Sens 19(14):2753–2774
Sobrino JA, Raissouni N (2000) Toward remote sensing methods for land cover dynamic monitoring: application to Morocco. Int J Remote Sens 21(2):353–366
Song J, Lu Y, Zhao Q, Zhang Y, Yang X, Chen Q et al (2022) Effect modifications of green space and blue space on heat–mortality association in Hong Kong, 2008–2017. Sci Total Environ 838:156127
Talukdar S, Pal S (2017) Impact of dam on inundation regime of flood plain wetland of Punarbhaba river basin of Barind tract of Indo-Bangladesh. Int Soil Water Conserv Res 5(2):109–121
Talukdar S, Pal S (2019) Effects of damming on the hydrological regime of Punarbhaba river basin wetlands. Ecol Eng 135:61–74
Talukdar S, Pal S (2020) Modeling flood plain wetland transformation in consequences of flow alteration in Punarbhaba river in India and Bangladesh. J Clean Prod 261:120767
Van de Griend AA, Owe M (1993) On the relationship between thermal emissivity and the normalized difference vegetation index for natural surfaces. Int J Remote Sens 14(6):1119–1131
Völker S, Baumeister H, Claßen T, Hornberg C, Kistemann T (2013) Evidence for the temperature-mitigating capacity of urban blue space—a health geographic perspective. Erdkunde 355–371
Weng Q, Fu P (2014) Modeling annual parameters of clear-sky land surface temperature variations and evaluating the impact of cloud cover using time series of Landsat TIR data. Remote Sens Environ 140:267–278
Xu T, Weng B, Yan D, Wang K, Li X, Bi W et al (2019) Wetlands of international importance: status, threats, and future protection. Int J Environ Res Public Health 16(10):1818
Xue Z, Hou G, Zhang Z, Lyu X, Jiang M, Zou Y et al (2019) Quantifying the cooling-effects of urban and peri-urban wetlands using remote sensing data: case study of cities of Northeast China. Landsc Urban Plan 182:92–100
Yu X, Ding S, Zou Y, Xue Z, Lyu X, Wang G (2018) Review of rapid transformation of floodplain wetlands in northeast China: roles of human development and global environmental change. Chin Geogr Sci 28:654–664
Zhao W, Li A, Bian J, Jin H, Zhang Z (2014) A synergetic algorithm for mid-morning land surface soil and vegetation temperatures estimation using MSG-SEVIRI products and TERRA-MODIS products. Remote Sens 6(3):2213–2238

Assessment of Ecological Conditions on Loktak Lake After Construction of Ithai Barrage

5

Md. Rejaul Islam, Swapan Talukdar, Ishita Afreen Ahmed, Mohd Rihan, Mohd Waseem Naikoo, Asif, and Ahmad Mujtaba Siddiqui

Abstract

The construction of the Ithai Barrage on the Manipur River has significantly altered the ecological balance of Loktak Lake, prompting a need to assess its impact on the lake's hydrology and ecology. This study systematically analyzes the spatiotemporal changes in surface water, land use/land cover (LULC), and ecological water conditions in Loktak Lake from 1976 to 2022, focusing on the influence of the barrage. Methodologically, the study employed Normalized Difference Water Index (NDWI) and Modified NDWI (MNDWI) for surface water mapping, utilizing thresholding techniques. LULC changes were assessed using random forest supervised classification. Additionally, water presence frequency was used to evaluate ecological water conditions, and statistical analyses were conducted to quantify the barrage's impact. LULC dynamics shows an increase in agricultural ponds from 6.10 km^2 in 1976 to 29.22 km^2 in 2022 and an expansion of built-up areas from 2.28 to 5.41 km^2 over the same period. Phumdi coverage exhibited variability, with thick vegetation areas peaking at 58.76 km^2 in 2007 before reducing to 19.96 km^2 by 2022. Ecological water assessment indicates an increase in 'Low Ecological Water' areas from 63.23 km^2 in 1976 to 76.16 km^2 in 2022, and in 'Ecological Water' areas from 6.10 to 29.22 km^2, suggesting shifts in ecological conditions. Statistical analysis of the barrage's impact showed an R^2 of 0.5008 and 0.1786 for NDWI and MNDWI, implying a moderate correlation with time respectively. The LULC R^2 values highlighted varied impacts across different land cover types. Therefore, the Ithai Barrage significantly influences Loktak Lake's hydrological and ecological dynamics, underscoring the need for integrated management to mitigate disruptions while considering socio-economic benefits.

M. R. Islam · I. A. Ahmed · M. Rihan · M. W. Naikoo · Asif
Department of Geography, Faculty of Sciences, Jamia Millia Islamia, New Delhi 110025, India
e-mail: mdrejaulislam23@gmail.com

Asif
e-mail: asif@jmi.ac.in

I. A. Ahmed
Department of Civil Engineering, Indian Institute of Technology Delhi, New Delhi 110016, India

M. W. Naikoo
Department of Geography and Disaster Management, University of Kashmir, Hazratbal, Srinagar, Jammu and Kashmir 190006, India

A. M. Siddiqui
Department of Geography, Faculty of Sciences, Aligarh Muslim University, Aligarh, Uttar Pradesh 202002, India

S. Talukdar (✉)
Department of Geography, Asutosh College, University of Calcutta, Kolkata 700026, India
e-mail: swapan.talukdar@asutoshcollege.in

S. Talukdar et al. (eds.), *Water Resource Management in Climate Change Scenario*, GIScience and Geo-environmental Modelling, https://doi.org/10.1007/978-3-031-61121-6_5

Keywords

Ecological water assessment · Hydrological dynamics · Spatiotemporal analysis · Land use and land cover (LULC) · Loktak Lake · Ithai Barrage

5.1 Introduction

Wetlands, encompassing roughly 6% of the Earth's land surface, are critical ecosystems offering diverse ecological service (Junk et al. 2016). These biologically rich zones, especially subtropical floodplain wetlands adjacent to water bodies, support unique aquatic biodiversity and are fundamental to the livelihoods and nutrition in developing countries (Davidson 2014; Mallick et al. 2021). Yet, these wetlands face global decline due to human activity and climate change, particularly in Asia and various mid- and high-latitude regions (Zedler and Kercher 2005). Factors, such as temperature changes, altered rainfall patterns, and extreme climatic events like droughts and floods have transformed these floodplains, often repurposed for intensive agriculture (Shahfahad et al. 2022; Furtak and Wolińska 2023). This transformation has paralleled shifts in freshwater ecosystems, impacting fish diversity and fisheries, contributing to the loss of up to 87% of the original global wetland area since 1700.

In India, wetlands are integral to rural economies, social structures, and cultures. However, they are endangered by environmental degradation, both from natural and human-induced factors (Laishram et al. 2022). The ongoing degradation, marked by ecosystem imbalance, water area shrinkage, biodiversity loss, and reduced fish production, is alarming. This situation is exacerbated by inconsistent policies in economics, environmental protection, and development planning, alongside inadequate governance in scientific management (Kangabam et al. 2018, 2019). Specifically, the construction of structures like the Ithai Barrage on the Manipur River has significantly altered the ecological balance of Loktak Lake, one of India's largest inland floodplain wetlands (Mahato et al. 2023). This barrage, built for hydroelectric power generation, has modified the natural hydrological regimes, leading to increased water levels, backwater effects, and consequent ecological changes. These changes have contributed to the transformation of wetland landscapes, affecting biodiversity and the livelihoods of local communities.

Despite numerous studies on land use changes and hydrological disturbances, there is a dearth of comprehensive research on long-term, spatiotemporal changes in floodplain wetland resources in the face of ecological degradation, anthropogenic activities, and impacts of structures like the Ithai Barrage. The potential of Geographic Information Systems (GIS) in managing wetland fisheries also remains underexplored in India. This study aims to fill these gaps by providing a detailed analysis of the spatiotemporal dynamics of Loktak Lake, with a focus on the impact of the Ithai Barrage. Employing advanced remote sensing techniques and GIS, the study quantitatively assesses changes in surface water, land use/land cover, and ecological water conditions over nearly five decades, offering insights into wetland management amid climatic and anthropogenic challenges (Paonam and Chatterjee 2023).

Despite extensive studies on aquatic ecosystems, there remains a significant research gap in understanding the long-term, comprehensive impact of human-made structures like barrages on complex wetland systems (Kangabam et al. 2017). Previous studies have often focused on short-term hydrological changes or specific ecological aspects, leaving a gap in holistic, long-term analyses that integrate surface water dynamics, land use and land cover changes, and ecological water conditions. Additionally, the socio-economic implications of such environmental changes, particularly in ecologically sensitive areas like Loktak Lake, are not sufficiently explored. This gap is critical as it hinders the development of effective, sustainable management strategies that can balance ecological conservation with local community needs (Kangabam et al. 2015).

The primary objective of this study was to fill these gaps by providing a comprehensive, long-term analysis of the impact of the Ithai Barrage on Loktak Lake. The study aimed to quantitatively assess changes in surface water extent, land use and land cover, and ecological water conditions over a span of nearly five decades, using advanced remote sensing techniques and statistical analyses. This approach is novel in its integration of various environmental indicators to provide a holistic understanding of the barrage's impact. Additionally, the study introduces a new perspective by linking these environmental changes with potential socio-economic consequences, paving the way for future research in this area. The novelty also lies in its methodological approach, combining NDWI and MNDWI for surface water analysis, random forest classification for LULC, and water presence frequency for ecological assessment, enriched with statistical trend analysis. This comprehensive, multidisciplinary approach contributes significantly to the field of environmental science and wetland management, offering insights that can inform policy-making and sustainable management practices.

5.2 Materials and Methods

5.2.1 Study Area

Loktak Lake in Manipur, India, stands as the largest freshwater lake in the northeastern region, renowned for its ecological and socio-economic significance (Fig. 5.1). Geographically positioned between latitudes 24° 25″ to 24° 41″ North and longitudes 93° 46″ to 93° 53″ East, it encompasses approximately 287 km^2 at an elevation of 768.5 m above mean sea level. Notable for its oval shape, with a maximum length of 32 km and width of 13 km, Loktak Lake is relatively shallow, with water depths ranging from 0.5 to 4.6 m. Its average depth is about 2.7 m. The lake is distinguished by the presence of phumdis, floating masses of soil, vegetation, and organic matter, covering a significant area of the lake. The southern part of the lake houses the Keibul Lamjao National Park (KLNP), spanning 40 km^2, which is the only floating wildlife sanctuary in the world and the natural habitat of the endangered brow-antlered deer, Cervus eldi eldi.

Loktak Lake's unique geology and geomorphology are integral to its character. It lies within the Manipur-Mikir Hills of the Himalayan orogenic belt, mainly consisting of sedimentary rock formations like sandstones, shales, and conglomerates, with occasional limestone deposits from the Mesozoic era. The lake features several small islands, some artificial, created for fishing and agriculture. Its water level undergoes seasonal fluctuations, particularly during the monsoon. The drainage network includes the Manipur River and its tributaries, feeding the lake and contributing significant silt. The climate of the region is tropical monsoon, with temperatures ranging from 0 to 35 °C and an average annual rainfall of 1183 mm. Loktak Lake boasts a rich biodiversity, with over 233 species of aquatic macrophytes and a diverse range of fauna, including the rare Indian python, sambhar, and barking deer. The lake's soils are predominantly alluvial, with patches of red ferruginous soil on slopes, characterized by poor drainage and high acidity.

Socio-economically, Loktak Lake is the lifeline for the surrounding population. It supports the livelihoods of a significant portion of Manipur's inhabitants, with 66% of the population residing in the catchment area. The central valley, including 30 towns and 1429 villages, houses 71% of the catchment's population, with Imphal as the largest town. The lake's proximity includes 53 settlements with a combined population of 200,000. Agriculture is a primary occupation, covering 15% of the catchment area, with rice being the dominant crop. The valley is known as the 'Rice Bowl of Manipur,' contributing significantly to the state's paddy production. However, intensified agricultural activities have led to increased fertilizer use, impacting the lake's ecology. The land use within the basin is divided into forests, agricultural areas, settlements, and wetlands. Phumdis cover a major part of the lake, and their presence is critical

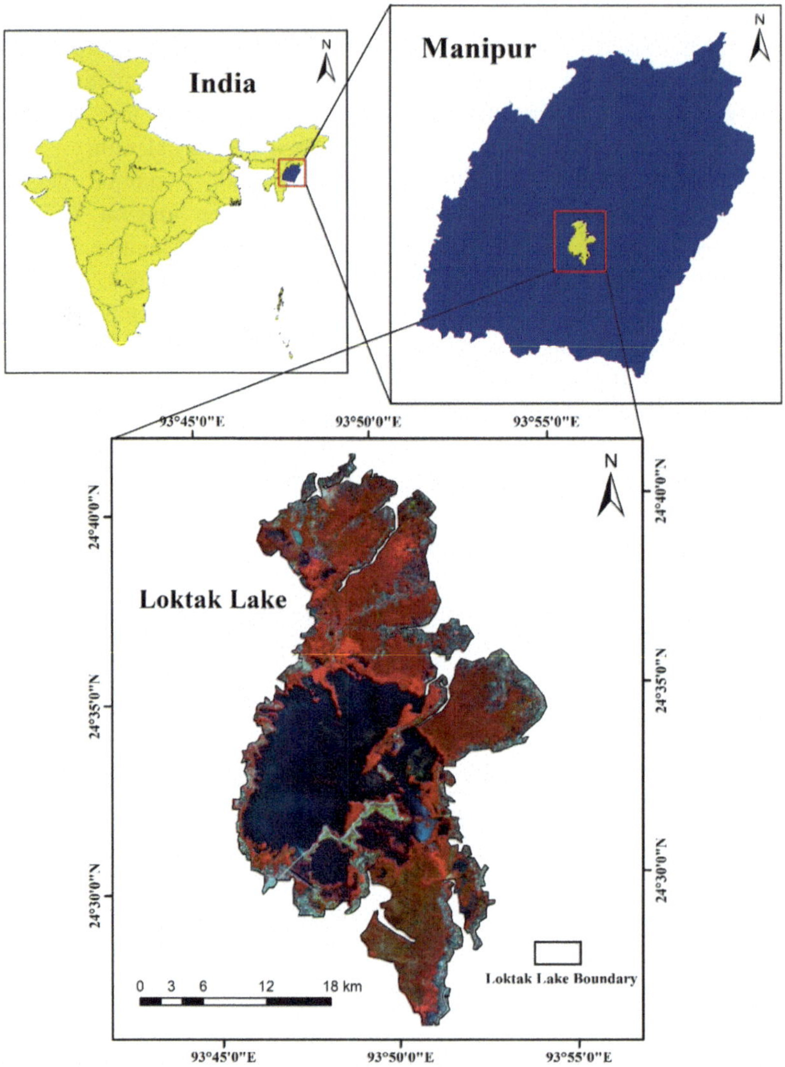

Fig. 5.1 Location of the study area: Loktak Lake, Manipur, India

for irrigation, drinking water supply, and as a food resource. These unique floating islands are key to the lake's ecosystem, providing habitat for numerous species and supporting local communities.

5.2.2 Materials

The study used a series of satellite images obtained from the United States Geological Survey (USGS) website and focused on data from the Landsat series to analyze the changes in Lake Loktak over time. Specifically, images from Landsat 2, Landsat 5, and Landsat 8 were used for the study, capturing different temporal snapshots of the area. The details of these images are listed in Table 5.1 and show the use of different sensors, including the Multispectral Scanner (MSS) for Landsat 2 and the Thematic Mapper (TM) for Landsat 5. For more recent observations, the operational land imager (OLI) and the thermal infrared sensor (TIRS) from Landsat 8 were used. These images cover the

Table 5.1 Details of satellite images use for the study

Satellites	Sensor	Path/raw	Date of acquisition	Spatial resolution
Landsat 2	MSS	145/43	01.03.1976	60
Landsat 5	TM	135/43	05.02.1988	30
Landsat 5	TM	135/43	01.03.1997	30
Landsat 5	TM	135/43	24.01.2007	30
Landsat 8	OLI_TIRS	135/43	06.03.2017	30
Landsat 8	OLI_TIRS	135/43	30.11.2022	30

path/line coordinates 145/43 and 135/43, with different acquisition dates between March 1, 1976 and November 30, 2022. The spatial resolution of the images was also different. The Landsat 2 MSS images had a resolution of 60 m, while the Landsat 5 TM and Landsat 8 OLI_TIRS images had a finer resolution of 30 m. This diversity in temporal coverage and spatial resolution enabled a comprehensive analysis of the physical and ecological changes in the Loktak Lake region over almost five decades (Garg 2015).

5.2.3 Methods for Surface Water of Wetland Mapping Using Indices and Thresholding

The methodology for mapping the surface water of the wetlands in the Loktak Lake region involved the use of pre-monsoon satellite imagery from 1976, 1988, 1997, 2007, 2017, and 2021 to study the impact of the Ithai Dam. This time frame was chosen because there is considerable variation in wetland coverage due to high monsoon rainfall in the study area. Extracting the area of Loktak Lake during these particular pre-monsoon seasons was crucial to identify the changes over the decades. To achieve this, several satellite-derived indices were used, including the Modified Normalized Difference Water Index (MNDWI) (Xu 2006) and the Normalized Difference Water Index (NDWI) (Gao 1996; McFeeters 2013). These indices were selected taking into account the uniqueness of Lake Loktak and the local environmental conditions, which play an important role in distinguishing wetlands from other land areas.

Two specific spectral indices for wetland extraction were used in the study: NDWI AND MNDWI. These indices were applied to satellite images obtained with the Landsat sensors MSS, TM, and OLI. In particular, the MNDWI separates water from non-water features and generally shows positive values for water areas and negative values for vegetated areas. NDWI, derived from the near infrared (NIR) and red spectral bands, helps identify open bodies of water by utilizing the different absorption and reflection properties of water in these bands. High NDWI values indicate a strong presence of water, while lower or negative values indicate the absence of water. This makes the NDWI a valuable tool for mapping and monitoring water-related features.

The selection of appropriate thresholds was a critical aspect of the study, necessary for distinguishing between water and non-water features in the satellite imagery. These thresholds varied depending on the scene and specific local conditions. A manual selection process was used to determine these values for each index, guided by visual comparisons with reference maps and Google Earth Pro imagery from corresponding time periods. False color composite (FCC) images of the study area served as the basis for this thresholding process. The maps with the indices were categorized into 30 different classes, of which water and non-water feature classes were identified. This methodological approach enabled a precise and customized analysis of the changes in the surface of Lake Loktak over the period studied, reflecting the dynamic nature of this unique wetland.

5.2.4 Methods for Land Use Land Cover Mapping of Wetland

The land use and land cover mapping of the Loktak Lake wetland was methodically carried out to categorize the area into five different types: open water areas, agricultural areas, phumdis with dense vegetation, phumdis with sparse vegetation, and settlements. This categorization required the selection of specific training patterns, i.e., the creation of polygons around locations that were representative of each land use type. The geo-referenced satellite images were then interpreted using supervised classification techniques, in particular using random forest classifiers (Belgiu and Drăguţ 2016). This approach is known for its effectiveness in processing large datasets and its ability to handle the complexity of multiple land use classes. Classification was facilitated by using a multispectral composition of bands red, blue, and green, using a false color composite to select regions of interest (ROIs) for the identified features. The spectral signatures for each land use class were carefully determined from the imagery using ArcGIS 14.2 software to ensure the accuracy of the classification process.

The accuracy of the classified land use and land cover maps was critically evaluated to ensure the reliability of the data. This accuracy assessment included a comparison of the classified images with high-resolution images and existing classified imagery that serve as baseline data. This process is crucial as it confirms the authenticity of the remote sensing data and ensures the quality and accuracy of the information. Classified maps from 1976, 1988, 1997, 2007, 2017, and 2022 were created using satellite imagery from Landsat 5 (Thematic Mapper) and Landsat 8 (OLI_TIRS). The accuracy for each year was quantified using a confusion matrix that organizes the classification results and allows the calculation of the user's accuracy and the producer's accuracy for each class. This matrix compares the reference dataset (ground truth) with the classification result and gives the number of correctly and incorrectly classified samples for each class. The overall accuracy of the classification results from the ratio between the sum of the elements along the diagonal of the confusion matrix and the total number of reference pixels. In addition, the kappa coefficient was used to obtain a statistical measure of inter-rater agreement on categorical elements to further validate the accuracy and reliability of the classification and identify areas for improvement.

5.2.5 Method for Ecological Water Condition Modeling

The methodology for ecological water condition modeling in the context of Loktak Lake focused on calculating the Water Presence Frequency (WPF) to understand the temporal dynamics of water availability and its ecological implications. The WPF is a crucial metric in assessing the frequency at which water is present in a given area over a specified time frame, thus reflecting the temporal pattern of water availability, such as seasonal variations or long-term changes (Mueller et al. 2013).

To calculate the WPF, the first step involves determining the total number of opportunities for water presence, referred to as 'N.' This could represent the total number of observations or instances where water presence is detectable. The next step is to count the number of occurrences where water presence was indeed detected, denoted as 'X.' The WPF is then calculated using Formula (5.1)

$$\mathrm{WPF} = \frac{X}{N} \times 100 \tag{5.1}$$

For instance, if water presence was observed in 25 out of 100 opportunities, the WPF would be 25%. This quantitative approach enables a clear and measurable understanding of water presence patterns over time.

The relationship between WPF and ecological surface water is crucial for ecosystem management and conservation. Ecological surface water refers to water that interacts directly with the environment, supporting various ecological processes and habitats (Talukdar and Pal 2017). Understanding this relationship is vital for identifying potential impacts on ecological surface water, predicting ecological responses, and guiding conservation efforts. This knowledge is also essential for developing sustainable water resource management strategies to maintain the ecological integrity of aquatic ecosystems.

For the mapping of inundation presence frequency, the MNDWI index from 1976 to 2022 was used, with observations made at regular 10-year intervals. Multi-date images from this period were integrated separately using the WPF approach, representing the average state of the dynamic wetlands. The generated maps of inundation frequency were classified into three categories: low (< 33%), moderate (33–66%), and high (> 66%) inundation frequency. Formula (5.2) was employed to compute it.

$$\mathrm{WPF}_j = \frac{I_n}{n} \times 100 \tag{5.2}$$

where WPF_j represents the frequency of water presence for the *j*th pixel over a specific period, I_j denotes the *j*th pixel containing water in the selected NDWI images, and n is the total number of images.

The value of WPF_j ranges from 0 to 100%, where values closer to 100% indicate a high frequency and regular appearance of water bodies, while values near 0 indicate a low frequency and irregular appearance. This approach was pivotal in quantifying permanent and seasonal lake water areas in pre- and post-dam periods, offering valuable insights into the changing ecological water conditions of Loktak Lake.

5.2.6 Impact Assessment Using Statistics

For assessing the impact on surface water areas, land use and land cover classes, and different zones of ecological water in Loktak Lake, a statistical approach using linear regression was employed. Linear regression is a statistical method that models the relationship between a dependent variable and one or more independent variables using a linear equation. In this context, the dependent variable could be the extent of surface water areas, the distribution of various land use and land cover classes, or the state of different ecological water zones. The linear regression model was applied to determine the trend in surface water areas over time. In this model, the extent of surface water areas (measured in square kilometers or as a percentage of total area) was set as the dependent variable, while time (in years) was the independent variable. This method allowed for the quantification of changes in surface water areas, highlighting whether there has been a significant increase or decrease over the study period. Similarly, linear regression was used to analyze trends in different land use and land cover classes within the Loktak Lake region. Each land use category (such as agricultural areas, phumdis with varying vegetation density, and settlements) was treated as a separate dependent variable in individual regression models, with time as the independent variable. This approach helped in understanding how different land use practices have evolved over the years and the rate at which these changes have occurred. For different zones of ecological water (such as areas of high, moderate, and low water presence frequency), linear regression models were created to examine the trends in these zones. The extent of each zone, defined by its water presence frequency or other relevant ecological parameters, was used as the dependent variable, against the time factor as the independent variable.

In each of these applications, the linear regression analysis provided insights into the nature of changes occurring in the Loktak Lake area. The statistical significance of the trends was determined using *p*-values and *R*-squared values, where a low *p*-value (typically <0.05) indicates a statistically significant trend, and the *R*-squared value shows the proportion of variance in the dependent variable that is predictable from the independent variable. This impact assessment method offered a systematic and quantitative way to understand and project the implications of environmental changes in the Loktak Lake ecosystem.

5.3 Results and Discussion

5.3.1 Spatiotemporal Surface Water of Wetland Mapping

The spatiotemporal analysis of surface water in the Loktak wetland, using thresholding techniques for water and non-water body identification, reveals significant changes over the years. The analysis utilized two water indices, NDWI and MNDWI, with specific threshold values for different years as listed in Table 5.2. These thresholds played a crucial role in accurately delineating water bodies from land or other features.

From 1976 to 1997, both NDWI and MNDWI analyses showed a steady increase in the surface area of water bodies, indicating an expansion phase in the Loktak Lake. The spatial coverage of open water bodies increased from 52.6392 km^2 in 1976 to 94.1004 km^2 in 1997 based on NDWI and from 52.6392 to 92.9601 km^2 based on MNDWI, as detailed in Table 5.3. However, the period between 1997 and 2007 marked a notable reduction in water bodies, as reflected in the decrease to 54.5625 km^2 (NDWI) and 71.4915 km^2 (MNDWI). Figures 5.2 and 5.3a–f illustrate these dynamics, showcasing the change in Loktak Lake's open water areas over 46 years. The period from 2007 to 2022 exhibited a more complex pattern. NDWI measurements indicated a slight increase in water bodies, reaching 85.1157 km^2 in 2022, while MNDWI measurements fluctuated, peaking in 2017 at 103.3578 km^2 before decreasing to 88.4259 km^2 in 2022. This mixed pattern suggests an oscillation in water body dynamics, potentially influenced by multiple factors, including climatic variations and human-induced changes.

The construction of the Ithai Barrage is a significant factor to consider in this context. The barrage, by manipulating water flow and levels, has likely contributed to the backwater effects observed in Loktak Lake. The increased backwater due to the barrage could explain the expansion of water bodies observed from 1976 to 1997. However, subsequent reductions and fluctuations in water body extent could be attributed to changes in barrage management, alongside other environmental factors. In summary, the study's quantitative approach, grounded in thresholding and NDWI/MNDWI analysis, demonstrates the dynamic nature of Loktak Lake's water bodies over nearly five decades. The impact of the Ithai Barrage, coupled with natural

Table 5.2 Threshold values for the water and non-water bodies

Indices	Land-water threshold 1976	Land-water threshold 1988	Land-water threshold 1997	Land-water threshold 2007	Land-water threshold 2017	Land-water threshold 2022
NDWI	0.036	0.062	– 0.007	0.0002	0.043	0.005
MNDWI	0.02	0.02	0.01	0.05	0.007	0.019

Table 5.3 Spatial coverage of surface water based on NDWI and MNDWI

Year	1976	1988	1997	2007	2017	2022
NDWI Water bodies (km^2)	52.6392	76.4721	94.1004	54.5625	79.6032	85.1157
MNDWI Water bodies (km^2)	52.6392	76.6017	92.9601	71.4915	103.3578	88.4259

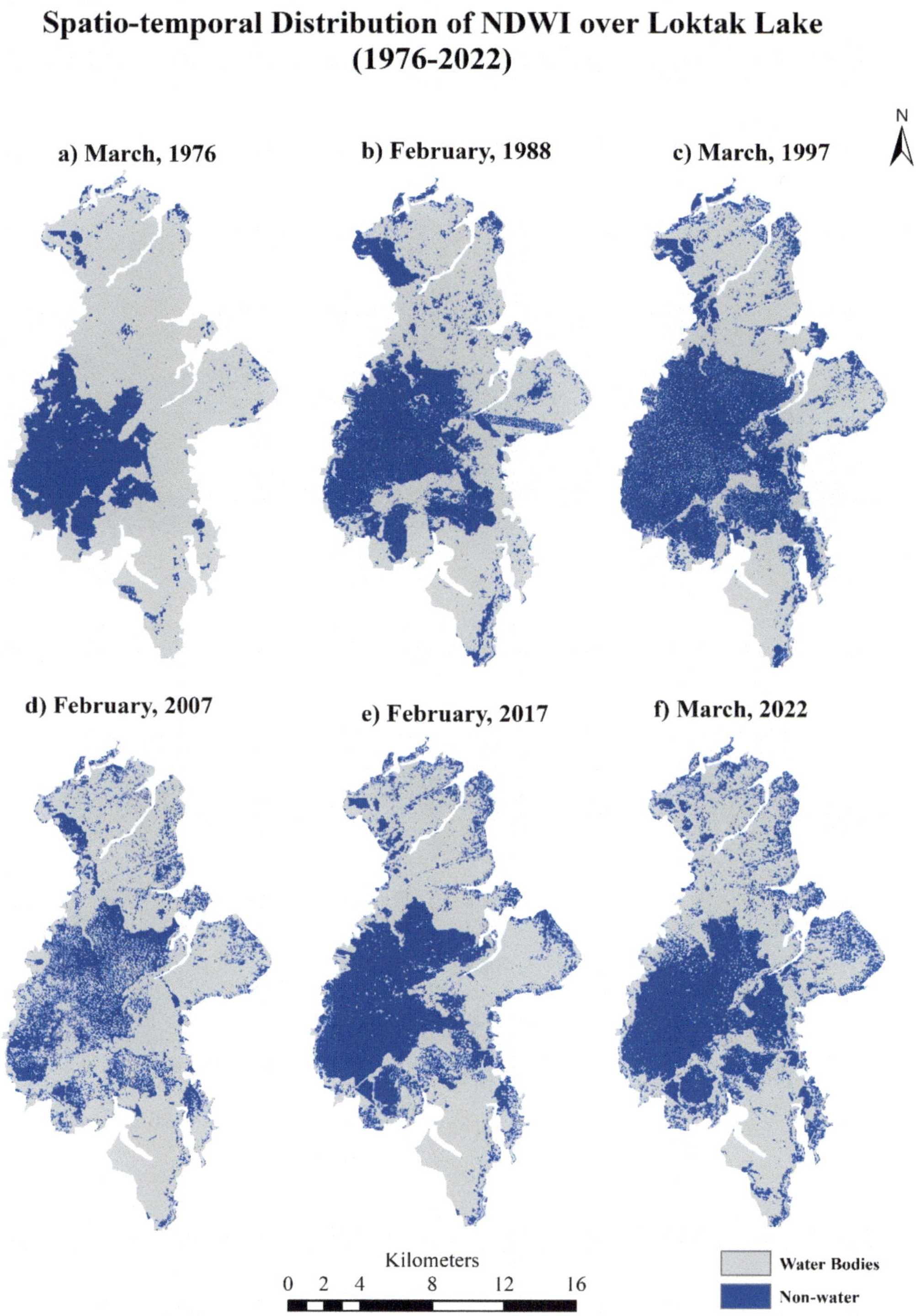

Fig. 5.2 Surface water bodies of Loktak Lake for pre-Ithai and post-Ithai using NDWI

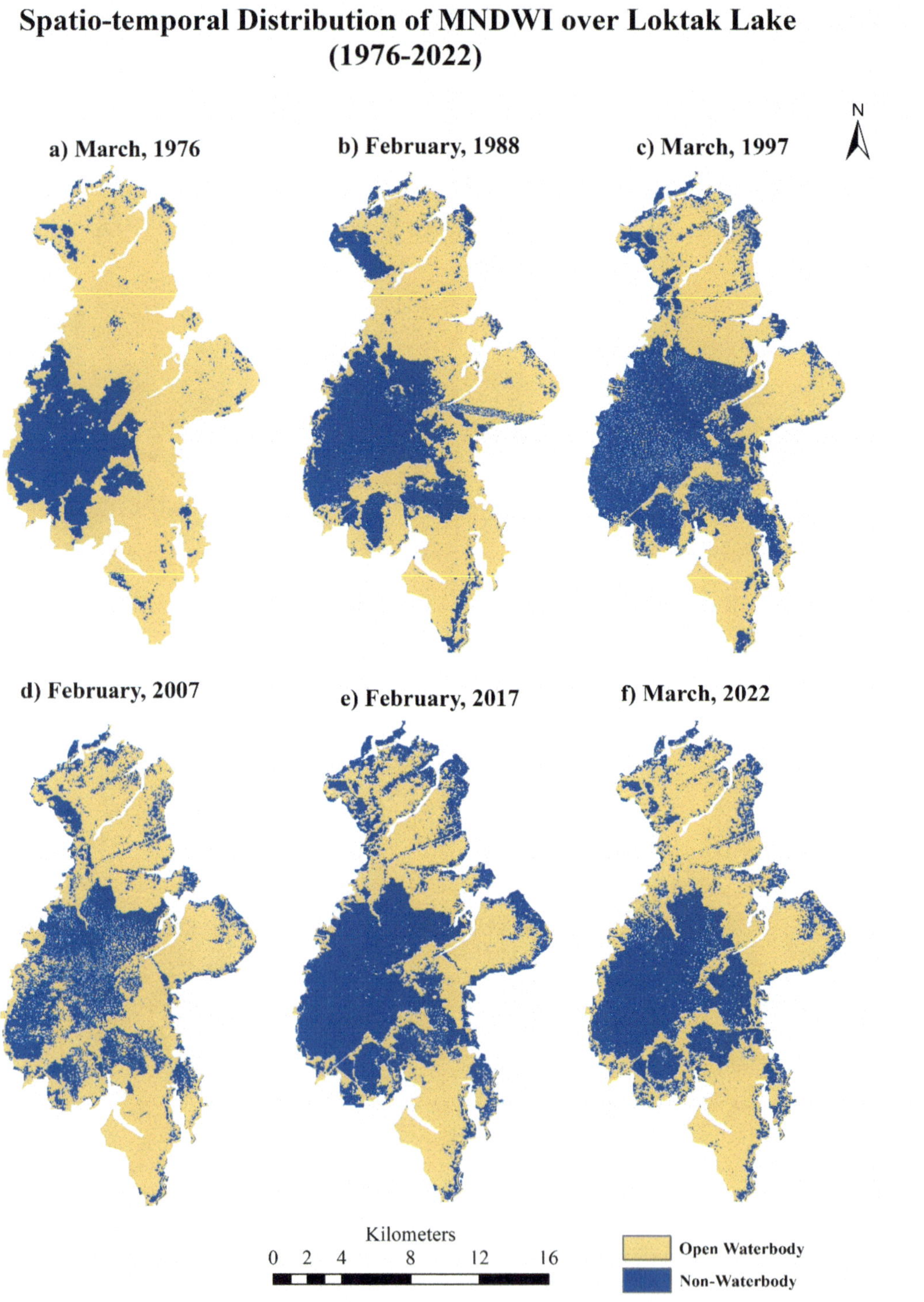

Fig. 5.3 Surface water bodies of Loktak Lake for pre-Ithai and post-Ithai using MNDWI

environmental factors, presents a complex picture of the lake's changing hydrology, with implications for ecological and human systems reliant on this vital wetland.

5.3.2 Spatiotemporal LULC Mapping of Wetland

The spatiotemporal Land Use and Land Cover (LULC) mapping of the Loktak wetland reveals significant changes in land cover types over a span of nearly five decades, as detailed in Table 5.4. The LULC categories analyzed include open water bodies, agricultural land, phumdis with thick vegetation, phumdis with thin vegetation, and built-up areas (Fig. 5.4). The spatial coverage data for these categories from 1976 to 2022 offers valuable insights into the evolving landscape around Loktak Lake. Open water bodies, a crucial component of the wetland ecosystem, displayed fluctuating trends. Initially, the area was 63.23 km^2 in 1976, peaked at 73.89 km^2 in 1997, dropped to 62.68 km^2 in 2007, and then saw a gradual increase to 76.16 km^2 by 2022. This fluctuation might be attributed to varying hydrological conditions, influenced by factors like rainfall patterns and the construction of structures like the Ithai Barrage. Agricultural land showed a consistently increasing trend, from 6.10 km^2 in 1976 to 29.22 km^2 in 2022. This increase reflects the intensification of agricultural practices and possibly the adaptation strategies of local communities to changing environmental conditions. The extent of phumdis, floating biomass, with thick vegetation, exhibited notable variations. Starting from 18.72 km^2 in 1976, it expanded dramatically to 46.84 km^2 in 1988, followed by a decrease and another increase, peaking at 58.76 km^2 in 2007. However, there was a substantial decrease to 19.96 km^2 by 2022. These fluctuations could be linked to environmental factors affecting phumdi growth and decay cycles, as well as human interventions for phumdi management. Phumdis with thin vegetation also showed fluctuations over the years. Initially, covering 111.49 km^2 in 1976, there was a decrease to 74.91 km^2 in 1997, with the coverage remaining relatively stable between 66 and 78 km^2 from 2007 to 2022. This trend suggests a dynamic balance between phumdi growth and degradation processes. The built-up area, indicative of human settlements and infrastructure development, exhibited a general increase, from 2.28 km^2 in 1976 to 5.41 km^2 in 2022. This steady growth reflects the increasing human footprint in the Loktak Lake region, which could have implications for the wetland ecosystem. Overall, the LULC analysis, as evidenced by the data and further illustrated in relevant figures, portrays a complex and dynamic landscape around Loktak Lake. The observed trends indicate not only natural fluctuations in the wetland ecosystem but also the impact of human activities. These changes highlight the need for sustainable management practices to balance ecological conservation with the socio-economic needs of the local communities dependent on this vital ecosystem.

5.3.3 Spatiotemporal Assessment of Ecological Water

The dynamics of surface ecological water in Loktak Lake from 1976 to 2022, as presented in Table 5.5, offer a detailed view of the changing

Table 5.4 Spatial coverage of LULC classes in Loktak Lake over time

Year	1976	1988	1997	2007	2017	2022
Open waterbodies	63.23	68.10	73.89	62.68	75.86	76.16
Agricultural ponds	6.10	6.75	19.69	12.80	28.37	29.22
Phumdies with thick vegetation	18.72	46.84	31.80	58.76	15.10	19.96
Phumdies with thin vegetation	111.49	78.79	74.91	66.15	76.17	74.87
Built-up	2.28	2.17	2.33	2.23	3.23	5.41

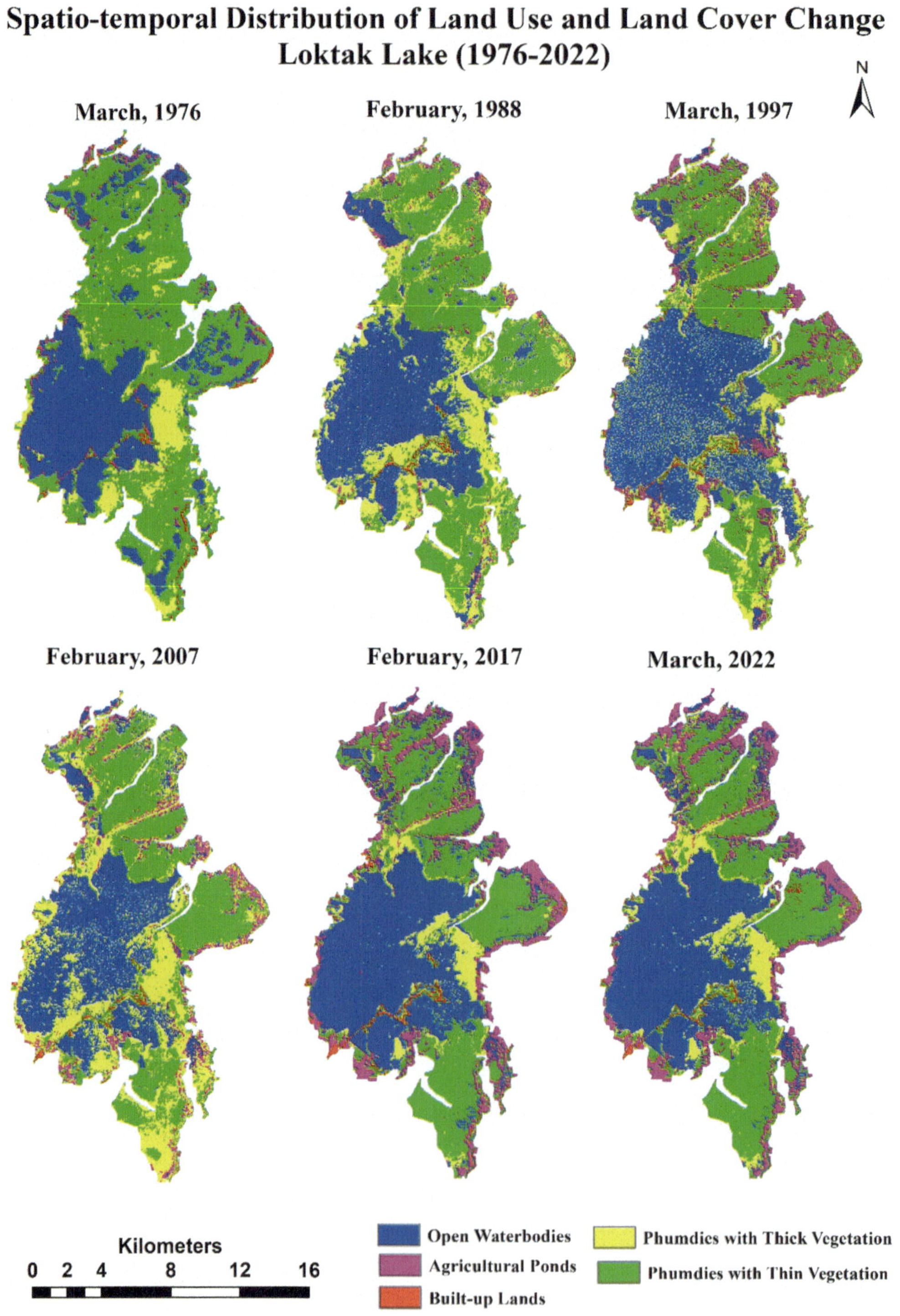

Fig. 5.4 LULC of Loktak Lake for pre-Ithai and post-Ithai Barrage

ecological conditions within the lake. The categories of 'Low Ecological Water,' 'Ecological Water,' and 'Beyond Ecological Water' reflect varying degrees of ecological significance and stress in the lake's ecosystem (Fig. 5.5). In 1976, the area categorized as 'Low Ecological Water' was 63.23 km^2, which indicates regions of lower ecological importance. This area showed a gradual increase over the years, reaching 76.16 km^2 by 2022. The increase in areas of low ecological significance could be associated with factors such as sedimentation, eutrophication, or changes in water quality. 'Ecological Water' areas, which represent zones within the optimal ecological range for the lake's ecosystem, have also shown fluctuations over the years. Notably, there was a significant increase from 6.10 km^2 in 1976 to 19.69 km^2 in 1997. However, this was followed by a decrease to 12.80 km^2 in 2007, before rising again to 29.22 km^2 by 2022. These fluctuations may be indicative of varying environmental conditions, management practices, and possibly the influence of climatic factors on the lake's ecology. The 'Beyond Ecological Water' category, which points to areas potentially experiencing ecological stress, also exhibited notable changes. The area increased from 18.72 km^2 in 1976 to a peak of 58.76 km^2 in 2007, and then decreased to 19.96 km^2 by 2022. The increase in areas beyond the ecological range could be a result of factors such as overexploitation of resources, invasive species, or altered hydrological regimes (Fig. 5.6).

The Ithai Barrage, constructed in 1983, has had a significant influence on the hydrological and ecological dynamics of Loktak Lake. The barrage has altered the natural flow and flushing characteristics of the lake, leading to increased water levels and backwater effects. This alteration may have contributed to changes in the distribution of ecological water categories. For instance, the increased water levels could have expanded the area of 'Ecological Water' temporarily but also led to more areas falling under 'Beyond Ecological Water' due to altered ecological conditions like water stagnation, reduced oxygen levels, and changes in sedimentation patterns. Therefore, the analysis of surface ecological water from 1976 to 2022 in Loktak Lake reveals a dynamic interplay between natural and human-induced factors, including the influence of the Ithai Barrage. The varying extents of different ecological water categories over time underscore the complexity of managing such a dynamic ecosystem, highlighting the need for adaptive and holistic management approaches to sustain the ecological integrity of this unique wetland system.

5.3.4 Impact Assessment of Ithai Barrage on Surface Water and Ecological Water of Loktak Lake

The impact assessment of the Ithai Barrage on the surface and ecological water of Loktak Lake from 1976 to 2022, as quantified using various indices and trend analyses, provides insightful observations about the changing dynamics of the lake's environment. The NDWI and MNDWI indices, crucial for assessing surface water extent, show different trends. The NDWI trend equation ($y = 6.7924x + 57.139$) with an R^2 of 0.5008 suggests that about half of the variability in NDWI is attributable to time, indicating a moderate positive trend in surface water extent over the years. This could reflect the impact of the Ithai Barrage, which by altering the natural water flow, could have contributed

Table 5.5 Dynamics of surface ecological water from 1976 to 2022

Year	1976	1988	1997	2007	2017	2022
Low ecological water (km^2)	63.23	68.10	73.89	62.68	75.86	76.16
Ecological water (km^2)	6.10	6.75	19.69	12.80	28.37	29.22
Beyond ecological water (km^2)	18.72	46.84	31.80	58.76	15.10	19.96

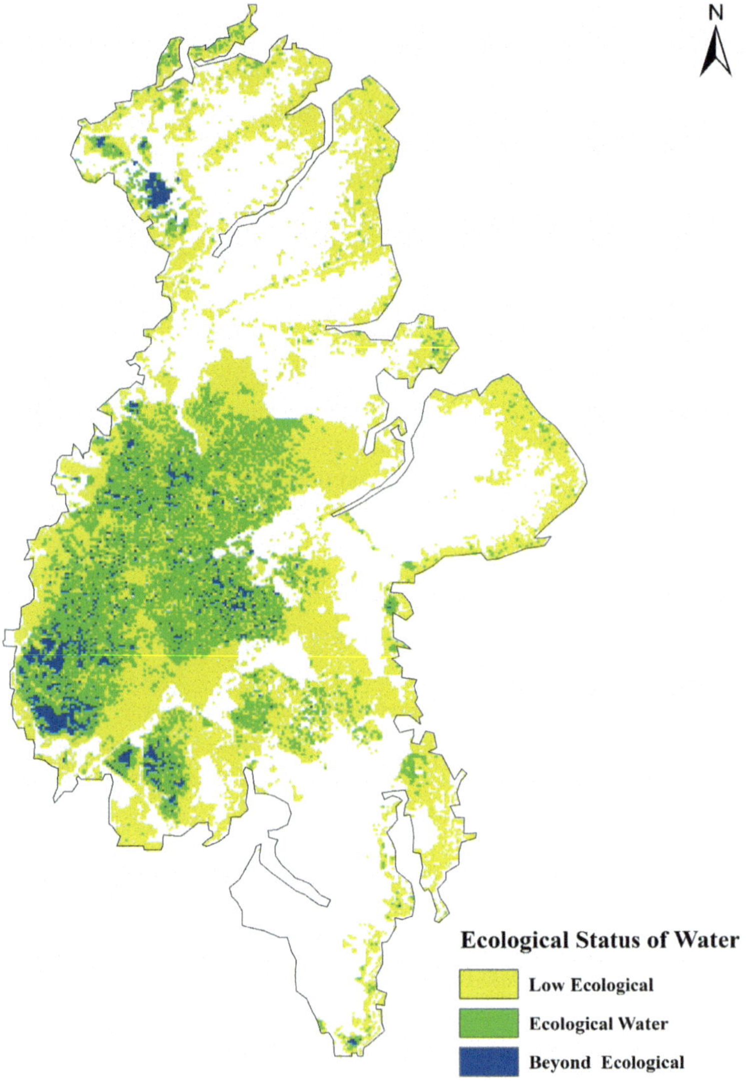

Fig. 5.5 Ecological status of water based on water or inundation presence frequency (WPF/IPF) from 1976 to 2022

to increased water retention in the lake. In contrast, the MNDWI ($y = 3.7782x + 60.525$) with an R^2 of 0.1786 implies a weaker correlation, suggesting less pronounced changes in water body dynamics over the same period. The trend equations for different LULC categories show varying degrees of changes. For open water bodies ($y = 2.192x + 62.315$; $R^2 = 0.4415$), there's a modest increasing trend, potentially influenced by the barrage's regulation of water levels. Agricultural ponds show a strong increasing trend ($y = 4.9591x - 0.202$; $R^2 = 0.8163$), reflecting changes in land use due to altered hydrological patterns post-barrage. Phumdis with thick vegetation and thin vegetation display decreasing trends, suggesting changes in vegetative cover possibly due to water level stabilization and other ecological impacts of the barrage. Built-up areas ($y = 0.5351x + 1.0687$; $R^2 = 0.6195$) show a significant increase, indicating expanding human settlement or infrastructure development, which could be a response to the altered landscape post-barrage. The category 'Low Ecological Water' shows a strong positive trend ($y = 3.2348x + 1.1249$; $R^2 = 0.844$), indicating an increase in areas with lower ecological significance, possibly due to persistent high-water levels caused by the

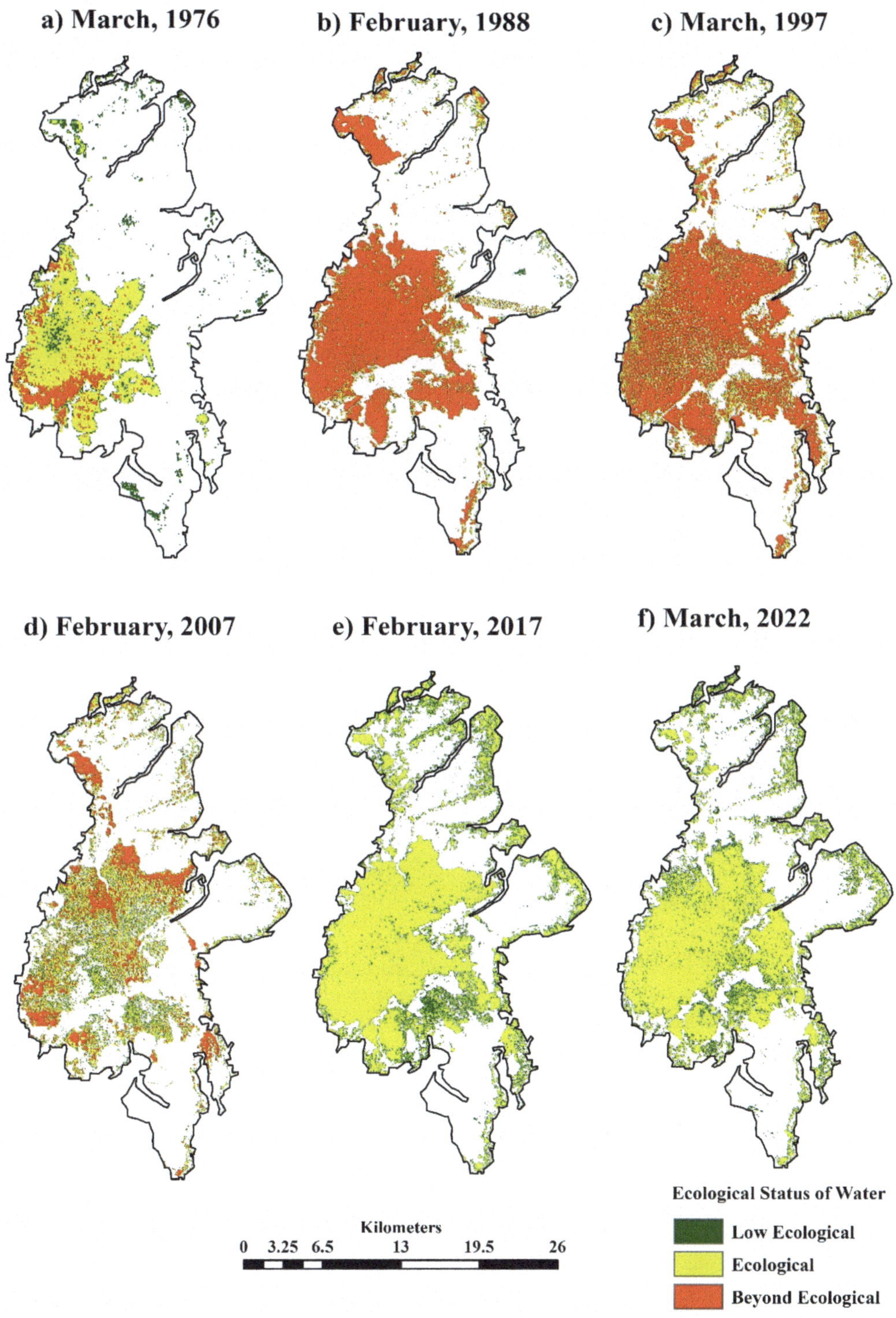

Fig. 5.6 Map showing the dynamics of surface ecological water of Loktak Lake over time

barrage. The 'Ecological Water' category also shows an increasing trend ($y = 11.472x + 0.4532$; $R^2 = 0.5536$), suggesting an expansion of areas within optimal ecological conditions. In contrast, the 'Beyond Ecological Water' category ($y = -7.555x + 51.96$; $R^2 = 0.231$) exhibits a negative trend, indicating a reduction in areas experiencing ecological stress.

Overall, the analysis suggests that the construction and operation of the Ithai Barrage have significantly influenced the hydrological and ecological dynamics of Loktak Lake. While it has led to an increase in surface water and areas falling within optimal ecological conditions, it has also contributed to ecological stresses, as evidenced by the increase in low ecological water areas and changes in LULC patterns. These findings underscore the need for careful management and monitoring of the barrage's operations to balance developmental needs with ecological sustainability.

5.4 Conclusion

This study provides a comprehensive quantitative analysis of the impact of the Ithai Barrage on Loktak Lake's hydrological and ecological dynamics. Utilizing NDWI and MNDWI, significant fluctuations in the surface water extent were observed, with an increase from 52.6392 to 94.1004 km^2 (NDWI) between 1976 and 1997, followed by a decrease and subsequent variability. Land Use and Land Cover (LULC) changes, assessed through random forest classification, revealed an increase in agricultural ponds from 6.10 to 29.22 km^2 and expansion of built-up areas from 2.28 to 5.41 km^2 over the study period. The ecological water assessment highlighted shifts in low ecological and ecological water categories, indicating altered ecological conditions. The study contextualizes the significant role of human interventions, particularly the Ithai Barrage, in altering natural ecosystems. It underscores the barrage's multifaceted impacts, from influencing surface water spread to affecting land use patterns and ecological water conditions. The research contributes to a deeper understanding of the complex interplay between human activities and aquatic ecosystems, emphasizing the need for integrated management strategies. However, the study has limitations, including reliance on remote sensing data and indices which might not capture all aspects of ecological changes. Future research should incorporate ground-based observations, more detailed hydrological modeling, and socio-economic aspects to provide a more holistic understanding of the barrage's impact. Additionally, exploring the long-term ecological and socio-economic impacts of the barrage could offer insights for better management and policymaking, aiming to balance developmental needs with environmental conservation and community welfare.

References

Belgiu M, Drăguţ L (2016) Random forest in remote sensing: a review of applications and future directions. ISPRS J Photogramm Remote Sens 114:24–31. https://doi.org/10.1016/j.isprsjprs.2016.01.011

Davidson NC (2014) How much wetland has the world lost? Determining global wetland loss and restoration needs. Proc Natl Acad Sci 111(13):4130–4137

Furtak K, Wolińska A (2023) The impact of extreme weather events as a consequence of climate change on the soil moisture and on the quality of the soil environment and agriculture—a review. CATENA 231:107378

Garg JK (2015) Wetland assessment, monitoring and management in India using geospatial techniques. J Environmanag 148:112–123

Gao BC (1996) NDWI—A normalized difference water index for remote sensing of vegetation liquid water from space. Remote Sens Environ 58(3), 257–266

Junk WJ, Annegarn H, Bonetto AA, Cavalcante RS, Crawter SJ, Noto LO et al (2016) The significance of wetlands for biodiversity conservation: a global perspective. Front Ecol Environ 14(7):436–447

Kangabam RD, Boominathan SD, Govindaraju M (2015) Ecology, disturbance and restoration of Loktak Lake in Indo-Burma Biodiversity Hotspot-an overview. NeBIO-An Int J Environ Biodivers 6(2):9–15

Kangabam RD, Bhoominathan SD, Kanagaraj S, Govindaraju M (2017) Development of a water quality index (WQI) for the Loktak Lake in India. Appl Water Sci 7(6):2907–2918. https://doi.org/10.1007/s13201-017-0579-4

Kangabam RD, Selvaraj M, Govindaraju M (2018) Spatio-temporal analysis of floating islands and their behavioral changes in Loktak Lake with respect to biodiversity using remote sensing and GIS techniques. Environ Monit Assess 190:1–14

Kangabam RD, Selvaraj M, Govindaraju M (2019) Assessment of land use land cover changes in Loktak Lake in Indo-Burma biodiversity hotspot using geospatial techniques. Egypt J Remote Sens Space Sci 22(2):137–143. https://doi.org/10.1016/j.ejrs.2018.04.005

Laishram RJ, Yumnam G, Alam W (2022) Assessment of ecohydrogeochemical status of freshwater Loktak Lake of Manipur, India. Environ Monit Assess 194(9):659. https://doi.org/10.1007/s10661-022-10336-w

Mahato S, Pukhrambam G, Joshi PK (2023) Damming effects on hydrological abundance and eco-hydrological alteration in upstream wetlands of Eastern Himalaya. J Clean Prod 418:138089

Mallick J, Talukdar S, Pal S, Rahman A (2021) A novel classifier for improving wetland mapping by integrating image fusion techniques and ensemble machine learning classifiers. Eco Inform 65:101426

McFeeters SK (2013) Using the normalized difference water index (NDWI) within a geographic information system to detect swimming pools for mosquito abatement: a practical approach. Remote Sens 5:3544–3561. https://doi.org/10.3390/rs5073544

Mueller S, Wang D, Fox MD, Yeo BT, Sepulcre J, Sabuncu MR, ... Liu H (2013) Individual variability in functional connectivity architecture of the human brain. Neuron 77(3):586–595

Paonam J, Chatterjee S (2023) Evaluating the land use land cover dynamics of Loktak Lake, A ramsar wetland of international importance in north east India. Int J Ecology Environ Sci 49(5):459–470

Shahfahad, Talukdar S, Islam ARMT, Das T, Naikoo MW, Mallick J, Rahman A (2022) Application of advanced trend analysis techniques with clustering approach for analysing rainfall trend and identification of homogenous rainfall regions in Delhi metropolitan city. Environ Sci Pollut Res 1–19

Talukdar S, Pal S (2017) Impact of dam on inundation regime of flood plain wetland of punarbhaba river basin of barind tract of Indo-Bangladesh. Int Soil Water Conserv Res 5(2):109–121

Talukdar S, Pal S (2019) Effects of damming on the hydrological regime of Punarbhaba river basin wetlands. Ecol Eng 135:61–74

Xu H (2006) Modification of normalized difference water index (MNDWI) to enhance open water features in Landsat imagery. Remote Sens Environ 105(4):404–426. https://doi.org/10.1016/j.rse.2006.06.007

Zedler JB, Kercher S (2005) Wetland resources: status, trends, ecosystem services, and restorability. Annu Rev Environ Resour 30:39:74

Integration of Machine Learning Models with Game Theory for Understanding Water-Induced Soil Erosion in an Urban Watershed

6

Ishita Afreen Ahmed, Swapan Talukdar, Jasmine Sultana, Mirza Razi Imam Baig, Hoang Thi Hang, and Atiqur Rahman

Abstract

In recent years, soil erosion has become a growing global concern, exacerbated by factors like abrupt climate shifts, natural resource exploitation, and soil quality degradation, with urban areas facing significant challenges. The Guwahati urban watershed, located in northeastern India, is one such urban region grappling with water-induced soil loss. To address this environmental issue, this study employs a range of models and techniques. Firstly, the Revised Universal Soil Loss Equation (RUSLE) model is utilized to estimate potential soil erosion in the area. To enhance the accuracy of erosion prediction, a finely tuned random forest (RF) machine learning model is deployed. Additionally, a Deep Neural Network (DNN)-based game theory model sheds light on feature contributions to soil erosion. The findings of the RUSLE model categorize the six sub-watersheds into varying risk zones, highlighting areas of high susceptibility to erosion, particularly Bharalu, Silsako, and Foreshore. Furthermore, after rigorous optimization, the RF model achieved impressive accuracy, with a root mean square error (RMSE) of 1.66 and mean absolute error (MAE) of 1.1, indicating reliability. Precipitation emerges as the primary influencer of erosion, emphasizing the importance of monitoring precipitation patterns and intensity. Also, the DNN-based game theory highlights that factors such as basin shape (F_f) and area significantly impact soil erosion. The heightened sensitivity to rainfall, Shannon Diversity Index (SHDI), and patch cohesion index (COHES) emphasize the importance of continuous monitoring of these parameters. These findings provide valuable insights into the dynamics of soil erosion in the study area, emphasizing the need for effective management strategies.

Keywords

Soil erosion · Urban watershed · Behavioral assessment · Random Forest · Game theory

I. A. Ahmed · M. R. I. Baig · H. T. Hang · A. Rahman (✉)
Department of Geography, Faculty of Sciences, Jamia Millia Islamia, New Delhi 110025, India
e-mail: arahman2@jmi.ac.in

I. A. Ahmed
Department of Civil Engineering, Indian Institute of Technology, New Delhi 110016, India

S. Talukdar
Department of Geography, Asutosh College, Kolkata, West Bengal 700026, India

J. Sultana
Amity Institute of Geoinformatics and Remote Sensing, Amity University, Noida, Uttar Pradesh 201301, India

S. Talukdar et al. (eds.), *Water Resource Management in Climate Change Scenario*, GIScience and Geo-environmental Modelling, https://doi.org/10.1007/978-3-031-61121-6_6

6.1 Introduction

Soil erosion, increasingly recognized as a critical environmental issue, is intensified by rapid climate change, overuse of natural resources, and declining soil health. The extensive ecological consequences of soil erosion are becoming a matter of heightened global attention, as indicated in multiple research findings (Maja and Ayano 2021). Over recent decades, the escalating risk of soil erosion and its implications for long-term soil sustainability have become increasingly evident. This situation calls for the formulation of thorough soil management strategies and effective conservation measures, aimed at protecting soil against erosive elements (Khalil and Aslam 2022). A pivotal element influencing soil erosion is the presence of vegetative cover, which performs a vital task in shielding the soil against erosive forces. Vegetation, comprising leaves, stems, and root systems, effectively intercepts rainfall and controls the flow of surface runoff, functioning as a protective barrier for the preservation of soil and water resources (Ma et al. 2021; Bai et al. 2019).

However, recently, changes in land use and modifications in soil structure, mainly attributable to severe farming and desertification, have made soil more prone to erosion. Urban development and population increase have also contributed to the removal of vegetation and increased soil displacement (Zerbe 2022). As a result, the formulation of strategic, efficient management approaches is crucial to prevent rapid ecological deterioration and ensure the long-term productivity of soil. Soil degradation significantly contributes to the formation of sediments, resulting in increased sedimentation in watersheds. According to Pal (2016), approximately 1964.4 million hectares of soil worldwide have been deteriorated as a result of human activities, with water erosion responsible for the degradation of 1903 million hectares of this amount. India, a humid subtropical country, grapples with significant threats to its fertile topsoil, with approximately 45% of its land area, equivalent to 1.468 billion acres, at risk of soil erosion, primarily driven by surface runoff (ICAR-NBSS&LUP 2021). To combat further soil erosion and conserve natural resources, comprehensive area-specific studies are imperative, as underscored by recent research (Pal et al. 2021).

Researchers have used various empirical and physical modeling approaches for estimating soil erosion and sediment yield, but these models often face limitations in small to medium-sized watersheds, especially in developing countries like India, due to a lack of data. To address this issue, Wischmeier and Smith introduced the Universal Soil Loss Equation (USLE) in 1978, and its updated version, the Revised Universal Soil Loss Equation (RUSLE), was developed by Renard in 1997. The RUSLE model, a more streamlined iteration of the USLE, has gained widespread adoption for soil erosion prediction and demonstrated superior accuracy (Nearing et al. 2005; McCool et al. 1987).

Recently, machine learning algorithms (MLAs) have emerged as valuable tools in studies related to natural hazards and the environment (Pourghasemi et al. 2023). Of these machine learning algorithms (MLAs), the random forest (RF) model has gained substantial acclaim and has been widely employed in environmental research (Pham et al. 2019). However, while MLAs offer powerful predictive capabilities, their outputs are often challenging to interpret. Post hoc interpretation methods, such as the Shapley additive explanations (SHAP) method, have been developed to enhance the interpretability of complex MLAs and improve their accuracy (Murdoch et al. 2019). This research employs a novel method, integrating deep neural networks (DNN) with game theory to interpret soil erosion through RUSLE and evaluate the behavior of different parameters influencing soil erosion. The study seeks to recognize the most influencing factors of soil erosion utilizing hyper-tuned ML model and evaluate their behavioral patterns under varying conditions conducive to soil erosion. The hypothesis posits soil erosion in the Guwahati, and this study aims to study the underlying mechanisms of this phenomenon while adopting technological advancements for effective soil erosion management.

6.2 Research Methodology

6.2.1 Study Area

The urban watershed of Guwahati is selected as the study area, due to its distinctive growth and patterns of development, along with the rapid transformations in its land use. Over the years, the population of the Guwahati Metropolitan Area (GMA) has seen remarkable growth, surging from 500,000 in 1991 to 814,000 in 2001 and 963,000 in 2011, according to Census of India data. There are six sub-watersheds in the study area (Fig. 6.1). These sub-watersheds, which have experienced rapid urbanization, are now integrated into the main sewer network. This integration has escalated the issue of soil erosion and degradation, primarily due to the area's swift expansion and the introduction of various pollutants. Also, the watershed is marked by distinct stratification in its soil composition, with layers of fine and coarse-grained soils of different thicknesses.

6.2.2 Data Used

The land use and land cover (LULC) map for the study area was generated from USGS Landsat 8 satellite imagery. The DEM used to define sub-watersheds was created from SRTM data. Soil taxonomy information came from the USDA, and IMD provided high-resolution rainfall data (0.25° × 0.25°). These datasets were used to identify the soil erosion influencing factors. Table 6.1 is summarizing the data sources of the present study.

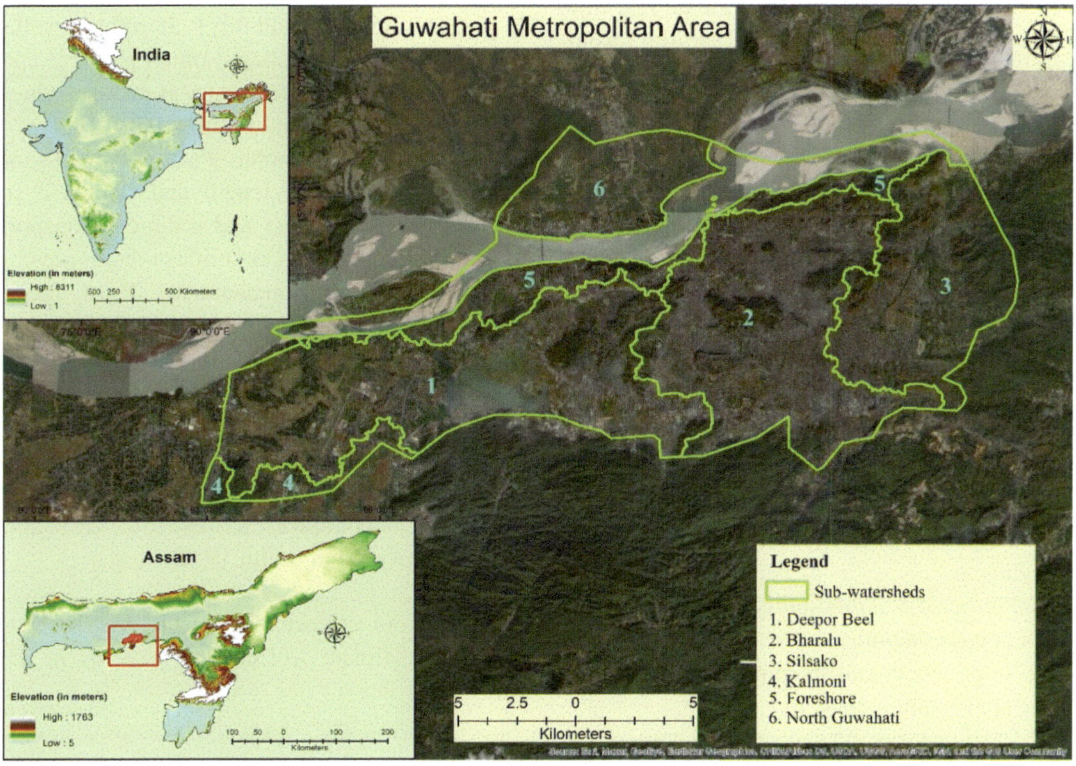

Fig. 6.1 Location map of the study area

Table 6.1 Dataset employed in the research

S. No.	Data source	Path/row	Year of acquisition	Spatial resolution (m)	Extracted information
	Landsat 8 (OLI/TIRS)	137/42	2020	30	LULC
1	SRTM-digital elevation model	137/42	2004	30	Aspect, slope, drainage, elevation, characteristics, etc.
2	Google Earth	–	2020	0.5	Guwahati city boundary and accuracy assessment
3	India Meteorological Department	–	1991–2020	–	Precipitation data
4	The US Department of Agriculture	–	2020	–	Soil data

6.2.3 Methodology

6.2.3.1 Methodology for Assessing Soil Erosion

RUSLE, known for its capacity to estimate soil erosion at pixel level, was employed in this study to acquire soil erosion data and conduct a detailed analysis. The RUSLE model is widely recognized for its capacity to simulate real-time soil erosion using spatial data of varying resolutions. As observed in previous studies (Kumar et al. 2022), RUSLE is particularly effective in analyzing streamflow variability and soil loss, considering factors such as rainfall, topology as well as land use information of a region.

RUSLE calculates annual soil loss based on five surface parameters using the following Equation (6.1):

$$\mathrm{SE} = R \times K \times LS \times C \times P \tag{6.1}$$

Here,

- SE: Predicted soil loss in tons per hectare per year.
- *R*: Regenerative index measured in MJ mm/ha/h/year.
- *K*: Predicted soil erodibility factor in tons per hectare per MJ/mm.
- *P*: Slope length and steepness, also known as the conservation element.
- *C*: Land use-specific management component.

The RUSLE model utilizes several factors for modeling soil erosion, including:

1. *R-factor* (Rainfall Erosivity): It quantifies precipitation's ability to cause soil erosion. Monthly IMD precipitation data and long-term rain gauge data from 1991 to 2020 were used. Empirical calculations based on annual rainfall data determined the *R*-factor.
2. *K-factor* (Soil Erodibility): It reflects soil's intrinsic erodibility based on its physical and chemical properties, using soil texture values from Wischmeier and Smith (1978).
3. *LS-factor* (Topographic Factors): LS-factor accounts for slope length and steepness in determining soil loss and sediment transport capacity, although it simplifies topography.
4. *C-factor* (Cover Management): The C-factor is derived from land use and land cover (LULC) maps, and it assesses soil loss differences between conservation practices and cultivation on slopes.
5. *P-factor* (Support Techniques): Zones are assigned according to values based on agricultural techniques, termed the 'P-factor' (Wischmeier and Smith 1978).

These factors collectively enable the RUSLE model to estimate soil erosion patterns in the study area.

6.2.3.2 Preparation of Soil Erosion Influencing Parameters

Numerous factors contribute to influencing soil erosion within a region. In consideration of the natural attributes of the study area, sixteen physical and anthropogenic factors have been identified, drawing from existing literature and available data. Table 6.2 provides a detailed description of these factors.

Table 6.2 Detailed description of the soil erosion influencing parameters

Basin properties	Predictors	Abbreviation	Descriptions
Basin morphology	Basin area	A	Sub-basin's area in square kilometers
	Basin length	L_b	Length of the sub-basin (km)
	Basin perimeter	P	The perimeter of the basin in kilometers (km)
	Form factor	F_f	The ratio of the basin's area (A) to the product of its circumference (P) and the square of its length (L_b)
Drainage network	Stream frequency	F_s	Ratio of the total number of channel segments per unit area for all stream orders
	Drainage density	D_d	Calculated total channel lengths per unit area, i.e., SL/A (km/km^2)
Topography	Slope gradient	S	Average % slope gradient of the sub-basin
	Basin relief	B_h	The difference between the highest and lowest relief points within the sub-basin, measured in meters (m)
	Elevation	DEM	Mean elevation of the sub-basin (m)
	Topographic wetness index	TWI	TWI = $\ln(\alpha/\tan(\beta))$, where α is the upslope area per unit contour length and $\tan(\beta)$ is the local slope
Climate	Mean annual precipitation	Rainfall	Mean annual precipitation (mm)
	Rainfall erosivity factor	R-factor	RUSLE erosivity factor (MJ mm ha^{-1} h$-^{1}$ yr^{-1})
Land use composition	Percentage of agricultural area	AGRI	Percentage of basin devoted to agricultural land use
	Percentage of built-up area	BUILT-UP	Percentage of basin devoted to built-up area
	Percentage of vegetation area	VEGETATION	Percentage of basin devoted to vegetation coverage
	Percentage of others (barren, open, waste, etc.)	OTHERS	Percentage of basin devoted to other lands (barren, open, waste, etc.)
	Percentage of grassland	GRASSLAND	Percentage of basin devoted to grassland
	Percentage of waterbody	WATERBODY	Percentage of basin devoted to waterbody
	Percentage of river	RIVER	Percentage of basin devoted to river
	Percentage of Riverbed	RIVERBED	Percentage of basin devoted to river bed
Land fragmentation	Patch density	PD	Number of patches per unit area (number per 100 ha)
	Edge density	ED	Total length of all edge segments per hectare for the considered landscape (m/ha)
	Shannon's diversity index	SHDI	Patch diversity in a landscape (unit = none)
	Patch cohesion index	COHES	Physical connectedness of the corresponding patch type (unit: none)

Land Use Composition and Land Fragmentation Factors

Landsat 8 (OLI/TIRS) imagery from Earth Explorer-USGS was utilized in the study to create the LULC map with eight categories. The map, dominated by built-up areas, underwent validation with a kappa coefficient of 0.86. Subsequently, FragStat software was employed to assess landscape fragmentation using patch density, edge density, Shannon Diversity Index (SHDI), and COHES. Eight LULC classes and four land parameters described the landscape composition. These metrics are commonly applied in studies on soil erosion patterns related to LULC (Zhang et al. 2021) (Fig. 6.2).

Basin Characteristics

Basin features, such as area, perimeter, and length, significantly influence watershed behavior. The size of the urban watershed, with an area of 339.25 km^2 and a perimeter of 922.16 km, is notably substantial for an urban area. Within the sub-watersheds, Deepor Beel is the largest (98.14 km^2), while Kalmoni covers a small area (8.75 km^2). Basin length, measured in a straight line from stream mouth to the farthest point on the divide, is 316.58 km for our urban watershed. Deepor Beel has the longest length at 10.50 km. Basin morphology strongly affects soil conditions and erosion risk, with larger basins generally posing an increased danger of soil erosion.

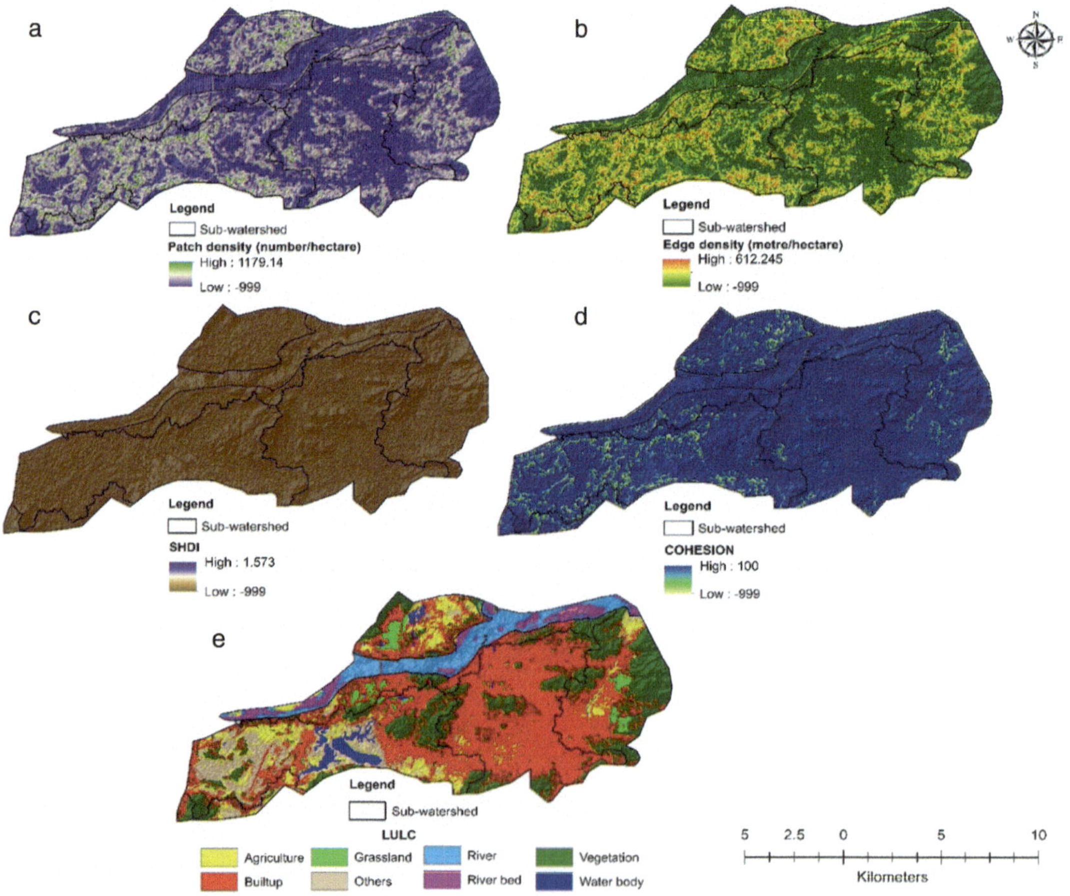

Fig. 6.2 LULC and land fragmentation: **a** PD, **b** ED, **c** SHDI, **d** COHES, and **e** LULC

The form factor (F_f), representing the ratio of basin area to the square of its length, characterizes basin shape (Horton 1932). Typically, F_f values are below 0.7854, indicating a round basin. Lower F_f values suggest more elongated basins. In our study, F_f ranges from 2.87 to 0.17, with Bharalu (2.87) and Silsako (1.48) sub-watersheds having higher F_f values, signifying a more rounded basin shape and potentially greater peak flows associated with soil erosion conditions (Fig. 6.3).

Drainage Characteristics

Stream Frequency (F_s): F_s is the ratio of the number of stream segments of all orders to the total watershed area (Horton 1945). This measure is affected by lithological factors and is crucial in understanding landscape development. In our research, the urban watershed's southern region, particularly the Kalmoni (3.89) and Foreshore (2.33) sub-watersheds, exhibited higher values. These increased values of Fs suggest reduced infiltration capacity, rendering these areas more prone to soil erosion.

Drainage Density (D_d): Horton (1945) notes that drainage density (D_d) is influenced by factors such as permeability of rocks, precipitation intensity, soil infiltration capacity, and slope gradient. D_d's value in a watershed has significant implications for vegetation, landforms, and the extent of soil erosion. In our research area, D_d values vary, with North Guwahati showing 0.84 km/km^2 and the Kalmoni sub-watershed reaching 1.43 km/km^2. These figures suggest the presence of rich vegetation and highly permeable surface materials in these areas (Fig. 6.4).

Topographic Characteristics

Slope: The slope angle, ranging from 0° to 46.5° in our study, crucially influences water flow speed and direction.

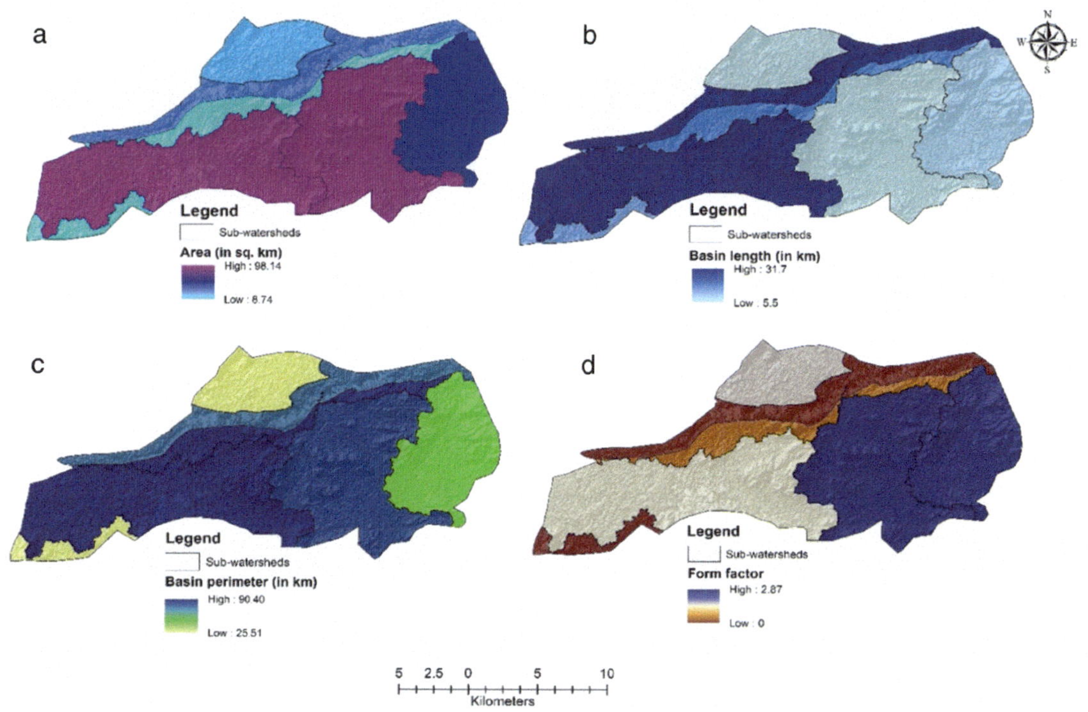

Fig. 6.3 Basin morphology: **a** area, **b** length, **c** perimeter, and **d** form factor

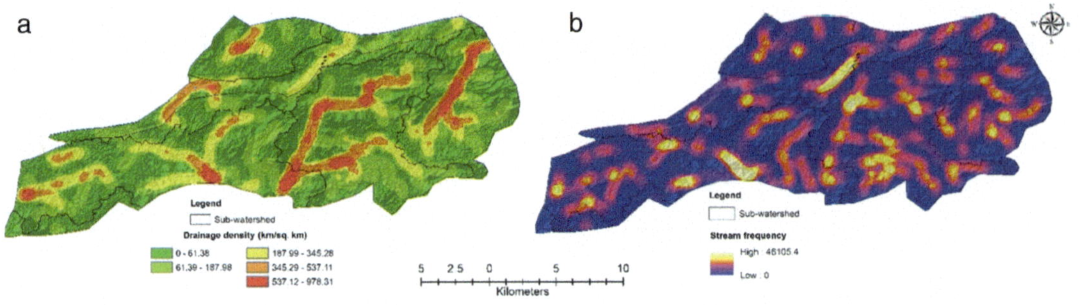

Fig. 6.4 Drainage network: **a** D_d, **b** F_s

Basin Relief (B_h): B_h, representing the maximum vertical distance within the watershed, varies from 28.17 to 83.33 in the Guwahati urban watershed. Higher Bh values indicate steeper relief and slopes, posing a higher risk of soil erosion.

Elevation: Elevation within the Guwahati urban watershed spans 0–40 m, signifying an undulating topography. This data was derived from SRTM dataset at a 30-m resolution.

Topographic Wetness Index (TWI): TWI characterizes watershed wetness based on spatial variations (Beven and Kirkby 1979). It ranges from −8.24 to 2.35 in our study, providing insights into wetness distribution across the Guwahati urban watershed (Fig. 6.5).

Rainfall

Figure 6.6 shows the precipitation patterns within the Guwahati urban watershed.

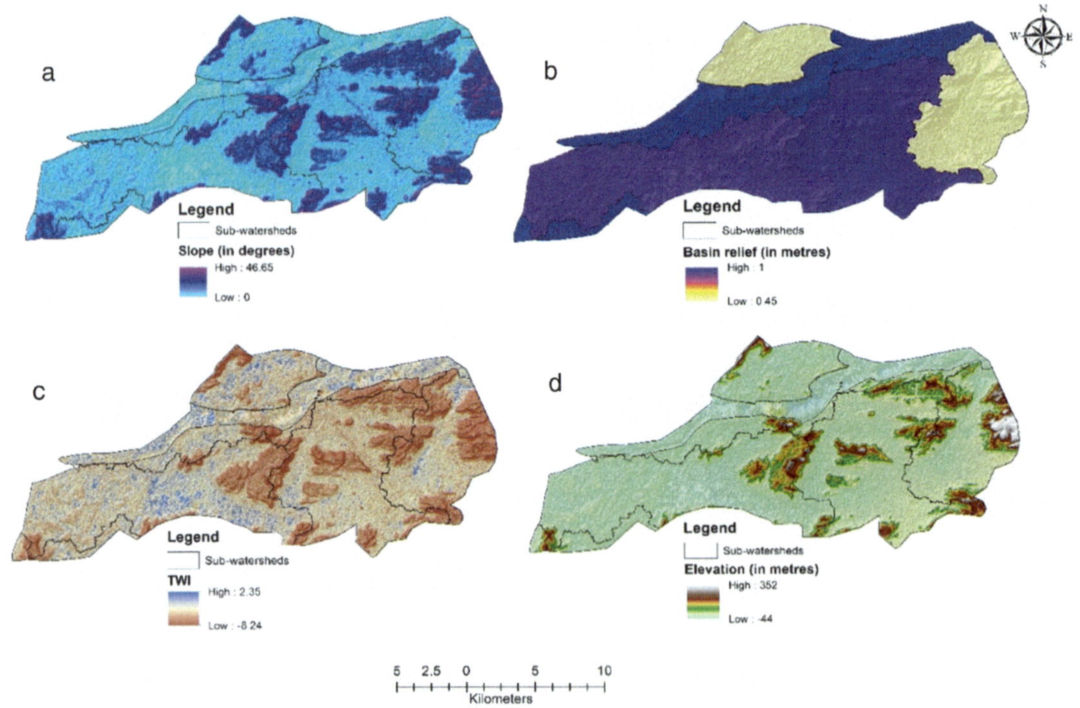

Fig. 6.5 Basin topography: **a** slope, **b** basin relief, **c** TWI, and **d** DEM

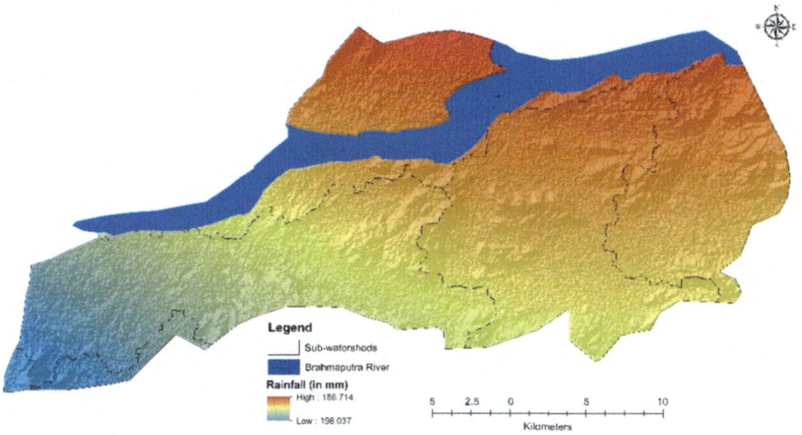

Fig. 6.6 Present rainfall scenario in the study area

Rainfall ranges with values varying from a peak of 186.7–156 mm in the current scenario. Numerous studies in climatology have explored the relationship between rainfall and soil erosion (Bagagiolo et al. 2018).

6.2.3.3 Analysis of Soil Erosion Factors Using Machine Learning Method

The present study uses the random forest (RF) machine learning technique to delve into the complex factors influencing soil erosion. To achieve this, a robust approach was taken by utilizing the H_2O platform in conjunction with the R programming language. The combination of these tools allowed for the implementation and refinement of machine learning models, with a particular focus on enhancing the performance of the random forest algorithm. The complete procedure was conducted using the H_2O framework, adeptly managing critical steps including the division of data for training and testing, building the model, fine-tuning hyperparameters via grid search, thorough validation, and ultimately applying the model to raster data.

The key innovation in this study lies in the use of a hyper-tuned random forest model. Unlike default models that rely on predefined settings, the hyper-tuned version meticulously optimizes multiple critical components. These components include the number of trees, the selection of variables at each split, tree depth, and other crucial parameters. This comprehensive optimization process ensures that the RF model performs at its peak, resulting in significantly improved accuracy when predicting soil erosion patterns. For model training, the *R* function 'h2o.randomForest()' was harnessed, enabling the machine to learn and adapt to the intricacies of the dataset. This approach not only boosts the accuracy of soil erosion predictions but also equips researchers and environmentalists with a powerful tool for understanding and managing this critical environmental issue.

6.2.3.4 Assessing Soil Erosion Parameters with Game Theory

The present work aimed to evaluate the influence of different parameters on soil erosion through the application of game theory. A deep neural network (DNN) model was employed to calculate Shapley values, which evenly distribute prediction differences among features, aiding in individual prediction explanations. DNNs, while more accurate than Artificial Neural Networks (ANNs), require additional computational resources and time due to their complexity. The approach involved two steps: developing a DNN model and integrating it into a framework for sensitivity analysis. After identifying the optimal DNN model, weights and structure were extracted, and parameter influence was assessed using the 'SensAnalysisMLP()' function. Finally, Shapley values were obtained with the SensFeaturePlot() method, providing insights into the relationship between outputs and inputs. This research enhances our understanding of soil erosion factors and prediction.

6.3 Result and Discussion

6.3.1 Soil Erosion Estimation

The RUSLE model is used in this study to evaluate potential soil erosion rates in the urban watershed of Guwahati. This method incorporates five key factors that are crucial in estimating soil loss of a region, i.e., *R*-factor, the *K*-factor, *C*-factor, LS-factor, and *P*-factor (Fig. 6.7). The results unveiled a spectrum of values for different factors influencing soil erosion. As depicted in Fig. 6.7a, the *K*-factors within the watershed exhibit a broad spectrum, spanning from 0.02 to 0.212 tons/ha/MJ/mm. The sub-watersheds like North Guwahati, Bharalu and Silsako, and Foreshore, stand out with exceptionally high *K*-factors, primarily due to their predominant sandy composition. In the study, the LS-factor varied significantly, ranging from −0.61 to 46.23, where higher numbers signaled a greater susceptibility to erosion, and lower figures

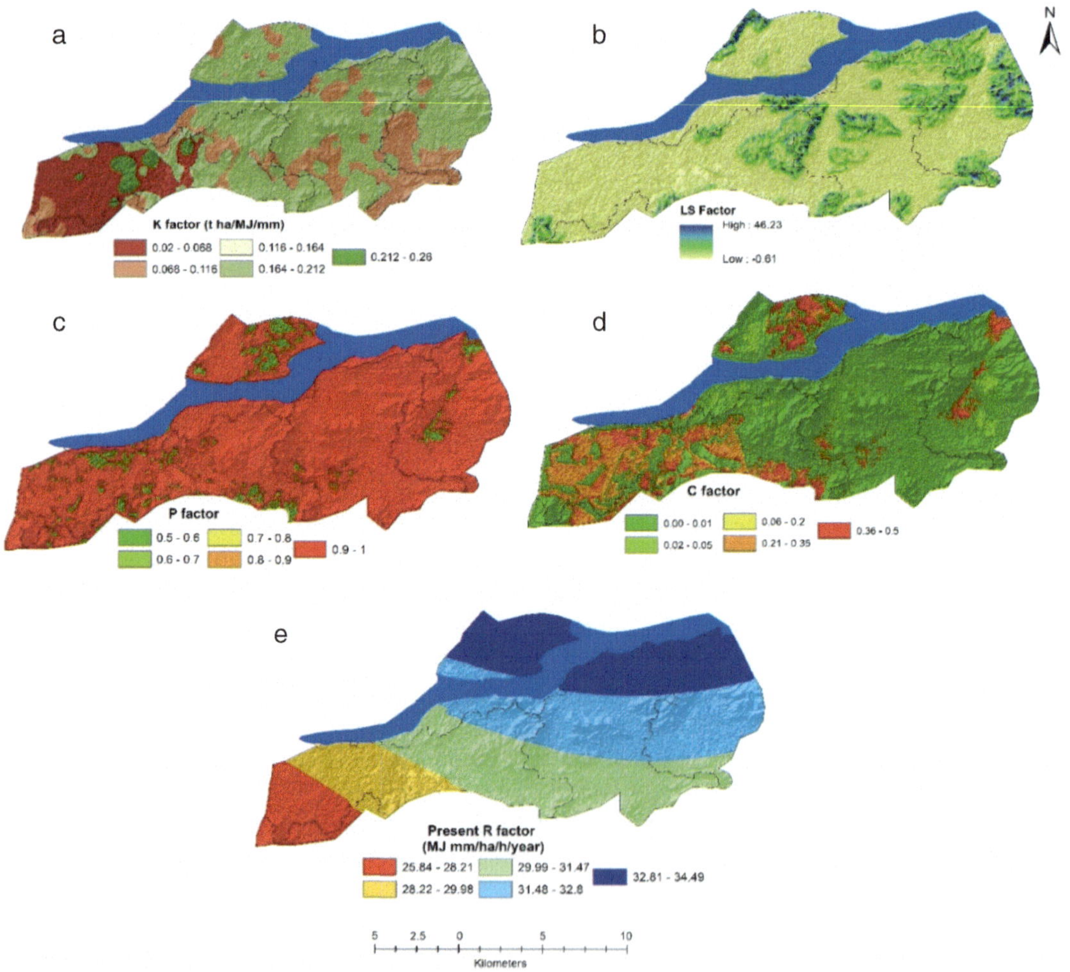

Fig. 6.7 RUSLE parameters: **a** K-factor, **b** LS-factor, **c** *C*-factor, **d** *P*-factor, **e** *R*-factor

indicated less vulnerability (Fig. 6.7b). The *P*-factor ranged from 0.5 to 1, with low estimates representing a greater erosion risk (Fig. 6.7c). Furthermore, Fig. 6.7e shows that the highest *R*-factor was observed in the North Guwahati area, with the northern regions of the Silsako and Bharalu sub-watersheds also exhibiting elevated *R*-factors. In a similar vein, the *C*-factor spanned from 0.01 to 0.35, indicating that lesser values implied a greater susceptibility to erosion caused by precipitation and runoff. These findings emphasize the pronounced risk of soil erosion in the Silsako and Bharalu sub-watersheds, primarily attributed to their consistently lesser values of *C*-factor (Fig. 6.7d).

The projected annual soil loss in the study area ranges from 166.96 to 188.41 tons per hectare (Fig. 6.7). The Bharalu, Foreshore, Silsako, and North Guwahati sub-watersheds displayed regions with significant soil loss, falling within the very high and high categories. Kalmoni and specific areas of Deepor Beel, however, showed lower erosion rates (Fig. 6.8). The removal of forest vegetation in hilly areas of Guwahati for urban development projects is a major factor contributing to soil erosion (Baruah and Naik 2020). This emphasizes the critical need for well-planned urban development and management strategies to mitigate soil erosion and its adverse consequences.

6.3.2 Determining Critical Factors Affecting Soil Erosion Using Machine Learning Models

This study aims to detect the key components initiating soil erosion using a finely tuned machine learning model, RF. While previous

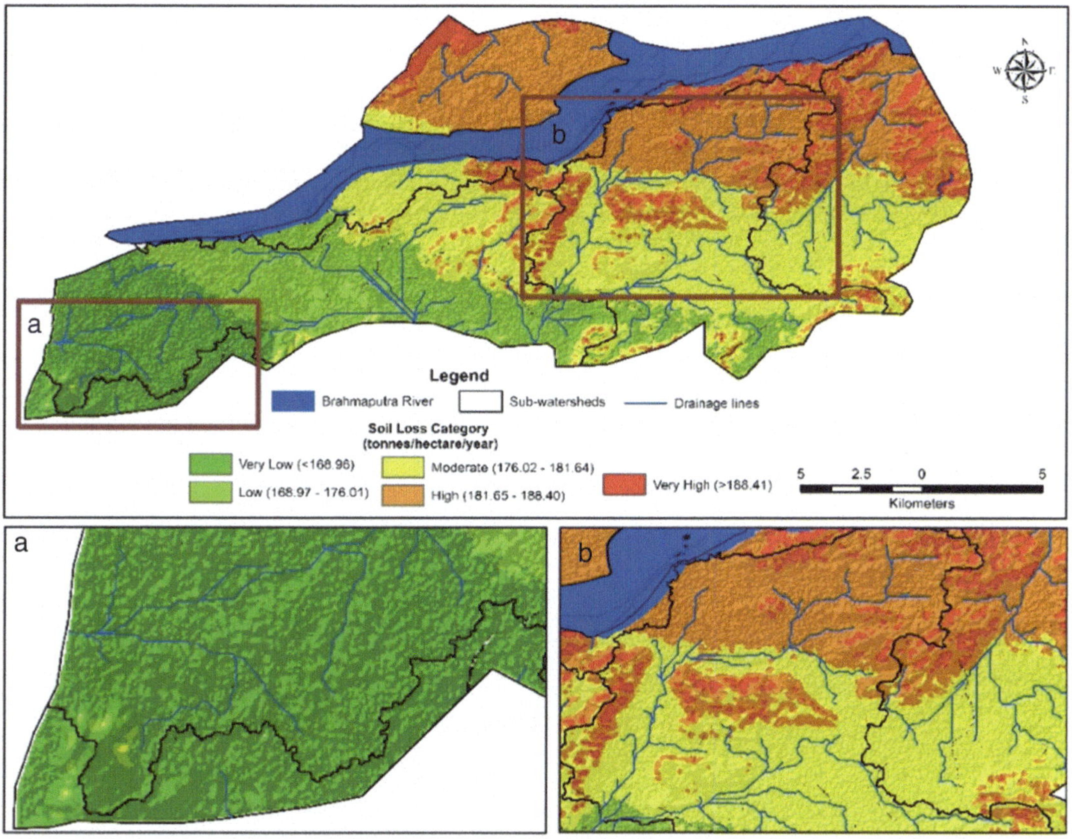

Fig. 6.8 Soil erosion condition: **a** low and very low soil erosion regions, **b** high and very high soil erosion region

research commonly relied on default RF models for predicting natural hazards, achieving accurate predictions requires precisely tuned models. The RF model consists of multiple components that necessitate optimization, a process that can be time-consuming to determine the ideal parameter combinations. In this study, an approach was adopted to identify optimal values for various RF model components, including the number of trees, variables sampled at each split, tree depth, minimum rows in terminal nodes, bins for continuous variables, and sample rate. Employing grid search, a set of 135 models was generated, ranked by accuracy in predicting test data (Fig. 6.9). The top-performing RF model was selected with optimized values: 250 trees, model size of 566,514 bytes, maximum depth of 21, mean depth of 14.54, minimum leaves of 157, maximum leaves of 201, and mean leaves of 175.60.

To review the model's accurateness efficiently, the optimal RF model was applied to predict new datasets (test data). The analysis involved various metrics, including mean squared error (MSE), root mean squared error (RMSE), mean absolute error (MAE), root mean squared logarithmic error (RMSLE), mean residual error, and coefficient of determination (R^2). The RF model demonstrated exceptional performance in predicting soil erosion. It achieved a low MSE of 2.74, indicating accurate predictions. The RMSE was 1.66, showcasing the model's precision. Additionally, the MAE of 1.11 and a small RMSLE of 0.01 further affirmed its reliability. The RF model's MSE of 2.739 and a high R^2 value of 0.98 indicated an effective fit to the data, emphasizing its effectiveness in soil erosion prediction. These findings suggest that the model excelled in accurately assessing the factors responsible for soil erosion and predicting it (Fig. 6.10).

The assessment highlights precipitation as the foremost factor impacting soil erosion, based on the RF model. This implies that changes in precipitation patterns and intensity influence the study area's soil erosion on a significant level. During periods of elevated precipitation, there is a greater likelihood of soil erosion due to increased surface runoff and the transport of soil particles. Additionally, the RF model identifies several other key parameters that contribute to soil erosion, such as land use (L_b), basin shape (F_f), size (area), perimeter, drainage density (D_d), slope steepness (slope), topographic wetness index (TWI), elevation, and basin relief (B_h) (Fig. 6.11). Together, these

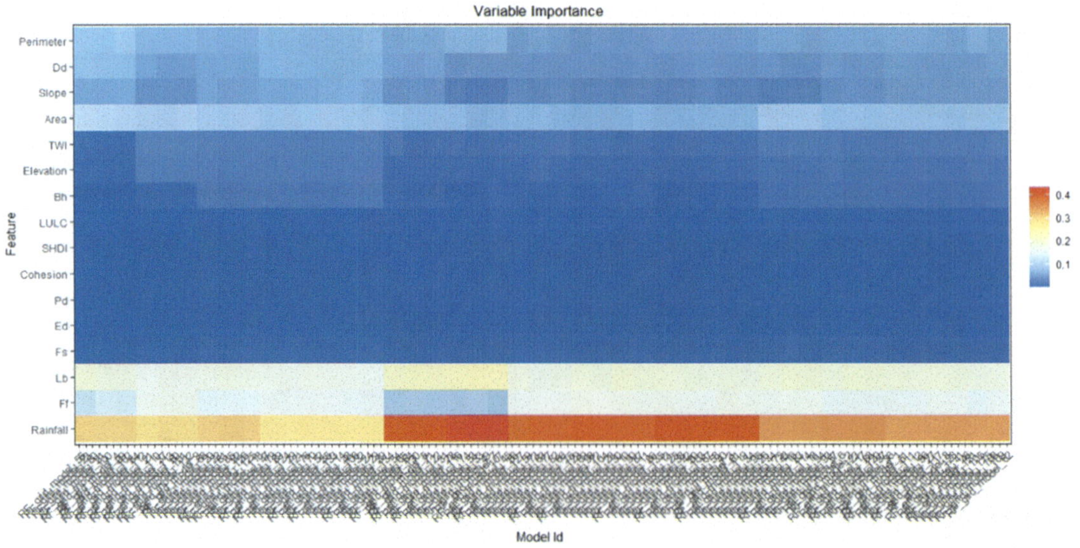

Fig. 6.9 Computation of importance for priority parameters for all RF models

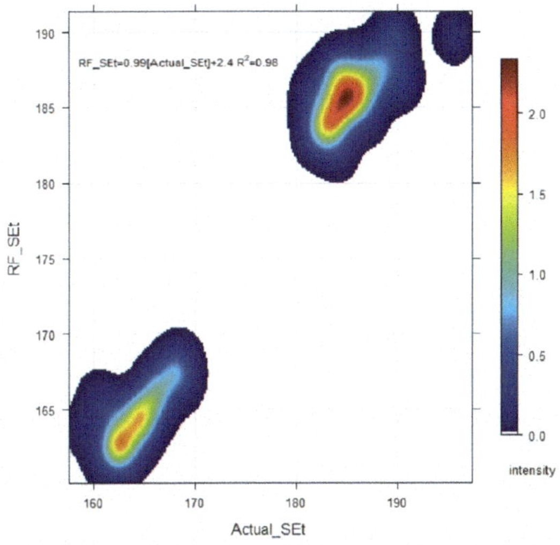

Fig. 6.10 Actual and predicted soil erosion

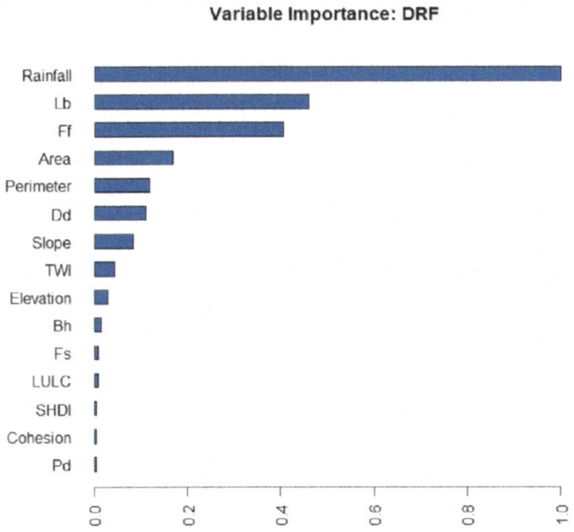

Fig. 6.11 Variable importance for soil erosion using RF

factors enhance the area's susceptibility to erosion. On the other hand, certain parameters like plant cover (P_d), cohesion, landscape diversity (SHDI), LULC, and stream frequency (F_s) were observed to exert a relatively minor influence on soil erosion, categorizing them as non-priority parameters. Therefore, this analysis provides significant comprehensions into the factors driving soil erosion, with precipitation being the most prominent, while others also play crucial roles in the erosion process within the studied region.

6.3.3 Evaluating the Behavior of Soil Erosion Influencing Factors Through Game Theory

The DNN-based game theory model generates a SHAP summary plot, offering insights into the impact of various features on soil erosion within the study area (Fig. 6.12). Summary or funnel plots serves as a visual tool to elucidate explanations for individual predictions. The presented plots demonstrate the amalgamation of feature significance and their corresponding influences. Additionally, the overlapping data points visually represent the scattering of Shapley values associated with each attribute. The base of the funnel plot represents the expected model output. Each line in the plot demonstrates how the positive or negative influence of particular features modifies the value from the predicted model output. This traversal extends from the background dataset to the model output for a specific estimate. Furthermore, the quantity of sample points exhibiting greater or lesser values reflects the sensitivity of the findings.

The study's findings revealed that both the F_f factor and the area factor had a notable impact on soil erosion. Additionally, the present work also highlighted the heightened sensitivity of soil erosion to greater values of SHDI, rainfall, and COHES. These results align with previous research, which has also examined soil erosion factors, revealing similar outcomes. F_f (Ahmed et al. 2023), elevation (Zhao et al. 2021), rainfall (Jia et al. 2022), slope (Guo et al. 2021), and LULC (Sooryamol et al. 2022) have consistently been recognized as key contributors to soil erosion in previous studies. Consequently, close and continuous monitoring of these parameters within the study area is strongly recommended to implement effective soil erosion management strategies.

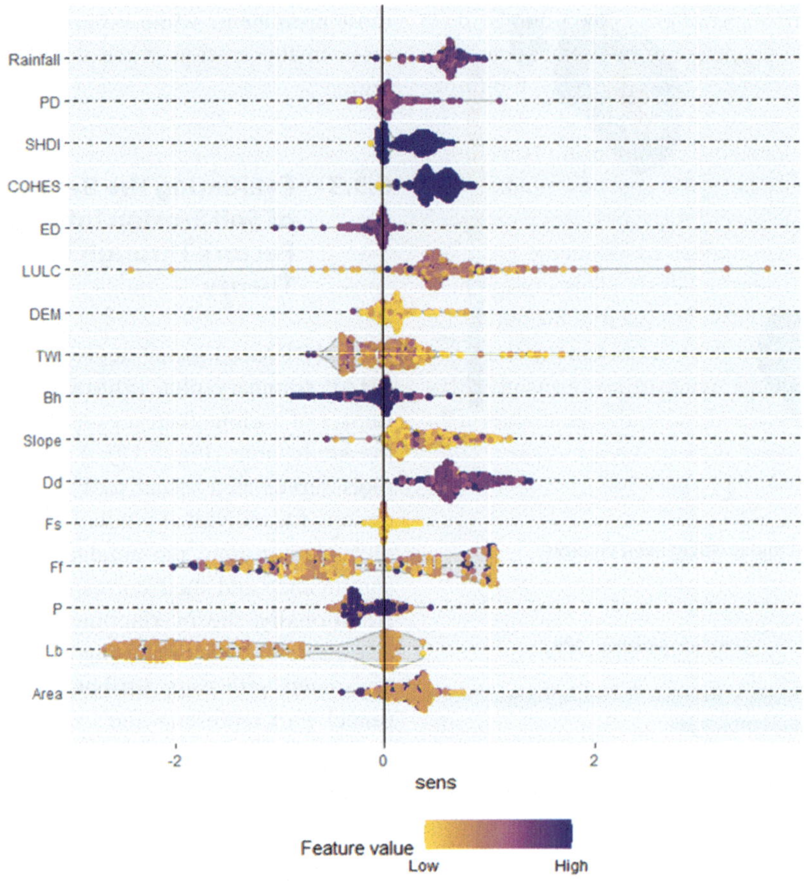

Fig. 6.12 Game theory application to study soil erosion patterns in individual samples

6.4 Conclusion

This study utilized RUSLE to assess the prospective soil erosion within the Guwahati urban watershed, considering a range of significant factors. The RUSLE model, incorporating *K*-factor, *R*-factor, *C*-factor, *P*-factor, and LS-factor, revealed a diverse range of values for these parameters. Notably, sub-watersheds with high *K*-factors, such as Foreshore, North Guwahati, Silsako, and Bharalu, were identified due to their sandy composition. LS-factor values indicated varying susceptibilities to erosion, with higher values signifying increased risk. The analysis further categorized the soil erosion into different zones, emphasizing the elevated risk in Foreshore, Silsako, North Guwahati, and Bharalu sub-watersheds. The study also underscored the importance of effective urban development planning and management in mitigating soil erosion caused by deforestation for construction projects in hilly areas.

To enhance soil erosion prediction, the study utilized a finely tuned machine learning model, random forest (RF), optimizing several model components. The top-performing RF model exhibited impressive accuracy, highlighting its effectiveness in predicting soil erosion within the urban watershed. Precipitation is the primary factor influencing erosion, emphasizing the need to monitor precipitation patterns and intensity. Additionally, other crucial parameters affecting soil erosion were identified, including land use, basin shape, size, perimeter, drainage density, slope steepness, topographic wetness index, elevation, and basin relief. These findings provide valuable insights into the influencing factors of soil erosion within the Guwahati urban watershed. Furthermore, the study employed a DNN-based game theory model that generated a SHAP funnel plot, which revealed the contributions of various features to soil erosion in the study area. Most significantly, F_f and area played significant roles in influencing soil erosion, along with heightened sensitivity to greater values of precipitation, SHDI, and COHES. These findings provide critical insights into the dynamics of soil erosion in the study area and emphasize the need for continuous monitoring of key parameters. Effective management strategies, informed by these results, are crucial for mitigating soil erosion's adverse effects on the region's environment and infrastructure. Moreover, the study underscores the importance of urban development planning that prioritizes vegetation cover preservation to combat soil erosion effectively. Policymakers should consider these findings when formulating strategies to address soil erosion in the region.

References

Ahmed IA, Talukdar S, Naikoo MW, Parvez A, Pal S, Ahmed S, Mosavi AH (2023) A new framework to identify most suitable priority areas for soil-water conservation using coupling mechanism in Guwahati urban watershed, India, with future insight. J Clean Prod 382:135363

Bagagiolo G, Biddoccu M, Rabino D, Cavallo E (2018) Effects of rows arrangement, soil management, and rainfall characteristics on water and soil losses in Italian sloping vineyards. Environ Res 166:690–704. https://doi.org/10.1016/J.ENVRES.2018.06.048

Bai J, Yang S, Zhang Y, Liu X, Guan Y (2019) Assessing the impact of terraces and vegetation on run-off and sediment routing using the time-area method in the Chinese Loess Plateau. Water 11(4):803. https://doi.org/10.3390/W11040803

Baruah P, Naik PRP (2020) Masterplan for Guwahati Metropolitan Area-2025 and sustainable environment. MuktShabd J 9:7400–7412

Beven KJ, Kirkby MJ (1979) A physically based, variable contributing area model of basin hydrology. Hydrol Sci Bull 24(1):43–69

Guo L, Liu R, Men C, Wang Q, Miao Y, Shoaib M, Wang Y, Jiao L, Zhang Y (2021) Multiscale spatiotemporal characteristics of landscape patterns, hotspots, and influencing factors for soil erosion. Sci Total Environ 779:146474. https://doi.org/10.1016/J.SCITOTENV.2021.146474

Horton RE (1932) Drainage-basin characteristics. Trans Am Geophys Union 13(1):350–361

Horton RE (1945) Erosional development of streams and their drainage basins; hydrophysical approach to quantitative morphology. Geol Soc Am Bull 56(3):275–370

ICAR-NBSS&LUP, Annual Report (2021) https://nbsslup.icar.gov.in/wpcontent/uploads/2021/10/Annual_Report/AnnualReport2021.pdf

Jia L, Yu KX, Li ZB, Li P, Zhang JZ, Wang AN, Ma L, Xu GC, Zhang X (2022) Temporal and spatial variation of rainfall erosivity in the Loess Plateau of China and its impact on sediment load. CATENA 210:105931

Khalil U, Aslam B (2022) Geospatial-based soil management analysis using novel technique for better soil conservation. Model Earth Syst Environ 8(1):259–275. https://doi.org/10.1007/S40808-020-01078-0/TABLES/6

Kumar M, Sahu AP, Sahoo N, Dash SS, Raul SK, Panigrahi B (2022) Global-scale application of the RUSLE model: a comprehensive review. Hydrol Sci J 67(5):806–830. https://doi.org/10.1080/02626667.2021.2020277

Ma S, Qiao YP, Wang LJ, Zhang JC (2021) Terrain gradient variations in ecosystem services of different vegetation types in mountainous regions: vegetation resource conservation and sustainable development. For Ecol Manage 482:118856. https://doi.org/10.1016/J.FORECO.2020.118856

Maja MM, Ayano SF (2021) The impact of population growth on natural resources and farmers' capacity to adapt to climate change in low-income countries. Earth Syst Environ 5(2):271–283. https://doi.org/10.1007/S41748-021-00209-6

McCool DK, Brown LC, Foster GR, Mutchler CK, Meyer LD (1987) Revised slope steepness factor for the universal soil loss equation. Trans ASAE 30(5):1387–1396

Murdoch WJ, Singh C, Kumbier K, Abbasi-Asl R, Yu B (2019) Interpretable machine learning: definitions, methods, and applications. arXiv preprint arXiv:1901.04592.

Nearing MA, Jetten V, Baffaut C, Cerdan O, Couturier A, Hernandez M, Le Bissonnais Y, Nichols MH, Nunes JP, Renschler CS, Souchère V, Van Oost K (2005) Modeling response of soil erosion and runoff to changes in precipitation and cover. CATENA 61(2–3):131–154

Pal S (2016) Identification of soil erosion vulnerable areas in Chandrabhaga river basin: a multi-criteria decision approach. Model Earth Syst Environ 2(1):1–11

Pal SC, Chakrabortty R, Arabameri A, Santosh M, Saha A, Chowdhuri I, Roy P, Shit M (2021) Chemical weathering and gully erosion causing land degradation in a complex river basin of Eastern India: an integrated field, analytical and artificial intelligence approach. Nat Hazards 110(2):847–879. https://doi.org/10.1007/s11069-021-04971-8

Pham BT, Prakash I, Singh SK, Shirzadi A, Shahabi H, Tran T-T-T, Bui DT (2019) Landslide susceptibility modeling using reduced error pruning trees and different ensemble techniques: hybrid machine learning approaches. CATENA 2019(175):203–218

Pourghasemi HR, Pouyan S, Bordbar M, Golkar F, Clague JJ (2023) Flood, landslides, forest fire, and earthquake susceptibility maps using machine learning techniques and their combination. Nat Hazards 116(3):3797–3816. https://doi.org/10.1007/S11069-023-05836-Y/FIGURES/9

Sooryamol KR, Kumar S, Regina M, David Raj A (2022) Modeling climate change impact on soil erosion in a watershed of north-western Lesser Himalayan region. J Sediment Environ 7(2):125–146

Wischmeier W, Smith D (1978) Predicting rainfall erosion losses—a guide to conservation planning. s. l.: USDA Agricultural Handbook No. 537

Zerbe S (2022) Global land-use development trends: traditional cultural landscapes under threat. Landscape Ser, 129–199. https://doi.org/10.1007/978-3-030-95572-4_4

Zhang Y, Wang N, Tang C, Zhang S, Song Y, Liao K, Nie X (2021) A new indicator to better represent the impact of landscape pattern change on basin soil erosion and sediment yield in the upper reach of Ganjiang, China. Land 10(9):990

Zhao Y, Liu L, Kang S, Ao Y, Han L, Ma C (2021) Quantitative analysis of factors influencing spatial distribution of soil erosion based on geo-detector model under diverse geomorphological types. Land 10(6):604

Part II
Water Resource Planning and Management

Challenges and Prospects in Using Very High-Resolution Optical Satellite Imagery for Urban Water Body Extraction—A Study Utilizing GeoEye Remote Sensing Data for Surface Water Management Perspective

7

Farhana Ferdos and Md. Rejaur Rahman

Abstract

The availability of high-resolution images has introduced a valuable method for utilizing remote sensing data to derive detailed insights from these images. However, the application of high-resolution images in urban settings presents certain challenges that have garnered significant attention recently. In urban areas, the presence of shadows cast by elevated structures like buildings, towers, and trees creates a form of interference when attempting to extract information from optical images, especially related to water features. Within urban environments, distinguishing between water and shadowed areas in remote sensing images poses a considerable challenge. Furthermore, the presence of vegetation canopies over bodies of water further complicates the accurate delineation of water boundaries. The accurate extraction of urban water bodies from remote sensing imagery is of paramount importance for effective water resource management and urban planning. Various techniques and methodologies have been proposed to tackle the extraction of surface water areas from remote sensing images. This study primarily addresses the issues encountered when employing commonly used methods such as the normalized difference water index (NDWI), object-based image classification (OBIC), and support vector machine (SVM) supervised classification with GeoEye remote sensing data to extract urban water areas. As a result, a combination of digitization, decision support, and field observation has been adopted to achieve a highly accurate identification and extraction of water bodies from the GeoEye's very high-resolution data. The investigation reveals that the NDWI, OBIC, and SVM supervised classification techniques are inadequate for achieving sufficient accuracy in identifying and extracting urban water bodies from GeoEye's very high-resolution satellite imagery due to the intricate characteristics of urban landscapes. Conversely, the manual digitization approach coupled with decision support and on-site observations proves effective in extracting water bodies, as it helps mitigate the complexities inherent in such environments. Nonetheless, there is a pressing need to develop advanced automated techniques tailored to efficiently extract water bodies from

F. Ferdos · M. R. Rahman (✉)
Department of Geography and Environmental Studies, University of Rajshahi, Rajshahi 6205, Bangladesh
e-mail: rejaur2001@yahoo.com; rejaur@ru.ac.bd

S. Talukdar et al. (eds.), *Water Resource Management in Climate Change Scenario*, GIScience and Geo-environmental Modelling, https://doi.org/10.1007/978-3-031-61121-6_7

extremely high-resolution images in complex urban areas. Additionally, the accuracy of the extracted water bodies has been validated through extensive field surveys and observations, ensuring that the information generated can be seamlessly integrated into urban water management planning efforts.

Keywords

Urban water body · High-resolution satellite data · Feature extraction · Geospatial analysis · Ponds quality

7.1 Introduction

Water resources hold significant importance within urban ecosystems, playing a pivotal role in comprehending urban ecological functions and ensuring sustainability. Consequently, the accurate identification and delineation of water bodies stand as a key concern for effective water resource management and planning in an urban area. The diverse land use and land cover patterns on Earth's surface encompass various features, with water bodies emerging as a dominant landscape element from an ecological perspective. These water bodies yield numerous benefits, including serving as habitats for distinct plant and animal species, facilitating surface and subsurface water retention, nutrient cycling, water purification, and groundwater replenishment. Furthermore, water bodies exert substantial influence on Earth's processes such as carbon cycling and methane emissions within urban systems (Mullen et al. 2023). The exponential growth of urban populations, urbanization, and land transformations have significantly impacted both the spatial distribution and quality of surface water bodies within urban regions, particularly evident in places like Bangladesh. Effectively managing the health and living environment necessitates a keen understanding of the spatial distribution and quality of surface water bodies within urban confines (Niemczynowicz 1999; Varis and Vakkilainen 2001).

Accurate extraction of water body data, encompassing metrics like size and distribution, serves as a vital resource for urban water resource analysis, determining conservation priorities, comprehending ecological processes, and formulating appropriate adaptation strategies (Lacaux et al. 2007; Viala 2008). Satellite remote sensing techniques have emerged as a potent tool for capturing Earth's surface phenomena, particularly water bodies. Remote sensing involves acquiring data or information about Earth's surface without physical contact, enabling remote observation. Remote sensing data has evolved into a pivotal resource for water resource monitoring and environmental research. Optical remote sensing data, renowned for their synoptic view and availability of varying spatial and temporal resolution images, are extensively employed in water body extraction, monitoring studies, and academic investigations across diverse scientific domains (Rahman 2006, 2023; Liao et al. 2014; Gong et al. 2018).

Over the past three decades, a multitude of techniques and methodologies have been proposed for the extraction, mapping, and monitoring of urban water bodies using remote sensing data (Gao 1996; Xu 2006; Feyisa et al. 2014; Santoro et al. 2015; Yang et al. 2017; Wu et al. 2018; Bie et al. 2020; Wang et al. 2021). Among these, several widely recognized methods include the single-band threshold approach (Jain et al. 2005), the Normalized Difference Water Index (NDWI) (McFeeters 1996; Kumari et al. 2022), modified NDWI (MNDWI) (Xu 2006), image classification methods (both supervised and unsupervised), the object-oriented extraction method (Chen et al. 2013; Yin et al. 2015), machine learning-based support vector machine (SVM) techniques (Sun et al. 2015; Gong et al. 2018; Talukdar et al. 2020), and convolutional neural networks (CNN) (Li et al. 2019; Qayyum et al. 2020; Gharbia 2023) were applied. Typically, these methods have been applied to low and medium spatial resolution images (> 30 m and 30 m, respectively) for water area extraction. In recent studies, advanced methods have been employed in conjunction with high spatial resolution (10 m) and very high-resolution (< 10 m) images (Nath and Deb 2010; Xie et al. 2016; Lai et al. 2019; Mullen et al. 2023). While these studies demonstrate some progress in achieving water area extraction, advanced remote sensing classification techniques like machine learning and deep learning exhibit enhanced accuracy with high and very high-resolution data compared to methods such as NDWI, threshold-based approaches, and even supervised classifications like maximum likelihood and support vector machines. However, it's worth noting that even these advanced methods have their limitations in achieving high accuracy in water body extraction. The spectral reflectance emanating from water bodies closely resembles that of shadows cast by elevated structures (such as buildings, towers, and trees). Additionally, the presence of vegetation canopy over water bodies poses a significant challenge in precisely delineating water body boundaries. Consequently, these factors collectively pose substantial challenges for accurately extracting water bodies from high- and very high-resolution remotely sensed data.

Given that the extent of surface water within urban settings is a crucial attribute, intricately tied to the richness and spatial diversity of the urban environment, as well as the quality of water it holds, coupled with the fact that freshwater ecosystems support a wealth of species diversity (Strayer and Dudgeon 2010), provide essential clean water sources for human populations (Ho and Goethals 2019), and significantly contribute to biogeochemical cycles (Wik et al. 2016; Kortelainen et al. 2020;), the accurate extraction of water surfaces using very high-resolution remote sensing data becomes imperative. Such precision is pivotal for effective urban water management. Consequently, this study aims to shed light on the limitations inherent in utilizing very high-resolution remote sensing data and prevalent methodologies. Simultaneously, it proposes strategies to achieve meticulous urban water body extraction, ensuring a heightened level of accuracy within the study area. Furthermore, the study seeks to evaluate the water quality of the identified urban water bodies, contributing valuable insights to the management of urban surface water resources.

7.2 Data Used and Methods

7.2.1 Study Area: Geographical Settings

The Rajshahi Metropolitan City in Bangladesh is situated within the geographical coordinates of 24°12′ to 24°42′ N latitude and 88°15′ to 88°50′ E longitude, as depicted in Fig. 7.1. Positioned in the north-western part of Bangladesh, this city holds a prominent status as one of the largest urban centres within the country. It covers an expanse of 48 km^2 and accommodates a populace of approximately 1.0 million individuals (BBS 2020). The city comprises four Police Stations, namely Boalia, Rajpara, Matihar, and Shah Makhdum, and under the purview of these stations, the urban layout is subdivided into 30 Wards. A significant geographical feature is the presence of the Padma River, coursing through the southern surroundings of the city. Historically, the city was originally referred to as Rampur Boalia. In 1876, its designation was altered to Rajshahi Pourashabha, and ultimately, in 1991, it was elevated to the status of Rajshahi City Corporation, obtaining metropolitan recognition (RMDP 2004; BBS 2013). The city finds itself enveloped by a tropical monsoon climate, with the long-term average temperature hovering around 24.5 °C and annual rainfall reaching approximately 1448 mm. Seasonal temperature variations are evident, with April and May witnessing maximum temperatures ranging from 35 to 44 °C, while minimum temperatures dip between 5 and 10 °C (BBS 2020, Kafy et al. 2020; Rahman and Rahman 2023).

This city finds its location within one of the hottest and driest regions of the nation, and recent decades have witnessed a notable rise in temperature trends (Rahman and Lateh 2017;

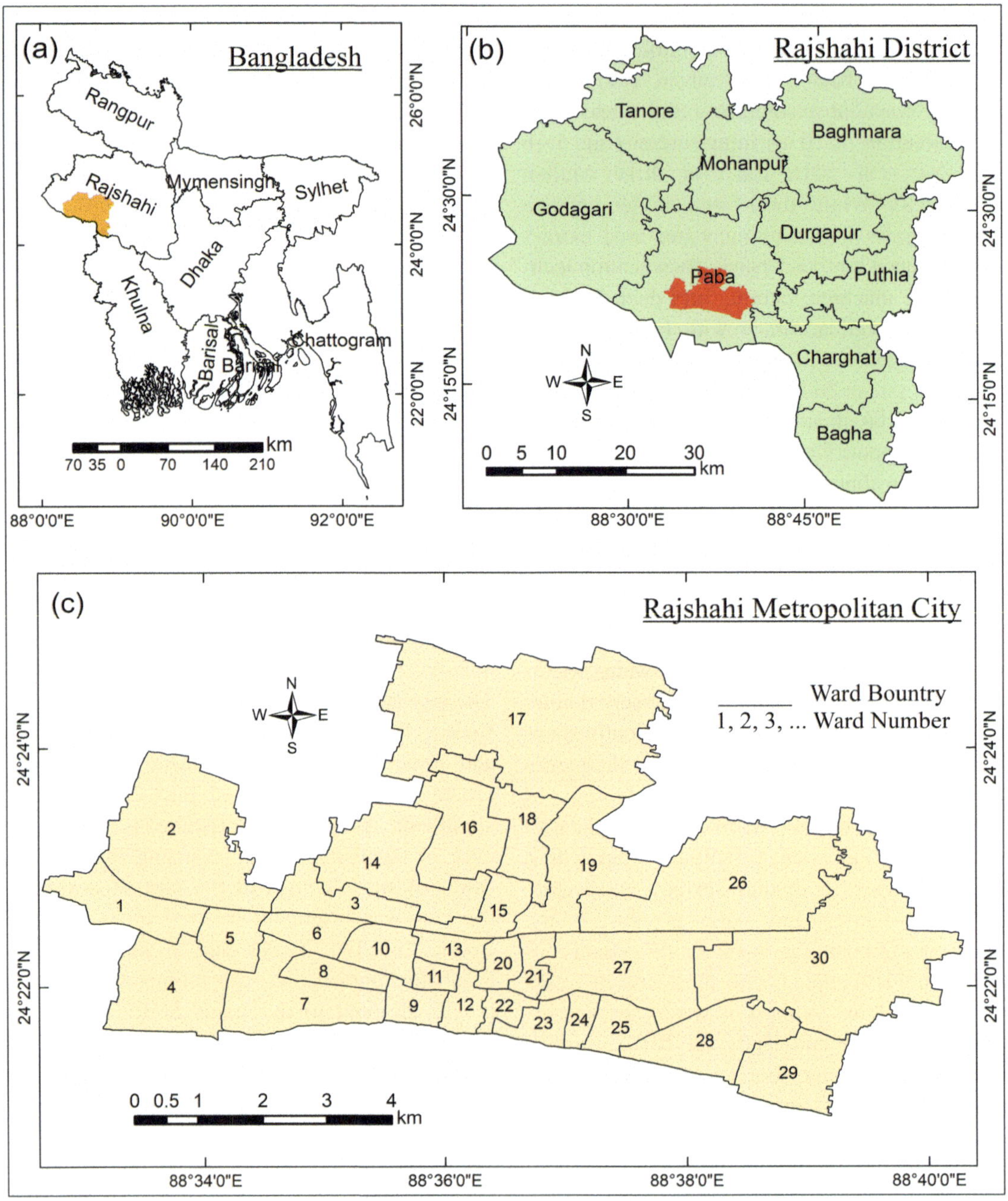

Fig. 7.1 Study area-geographic location of Rajshahi Metropolitan City

Rahman et al. 2018). The population surge, increasing from around 0.52 million in 1990 to 1.0 million in 2022, has fuelled rapid urbanization and consequently, drastic changes in land use and cover patterns, resulting in urban expansion. This urban growth has translated into the transformation of greenery and water bodies into impermeable surfaces within the cityscape. This conversion, primarily driven by the growth of built-up areas and urban infrastructure, accentuates alterations in land use and cover patterns, with a particularly pronounced impact on water bodies. This, in turn, diminishes the city's overall ecological resilience and protective capacity. Given these dynamics, the accurate identification and monitoring of urban water bodies play a pivotal role in effective urban planning and development. The advent of advanced geospatial technology, particularly remote sensing data, provides a valuable avenue for extracting detailed information from images, specifically pertaining to water bodies. However, as indicated in the introductory section, challenges persist in utilizing high and very high spatial resolution remote sensing data for accurate water body identification and extraction. This chapter endeavours to illuminate the constraints associated with such data and prevalent methodologies, while also proposing strategies to achieve precise extraction of urban water bodies, ensuring a high level of accuracy within the study area. The aim is to discern the spatial distribution, quantify areas, and assess the quality of water bodies in the designated region.

7.2.2 Data Used and Methods

In order to delineate water bodies within the designated study area, highly detailed optical remote sensing data known as GeoEye has been employed. This chapter delves into the meticulous extraction of water bodies from this exceptionally high-resolution remote sensing data, while also addressing the constraints associated with its application using diverse methodologies. The process entails an initial exploration of extraction techniques followed by an identification of limitations, challenges, and discrepancies inherent in the process. Ultimately, the chapter presents a refined methodology to accurately compute water area from the GeoEye image by surmounting encountered issues. GeoEye-1, launched in 2008, stands as a satellite offering remarkable spatial resolution. Its sensor captures imagery at 0.41-m panchromatic resolution and 1.65-m resolution for a 4-band multispectral dataset, comprising Blue, Green, Red, and Near-Infrared (NIR) spectral bands. This investigation utilized specific bands of GeoEye, including panchromatic (450–800 nm), Green (510–580 nm), Red (655–690 nm), and Near-Infrared (780–920 nm). The data was acquired from the Department of Geography and Environmental Studies at the University of Rajshahi, Bangladesh. The availability of this data is attributed to the support extended by the Higher Education Quality Improvement Project (HEQEP), Bangladesh. Supplementary spatial information sourced from Geographic Information Systems (GIS) includes city boundaries, ward boundaries, and ancillary data. These datasets have been procured from secondary sources, on-site surveys, and observations. The entirety of these datasets, including the remote sensing imagery, were meticulously prepared and subjected to analysis utilizing ArcGIS software. The chapter, in a systematic manner, navigates through the intricacies of extracting water bodies from GeoEye data, encompassing both the application of diverse techniques and the ensuing limitations. By addressing these complexities, the chapter culminates in a refined methodology, ensuring the precise determination of water area from the GeoEye image within the context of the Rajshahi Metropolitan City (Fig. 7.1).

Within the realm of image processing, a technique known as image fusion was applied to enhance the visibility and discernment of image objects. This technique involved the merging of geocoded panchromatic (offering high spatial resolution) and multispectral (providing high spectral information) bands. The execution of this image fusion approach was carried out using ERDAS Imagine, an image processing software, and the process entailed utilizing principal component analysis (PCA). The resultant fused image encapsulates a combination of both high spatial and high spectral details, effectively integrating the multispectral bands with a spatial resolution of 0.41 m. Image fusion, by its nature, amalgamates pertinent information from two or more images, resulting in a singular image that is more enlightening and informative than its individual components. Subsequently, the augmented fused image was employed for the extraction of water bodies. Initially, three well-established techniques were considered for this purpose:

normalized difference water index (NDWI), object-based image classification (OBIC), and support vector machine (SVM). These techniques are frequently employed for extracting water bodies from imagery due to their effectiveness in identifying water-related features.

The Normalized Difference Water Index (NDWI) is commonly employed to accentuate water bodies within satellite images, highlighting them in contrast to other features like soil, vegetation, and impervious surfaces. Initially introduced by McFeeters (1996), NDWI utilizes the Near-Infrared (NIR) and green bands of optical remote sensing data to identify water bodies. The resulting index yields higher values for water bodies while lower values are indicative of other features. The index values of NDWI are dimensionless and range from − 1 to + 1. Here, the NDWI is computed using Eq. 7.1, where B_2 and B_4 represent the Green and NIR bands, respectively, from the GeoEye sensor.

$$\text{NDWI} = \frac{B_2 - B_4}{B_2 + B_4} \tag{7.1}$$

The object-based image classification (OBIC) method is harnessed to categorize objects or features within an image by incorporating both spectral and spatial information. This technique takes into account not only the spectral attributes but also factors like shape, texture, and the spatial relationship of the identified object with its neighbouring pixels. The availability of very high-resolution remote sensing data has prompted the requirement for novel image analysis approaches. Traditional pixel-based algorithms often overlook the spatial coherence inherent in high-resolution imagery. The incorporation of object attributes such as scale, compactness, orientation, and texture, coupled with spectral properties, substantially widens the potential of remote sensing satellite image analysis. In the OBIC approach, the initial step involves the identification of image objects. This is achieved through the application of a multi-resolution segmentation technique. By implementing this segmentation procedure, the entire image is divided into segments, generating distinct image objects based on adjustable criteria of colour and shape homogeneity or heterogeneity (Rahman and Saha 2008; Roostaei et al. 2012). Rather than classifying individual pixels, this approach extracts cohesive image objects during the segmentation process. In remote sensing, the segmentation of an image is defined as the process of "searching for regions with uniform characteristics within an image, followed by their subsequent classification" (Mather 1999; Guo et al. 2007). Hence, a multi-resolution segmentation was carried out on a false colour composite (FCC) of the NIR, Red, and green bands. The segmentation classifier embedded in Idrisi TerrSet 2020 software (TerrSet 2020) was employed for this purpose. This method allows for a more comprehensive analysis of the image, capitalizing on both spectral and spatial intricacies to achieve a more accurate object classification.

Further, to extract and categorize the water bodies in the image, a technique called support vector machine (SVM) image classification was employed. SVM was chosen due to its ability to achieve high classification accuracy and robust generalization, as demonstrated in previous studies (Awad and Khanna 2015; Hong et al. 2016; Kranjcic et al. 2019). In the context of image classification using SVM, a supervised machine learning algorithm is employed, enabling the computer to learn autonomously from data rather than explicit programming. SVM classifies satellite images by identifying a hyperplane that best separates different classes. It focuses on support vectors, which are instances at the edge of the class distribution, in order to effectively distinguish classes and disregard other training instances (Brown et al. 2003). In this research, the SVM classification approach was followed, involving the creation of training and testing sites. These sites encompassed four land use and land cover categories: vegetation, water bodies, built-up areas, and bare land. Ground truth information played a role in selecting appropriate training and testing sites. The primary objective of the SVM classifier is to find a hyperplane that optimally separates two classes, as highlighted by Hong et al. (2016). The selection of a suitable kernel is crucial within the SVM framework, as it greatly influences training and classification accuracy. Notably, the radial basis function (RBF) kernel has demonstrated strong performance in data modelling (Kia et al. 2012). Thus, the SVM model in this study employed the RBF kernel to classify the GeoEye satellite image within the study area. For a deeper understanding of SVM and the mathematical details of the RBF kernel, references such as Mathur and Foody (2008), Kalantar et al. (2018), Wang et al. (2020), and He et al. (2021) provide relevant information. To specifically train the SVM for water body assessment and land use and land cover classification, the training dataset and the "Train Support Vector Machine Classifier" tool within ArcGIS 10.6 were utilized, facilitating the extraction and classification process.

The outcomes of employing the NDWI, OBIC, and SVM methods for accurately extracting water bodies from very high-resolution satellite data are detailed in the results section, highlighting their associated limitations and challenges. Given their inadequacy in achieving the desired level of precision, an alternative approach was adopted, which involved an on-screen digitizing procedure complemented by field observations and decision support mechanisms. This technique offers enhanced flexibility for extracting specific objects and areas with a heightened degree of accuracy. The on-screen digitizing process parallels traditional digitization, with users employing their mouse to generate a map layer on the screen while referencing pertinent data or employing an image as a background. Through this methodology, an object's attributes can be traced from a scanned map or image, and new layers or themes can be created. Since water bodies constitute area features, they were digitized as polygon layers in this study, utilizing ArcGIS software to extract their respective areas. Additionally, supplementary data regarding the quality of surface water sources was acquired through first-hand visits and observations in the field. These visits involved evaluations of the physical condition of ponds, water colour, pollution levels, suitability for diverse purposes, and water transparency. Thorough documentation of all relevant data was meticulously conducted during these field trips. Furthermore, the data layers comprising the city and ward boundary maps were obtained from secondary sources and utilized in different analyses. An overview of the comprehensive methodology is visually represented in Fig. 7.2.

7.3 Results

7.3.1 NDWI, OBIC, and SVM: Limitations and Consequences Towards Water Body Extractions

The primary goal of the research was to identify the water bodies in the study area by utilizing high-resolution optical remote sensing data from GeoEye. We employed three different techniques for this purpose, but we encountered certain limitations and accuracy issues with each of them. While NDWI, OBIC, and SVM are commonly employed techniques for water body extraction from remote sensing data, they may not always be appropriate for identifying water bodies in all types of remote sensing data. This is especially relevant for very high spatial resolution data, where accurate identification of water areas is crucial. When utilizing the NDWI technique to detect and extract water bodies from high-resolution images, we came across a challenge. The issue at hand arises from the fact that both shadows and water bodies have a similar dark signature in the image, leading to overlapping (Fig. 7.3a–d). It is worth noting that having a canopy cover above a pond can cause a decrease in the visible surface area of the water when using NDWI. This may result in an inaccurate estimation of the water area, which could have important consequences for applications that rely on precise measurements of water resources (Fig. 7.3e–f). Therefore, it is essential to take into account the impact of the canopy cover on the pond's surface area when analysing or evaluating the water resources in the vicinity.

On the other hand, within the realm of OBIC, segmentation has been demonstrated to be an

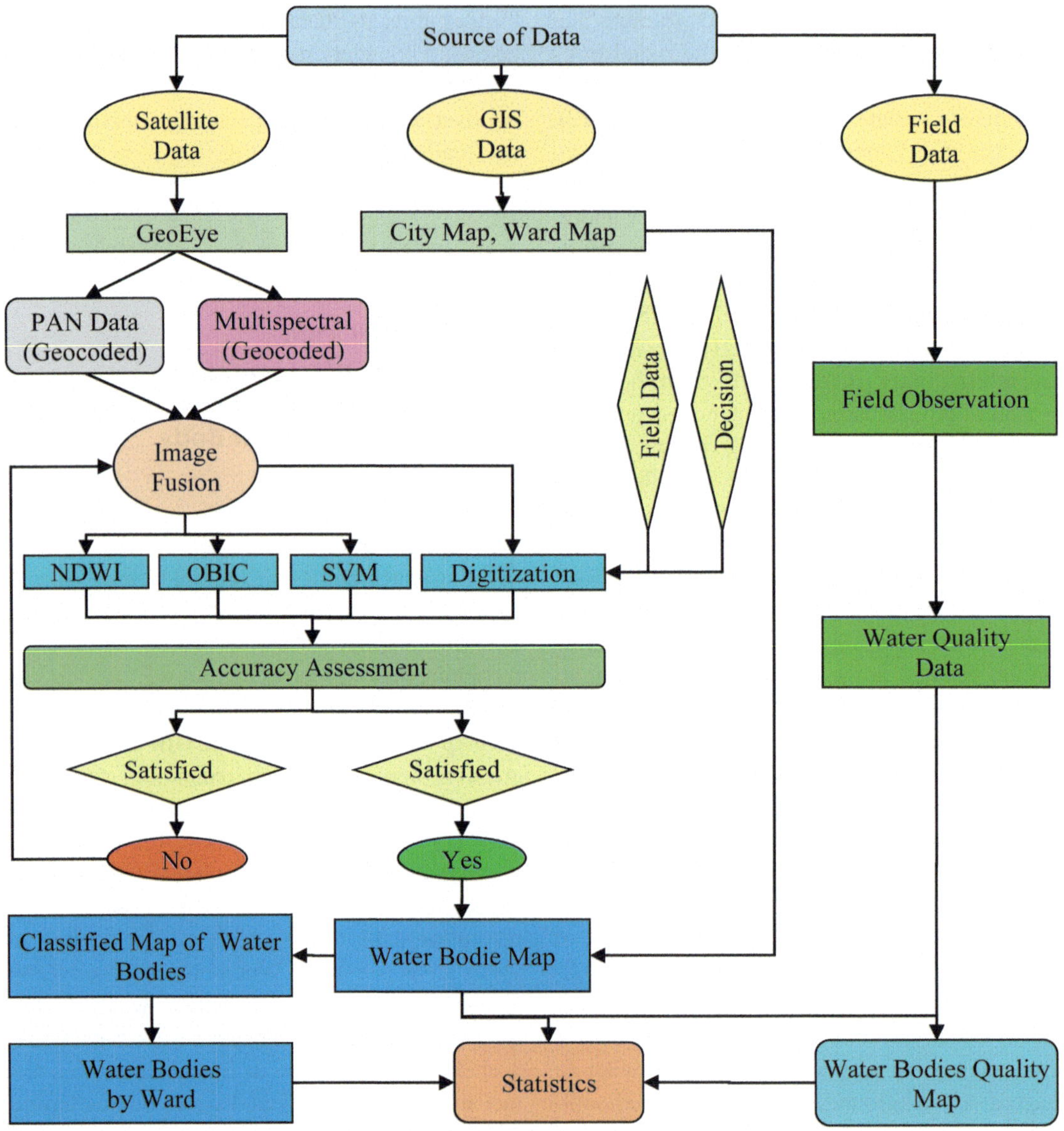

Fig. 7.2 Overall methodology of the study

effective methodology for identifying water bodies within specific areas, as is evidenced by Fig. 7.4a and b. It is, however, crucial to take note of the fact that inaccurate estimations may arise when shadowed regions overlap with water bodies, as evinced by Fig. 7.4c and d, as well as e and f. Thus, the process of identifying water bodies using digital classification methods, such as OBIC and segmentation, can prove to be a challenging task. This is particularly true for very high-resolution images captured in urban areas where tall buildings, trees, and other structures cast shadows that can create confusion. Consequently, distinguishing between water bodies and shadows using this method can be a complex and demanding process.

Similarly, when employing SVM for efficient water body extraction, we encountered analogous limitations and challenges. It's noteworthy that shadow regions were erroneously

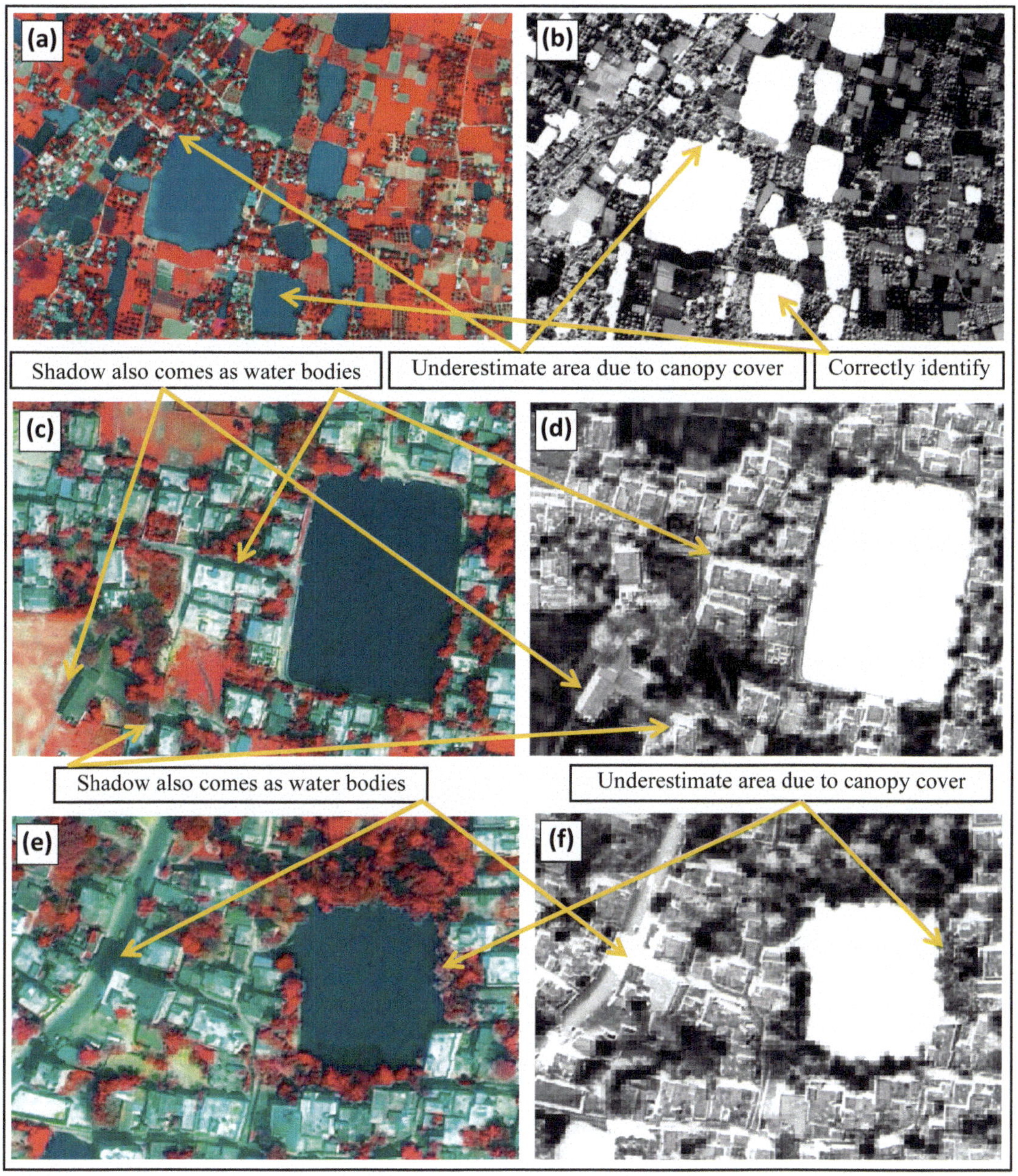

Fig. 7.3 Findings (limitations) based on NDWI

categorized as water bodies (depicted in Fig. 7.5a, b and e, f), and the presence of canopy cover over water bodies led to an underestimation of their extent (shown in Fig. 7.5c and d). Just like the other two methods, SVM demonstrated high accuracy when there were no obstacles such as shadows or canopy cover effects (illustrated in Fig. 7.5e and f). Through the utilization of the SVM technique, images can indeed be classified with a notable degree of accuracy to extract various objects. Nevertheless, when it comes to extracting urban water bodies from images with extremely high resolution, SVM demonstrates limited suitability in achieving the desired level of high accuracy.

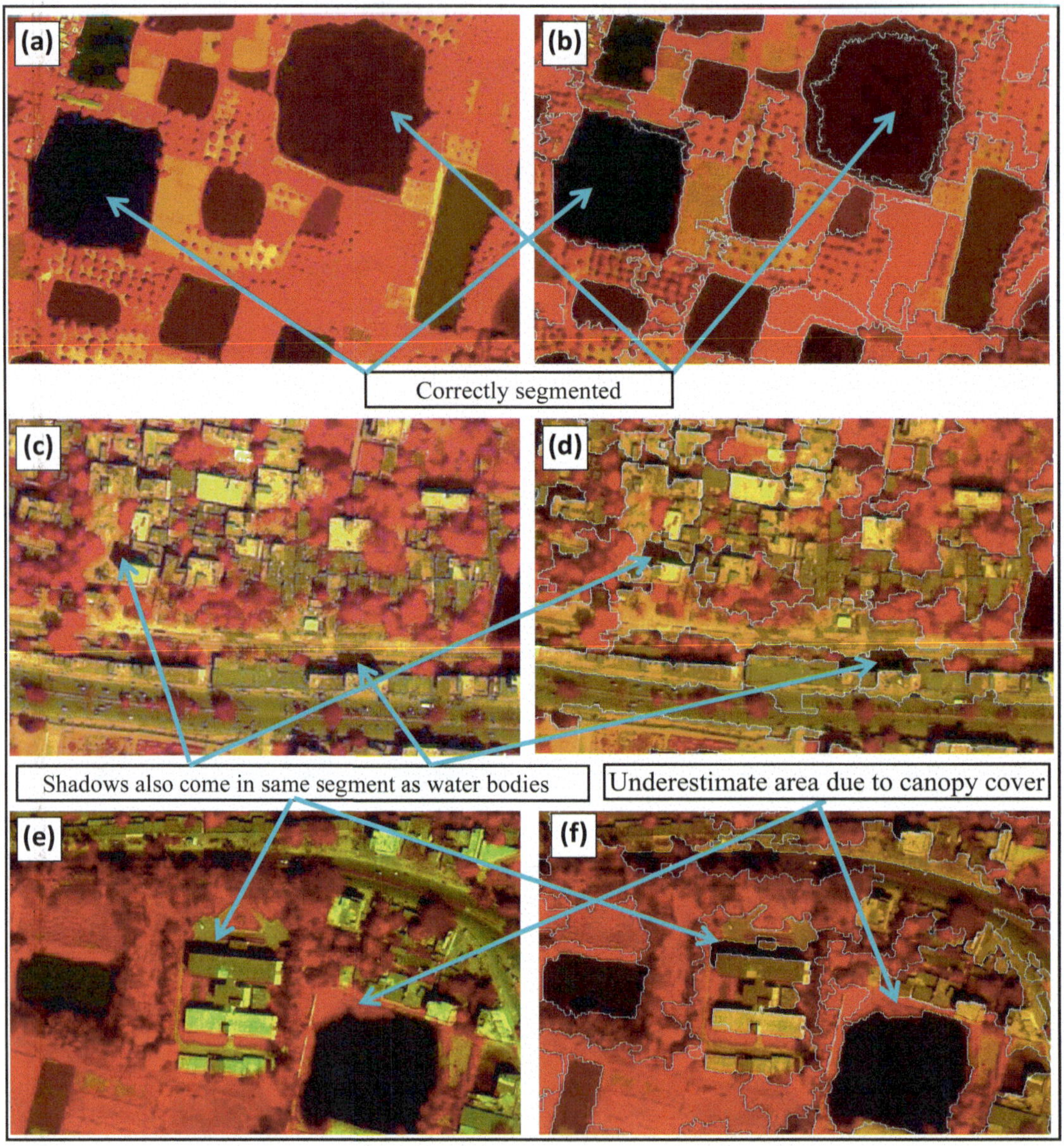

Fig. 7.4 Findings (limitations) based on OBIC

7.3.2 Digitization: Opportunities and Consequences Towards Water Body Extractions

During our study, we encountered challenges related to the accuracy of digital extraction methods, namely NDWI, OBIC, and SVM, due to the presence of shadows and canopy cover that interfered with the identification of water bodies. In order to address this issue, we employed an on-screen digitizing technique in conjunction with field visit data and decision support. This approach proved to be successful in extracting water bodies from the GeoEye image of the area, effectively mitigating the challenges previously encountered. The utilization of digitizing techniques allows for the extraction of specific areas of interest with increased flexibility. This approach takes into account potential obstacles, such as shadows, canopy cover,

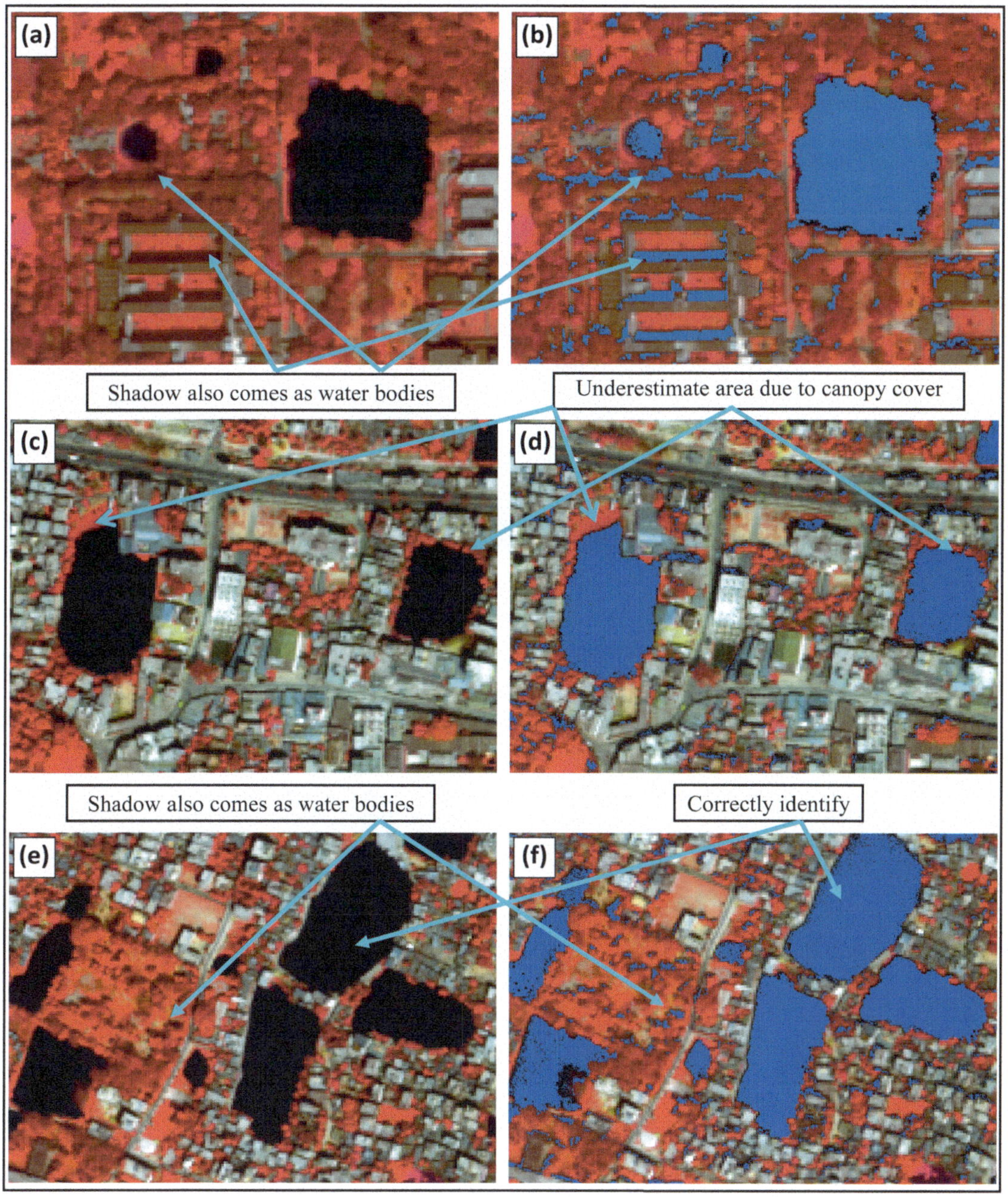

Fig. 7.5 Findings (limitations) based on SVM

and other factors that may hinder the process, particularly when dealing with water bodies. By leveraging this methodology, we are able to achieve a higher level of precision, which is particularly beneficial for the micro-level analysis and management of water bodies. Following the implementation of the digitization technique, we conducted a comprehensive evaluation to verify the accuracy of the measurements of water body areas. We achieved an overall 97% accuracy. Table 7.1 illustrates a range of testing sites along with their respective accuracy levels. The table demonstrates that automated techniques such as NDWI, OBIC, and SVM achieve substantial

Table 7.1 Accuracy level of different methods with pond conditions

Testing sites	Pond area (acre)	Extraction methods (area in acre)				Pond condition	Accuracy (%)			
		NDVI	OBIC	SVM	Digitizing		NDVI	OBIC	SVM	Digitizing
1	1.740	1.450	1.550	1.550	1.700	Canopy cover over pond	83	89	89	98
2	2.580	1.850	1.860	1.880	2.500	Canopy cover over pond	72	72	73	97
3	0.300	0.250	0.230	0.220	0.290	Canopy cover over pond	83	77	73	97
4	0.450	0.385	0.380	0.390	0.430	Canopy cover over pond	86	84	87	96
5	2.230	2.100	2.120	2.050	2.180	No canopy cover and shadow	94	95	92	98
6	0.240	0.215	0.210	0.210	0.235	No canopy cover and shadow	90	88	88	98
7	0.530	0.510	0.490	0.495	0.515	No canopy cover and shadow	96	92	93	97
8	0.150	0.125	0.131	0.130	0.145	No canopy cover and shadow	83	87	87	97
9	0.530	0.570	0.550	0.590	0.510	With shadow area	108	104	111	96
10	0.660	0.670	0.680	0.690	0.640	With shadow area	102	103	105	97
Total	9.410	8.125	8.201	8.205	9.145	–	86	87	87	97

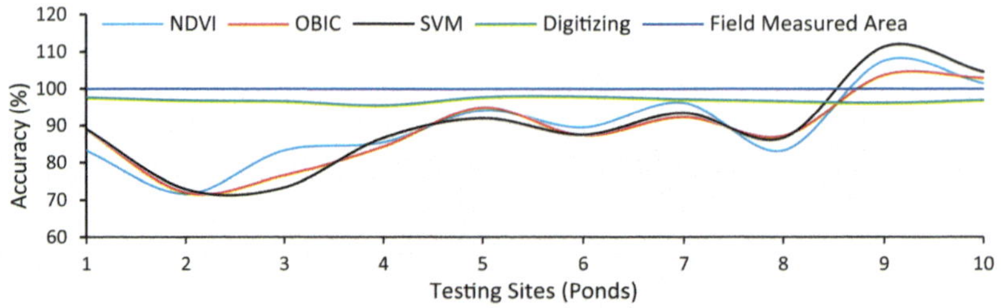

Fig. 7.6 Accuracy level of different methods

accuracy (ranging from 83 to 96%) when no hindrances like shadows or canopy cover are present over water bodies. However, when applying these three extraction methods under conditions of canopy cover or shadow influence, the achieved accuracy is comparatively lower (ranging from 72 to 89%). Notably, instances where both shadow and pond areas coexisted with these three methods led to an overestimation problem due to similar spectral signatures, resulting in accuracy values exceeding 100% (ranging from 102 to 111%) as shown in Table 7.1. However, the digitization approach displayed a distinct advantage by offering the flexibility to mitigate complexities and incorporate on-ground information and judgments. This flexibility translated into remarkably very high accuracy rates (ranging from 96 to 98%) across all intricate pond scenarios, as depicted in Table 7.1 and Fig. 7.6. Thus, digitization is the best way to identify water bodies in very high spatial resolution images, especially considering the current lack of appropriate digital methods due to the complex urban landscape.

7.3.3 Spatial Distribution and Pattern of Water Bodies (Ponds)

The spatial distribution of water bodies in the study area was obtained using digitizing techniques and water body extraction, as shown in Fig. 7.7a. The data indicates that the city hosts a significant number of ponds that are scattered throughout, with a higher concentration in the north-eastern, western, and northern parts of the city. Our analysis revealed a total of 1526 ponds in the city, which covers an area of 2.76 km^2 (276 ha), representing approximately 5.73% of the city's total area. The study of urban ponds involves the classification of their physical characteristics and patterns, with size and shape being the primary factors of interest. Our research has identified five distinct categories of ponds, classified as very big (greater than 8094 m^2), big (between 4047 and 8093 m^2), medium (between 1174 and 4046 m^2), small (between 607 and 4045 m^2), and very small (less than 607 m^2). It is essential for urban surface water

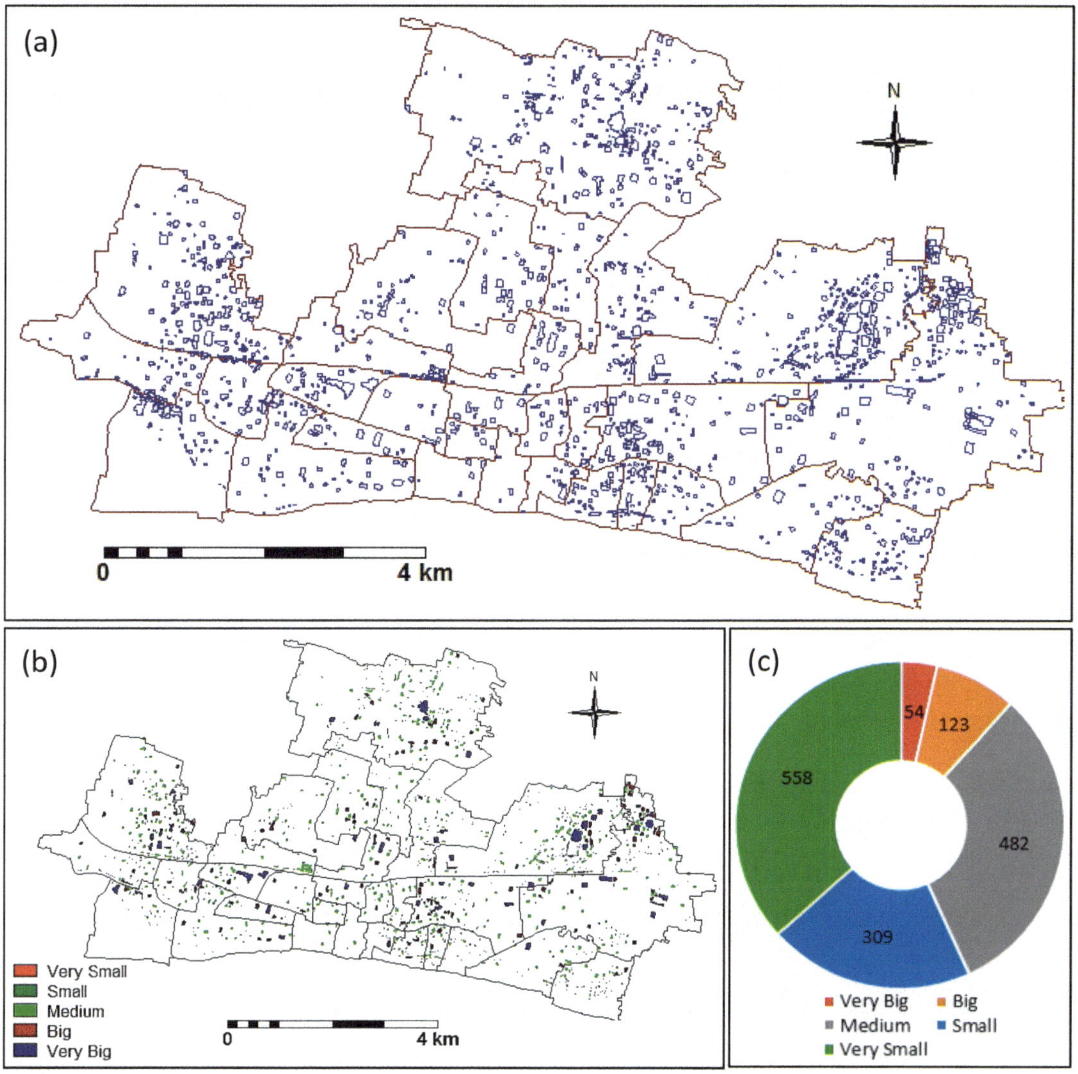

Fig. 7.7 Spatial pattern of water bodies: **a** spatial distribution; **b** spatial pattern of pond types; **c** statistics of pond types

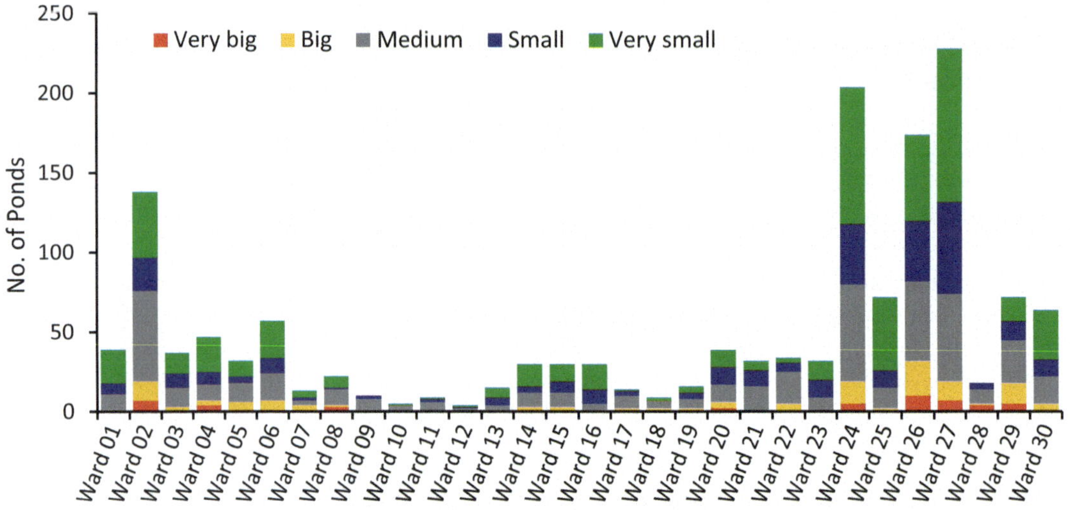

Fig. 7.8 Ward-wise distribution of ponds by types

planning and management to consider the size of water bodies, as the diversity of water surfaces has a significant impact on various aspects of the urban environment, including biodiversity, surface temperature, and comfort levels. The distribution patterns of different types of ponds are depicted in Fig. 7.7b.

There are total of 54, 123, 482, 309, and 558 very big, big, medium, small, and very small ponds are classified there in the city and occupied an area about 58, 61, 100, 29, and 27 hectares, respectively (Fig. 7.7c). An analysis of the distribution of very big ponds within the city revealed that these water bodies were primarily situated in the eastern, western, and partially northern and central parts. Conversely, the concentration of big ponds was found to be more prevalent in the western, southern, and some central areas. Medium, small, and very small ponds were dispersed throughout the city, with a slightly higher concentration observed in the eastern, northern, and western parts (Fig. 7.7b). Again, an analysis of the distribution of ponds by type across various wards has revealed intriguing insights. Based on Fig. 7.8, Ward 26 had the highest number of very big ponds, with a total of 10, followed by Wards 27, 2, and 29, which had 7, 7, and 5 big ponds, respectively. Ward 26 ranked first with 10 big ponds, while Wards 12 and 6 ranked second and third with 12 and 6 big ponds, respectively. Furthermore, Ward 24 had the highest number of medium-sized ponds, with a total of 61, followed by Wards 2, 27, and 26, which had 57, 55, and 50 medium ponds, respectively. The distribution of small and very small ponds was primarily concentrated in Wards 24, 25, 26, 27, and 2, as illustrated in Fig. 7.8.

7.3.4 Ponds Quality

The aim of this study was to evaluate the quality of ponds by considering factors such as their suitability for daily activities, physical condition, pollution levels, and water clearness. To achieve this objective, comprehensive field surveys and data collection efforts were carried out. For assessing the suitability of ponds for daily activities, factors such as bathing, washing, and other uses of pond water were taken into consideration. The outcomes reveal that out of the total ponds assessed, 371 ponds (14.22%) were categorized as suitable, 176 ponds (30.30%) were classified as moderately suitable, and 675 ponds (55.28%) were deemed unsuitable for everyday uses. Remarkably, a significant concentration of the suitable ponds was situated in Wards 17 and 47, as depicted in Fig. 7.9a. These findings yield valuable insights that can inform the

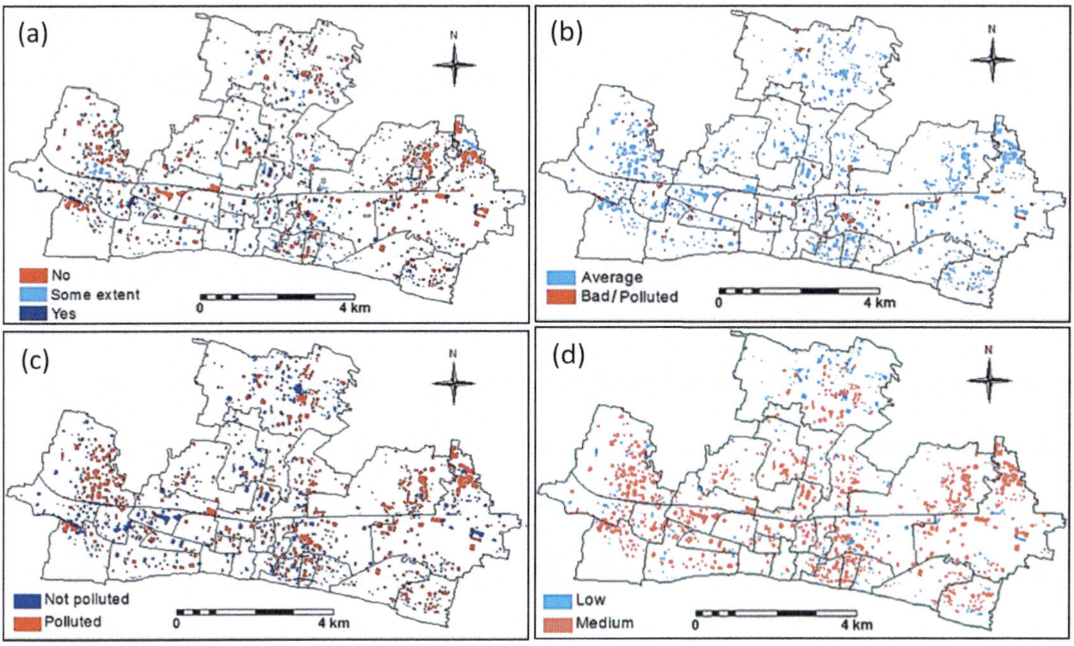

Fig. 7.9 Quality of ponds: **a** suitability for daily uses; **b** overall physical condition; **c** pollution condition; **d** water transparency

decisions made by policymakers and stakeholders involved in the management of both ponds and their surrounding environments. Secondly, we conduct an evaluation of the ponds' physical condition through an analysis of variables such as water volume, colour, dryness, and dirtiness. Our observations indicate that the ponds do not consistently manifest favourable conditions, as evidenced by 300 ponds (22.22%) exhibiting a state of deterioration (bad condition). The remaining ponds within the studied area, accounting for 77.78% of the total, exhibit an average condition. It is worth noting that a concentration of ponds in bad condition is notably prevalent in Wards 27 and 24 (Fig. 7.9b). These findings underscore the significance of ongoing surveillance and maintenance endeavours to safeguard the overall well-being and quality of the ponds.

Moreover, in order to assess the pollution status of ponds, we employ observational techniques that take into account various factors, including the presence of waste material, fish cultivation, odours, and instances of pond abandonment. Our findings reveal that, of the total number of ponds surveyed, 964 (71.35%) were found to be polluted, with the highest level of pollution recorded in Wards 30 and 2. Conversely, only 28.65% of the ponds were classified as unpolluted in the area (Fig. 7.9c). Besides, the transparency of the water was evaluated through the observation of pond water clarity. Generally, it is understood that higher water clarity or transparency correlates with good pond water quality, while the opposite holds true. In this context, water clarity was classified as low, medium, and high. Our observations indicated that a total of 967 ponds (71.70%) exhibited medium levels of water transparency. In contrast, 28.30% of the overall pond count displayed low water transparency. The greatest concentration of ponds with low transparency was observed in Wards 17 and 2 (Fig. 7.9d). No ponds with high water transparency were identified. Consequently, based on water transparency, the ponds in the study area are found to be in unfavourable conditions.

7.4 Surface Water Resource Management Perspectives

Urban aquatic features play a pivotal role in overall urban landscape and eco-environment settings. In effectively managing water resources within dynamically evolving urban areas, it is imperative to accurately identify and acquire information about these water bodies. Contemporary satellite technology now affords us the capability to access very high-resolution satellite images, offering intricate insights into terrestrial phenomena. Nevertheless, the extraction of elements such as water bodies from these images remains a formidable challenge due to the intricate reflectance patterns exhibited by diverse features. Given the wealth of detailed information present in the high-resolution image and its potential for extraction, this image holds significant value as a robust data and information source. This investigation achieved a notably high precision in delineating water bodies as ponds. The accuracy and depth of knowledge concerning urban water bodies stand as prerequisites for effective urban water planning and management, contributing to the enhancement of the urban milieu. Gaining accurate and comprehensive spatial data on the distribution and types of water bodies enables the subsequent collection of information concerning the conditions and quality of these water bodies. Integrating this comprehensive information becomes instrumental in endeavours like water body preservation, flood mitigation, and the strategic planning and administration of urban water features. The meticulous management of urban water resources necessitates a thorough strategy that addresses the distinctive obstacles presented by swiftly growing urban environments. The increase in urban population strains current water infrastructure, giving rise to challenges like insufficient water supply and uneven distribution trends. Additionally, the expansion of urban populations prompts the transformation of water features into impermeable surfaces like settlements and other structures, ultimately leading to the deterioration of urban ecosystems. Urbanization also adds to the contamination of surface water via runoff containing pollutants from roads, industries, and residences. Hence, geospatial data concerning water features, encompassing their scope, water quality, pollution status, physical state, and suitability for surface water utilization, plays a vital role in effective planning and management.

In this investigation, the spatial distribution of water bodies in the form of ponds, classification of ponds based on their dimensions (area), evaluation of water quality in terms of its suitability for daily uses, pollution levels, physical conditions, and water clarity (transparency) were assessed. These findings can be effectively integrated into water body management and planning strategies or decisions. Regarding the distribution and types of ponds, a fundamental strategy involves enforcing strict land use regulations and urban planning to prioritize the safeguarding of existing ponds and their surrounding areas against encroachment and pollution. Especially for very big ponds, it is important to preserve them while recognizing their distinctive attributes and significance within the urban environment. Similarly, the Wards which have minimum number of water bodies should be conserved and protected through specialized legislation. To address pollution, cleanliness, and suitability for daily activities, priority should be given to restoring ponds with poor water quality, dirtiness, and unsuitable uses to enhance their overall quality. Combating pollution necessitates a combination of measures like implementing optimal practices for managing runoff in surrounding regions, reducing the application of fertilizers and chemicals, preventing waste dumping, and establishing vegetated buffers. Enhancing water clarity can be attained through strategies such as controlling sediment accumulation, managing nutrient levels, regular dredging to prevent sediment growth, and raising awareness about responsible practices. Adapting pond utilization based on their conditions is vital; heavily polluted ponds might be best suited for ecological revitalization, while clearer ones can accommodate recreational activities like fishing or boating. Integrating both water quality testing and community involvement ensures that conservation endeavours align with each urban pond's specific requirements and potential, fostering a balanced coexistence between human endeavours and the ecological well-being of these crucial urban water bodies. Furthermore, collaborative initiatives involving local communities, government entities, and environmental organizations play a pivotal role, encouraging residents to actively participate in pond restoration, clean-up campaigns, and educational initiatives. By implementing these strategies, urban ponds can flourish as invaluable ecosystems, offering spaces for recreation, enhancing water quality, and promoting biodiversity within urban landscapes.

7.5 Conclusions

This study extensively investigated the challenges and opportunities inherent in utilizing very high-resolution optical satellite imagery for the purpose of identifying urban water bodies. This was exemplified through the comprehensive analysis of GeoEye remote sensing data, followed by the utilization of the extracted water body information for surface water management perspectives. The research unveiled the limitations and potential advantages associated with this technology-driven approach. It acknowledged the complexities of accurately recognizing and characterizing water bodies within intricate urban landscapes, particularly given factors such as spectral confusion and the intricate urban layout. Notably, the study underscored the persisting challenges in distinguishing the spectral signatures of water bodies and accounting for shadows cast by tall structures, trees, towers, and other elements. Additionally, the presence of a canopy cover over urban water bodies emerged as a significant hurdle, impacting the precision of water body extraction using conventional automated techniques designed for high spatial resolution imagery.

In this investigation, employing methods such as NDWI, OBIC, and SVM, none of these techniques were found to be suitable for accurately extracting water bodies from the GeoEye imagery. Due to the intricate nature of very high-resolution imagery, which provides detailed data, and its applicability to micro-level analyses with the capacity to enrich our comprehension of urban water dynamics, an approach involving on-screen digitization combined with field observations and decision support was employed. This technique proved notably suitable due to its adaptability in accurately extracting specific objects and areas with a notable degree of precision. However, despite its effectiveness, digitization emerged as a time-intensive procedure, and in this context, the study's accuracy was strengthened by extensive fieldwork. Such challenges hold particular significance in developing countries like Bangladesh, where data collection can prove burdensome and manual.

Thus, based on the outcomes of the study, it is recommended that advanced automated methodologies be devised, especially tailored for the complexities of very high-resolution imagery within complex urban environments. This is pivotal for streamlining and enhancing the extraction of features such as water bodies. Given the evolving nature of remote sensing technology and its potential to significantly influence urban water management, these findings underscore the necessity of continual innovation to tackle the unique challenges posed by urban water body identification in dynamic urban landscapes. By addressing challenges directly and capitalizing on the possibilities afforded by extremely detailed optical satellite imagery, we can forge a path towards an urban future that is both more resilient and environmentally mindful. In this trajectory, the conservation and adept management of urban water bodies assume a central role in shaping the development of our cities.

References

Awad M, Khanna R (2015) Support vector machines for classification. In: Efficient learning machines. Apress, Berkeley, CA, USA, pp 39–66

BBS (2013) District statistics, Rajshahi. Ministry of Planning, Government of the People's Republic of Bangladesh. Bangladesh Bureau of Statistics, Dhaka, Bangladesh

BBS (2020) Statistical yearbook Bangladesh-2019. Bangladesh Bureau of Statistics (BBS), Statistics & Informatics Division (SID), Ministry of Planning, Government of the People's Republic of Bangladesh Dhaka, Bangladesh

Bie W, Fei T, Liu X, Liu H, Wu G (2020) Small water bodies mapped from Sentinel-2 MSI (MultiSpectral Imager) imagery with higher accuracy. Int J Remote Sens 41(20):7912–7930

Brown M, Lewis HG, Gunn SR (2003) Linear spectral mixture models and support vector machines for remote sensing. IEEE Trans Geosci Remote Sens 38:2346–2360

Chen C, Chen JX, Xiang-Qian LI, Zhang Q, Zheng SQ (2013) Study on water body information extraction from high resolution remote sensing image based on object-oriented method. Yellow River 35(9):68–70

Feyisa GL, Meilby H, Fensholt R, Proud SR (2014) Automated water extraction index: a new technique for surface water mapping using Landsat imagery. Remote Sens Environ 140:23–35

Gao BC (1996) NDWI—a normalized difference water index for remote sensing of vegetation liquid water from space. Remote Sens Environ 58:257–266

Gharbia R (2023) Deep learning for automatic extraction of water bodies using satellite imagery. J Indian Soc Remote Sens 51:1511–1521. https://doi.org/10.1007/s12524-023-01705-0

Gong WF, Wang P, Wang SY, Zhou Y, Cao KH (2018) Methods of water body extraction in Boundary River based on GF-2 satellite remote sensing image of high resolution. J Eng Heilongjiang Univ 9(04):5–11

Guo Q, Kelly M, Gong P, Liu D (2007) An object-based classification approach in mapping tree mortality using high spatial resolution imagery. GIsci Remote Sens 44:24–47

He Y, Ma D, Xiong J, Cheng W, Ji H, Wang N, Guo L, Duan Y, Liu J, Yang G (2021) Flash flood vulnerability assessment of roads in China based on support vector machine. Geocarto Int. https://doi.org/10.1080/10106049.2021.1926560

Ho LT, Goethals PLM (2019) Opportunities and challenges for the sustainability of lakes and reservoirs in relation to the sustainable development goals (SDGs). Water 11(7):1462. https://doi.org/10.3390/w11071462

Hong H, Pradhan B, Bui DT, Xu C, Youssef AM, Chen W (2016) Comparison of four kernel functions used in support vector machines for landslide susceptibility mapping: a case study at Suichuan area (China). Geomat Nat Haz Risk, 1–26. https://doi.org/10.1080/19475705.2016.1250112

Jain SK, Singh RD, Jain MK, Lohani AK (2005) Delineation of flood-prone areas using remote sensing techniques. Water Resour Manage 19(4):333–347

Kafy AA, Rahman MS, Faisal AA, Hasan MM, Islam M (2020) Modelling future land use land cover changes and their impacts on land surface temperatures in Rajshahi, Bangladesh. Remote Sens Appl: Soc Environ 18:100314

Kalantar B, Pradhan B, Naghibi SA, Motevalli A, Mansor S (2018) Assessment of the effects of training data selection on the landslide susceptibility mapping: a comparison between support vector machine (SVM), logistic regression (LR) and artificial neural networks (ANN). Geomat Nat Haz Risk 9(1):49–69. https://doi.org/10.1080/19475705.2017.1407368

Kia MB, Pirasteh S, Pradhan B, Mahmud AR, Sulaiman WNA, Moradi A (2012) An artificial neural network model for flood simulation using GIS: Johor River Basin, Malaysia. Environ Earth Sci 67:251–264. https://doi.org/10.1007/s12665-011-1504-z

Kortelainen P, Larmola T, Rantakari M, Juutinen S, Alm J, Martikainen PJ (2020) Lakes as nitrous oxide sources in the boreal landscape. Glob Change Biol 26(3):1432–1445. https://doi.org/10.1111/gcb.14928

Kranjcic N, Medak D, Zupan R, Rezo M (2019) Machine learning methods for classification of the green infrastructure in city areas. Earth Environ Sci 362:012079

Kumari N, Srivastava A, Kumar S (2022) Hydrological analysis using observed and satellite-based estimates: case study of a lake catchment in Raipur, India. J Indian Soc Remote Sens 50(1):115–128

Lacaux JP, Tourre YM, Vignolles C, Ndione JA, Lafaye M (2007) Classification of ponds from high-spatial resolution remote sensing: application to Rift Valley Fever epidemics in Senegal. Remote Sens Environ 106(1):66–74

Lai Y, Zhang J, Song Y (2019) Surface water information extraction based on high-resolution image. IOP Conf Ser Earth Environ Sci 330:032013. https://doi.org/10.1088/1755-1315/330/3/032013

Li L, Yan Z, Shen Q, Cheng G, Gao L, Zhang B (2019) Water body extraction from very high spatial resolution remote sensing data based on fully convolutional networks. Remote Sens 11:1162. https://doi.org/10.3390/rs11101162

Liao A, Chen LJ, Chen J, Chaoying HE, Cao X, Chen J, Peng S, Sun FD, Gong P (2014) High-resolution remote sensing mapping of global land water. Sci China Earth Sci 57(10):2305–2316

Mather P (1999) Computer processing of remotely sensed images. Wiley, Chichester

Mathur A, Foody GM (2008) Multiclass and binary SVM classification: implications for training and classification users. IEEE Geosci Remote Sens Lett 5:241–245

McFeeters SK (1996) The use of the normalized difference water index (NDWI) in the delineation of open water features. Int J Remote Sens 17(7):1425–1432. https://doi.org/10.1080/01431169608948714

Mullen AL, Watts JD, Rogers BM, Carroll ML, Elder CD, Noomah J et al (2023) Using high-resolution satellite imagery and deep learning to track dynamic seasonality in small water bodies. Geophys Res Lett 50:e2022GL102327. https://doi.org/10.1029/2022GL102327

Nath RK, Deb SK (2010) Water-body area extraction from high resolution satellite images—an introduction, review, and comparison. Int J Image Process 3(6)

Niemczynowicz J (1999) Urban hydrology and water management-present and future challenges. Urban Water 1:1–14

Qayyum N, Ghuffar S, Ahmad H, Yousaf A, Shahid I (2020) Glacial lakes mapping using multi satellite PlanetScope imagery and deep learning. ISPRS Int J Geo Inf 9(10):560. https://doi.org/10.3390/ijgi9100560

Rahman MR (2006) Flood inundation mapping and damage assessment using multi-temporal RADARSAT and IRS 1C LISS III Image. Asian J Geoinf 6(2):11–21

Rahman MR (2023) River dynamics—a geospatial analysis of Jamuna (Brahmaputra) River in Bangladesh during 1973–2019 using Landsat satellite remote sensing data and GIS. Environ Monit Assess 195:96. https://doi.org/10.1007/s10661-022-10638-z

Rahman MR, Lateh H (2017) Climate change in Bangladesh: a spatio-temporal analysis and simulation of recent temperature and rainfall data using GIS and time series analysis model. Theor Appl Climatol 128:27–41. https://doi.org/10.1007/s00704-015-1688-3

Rahman MR, Saha SK (2008) Multi-resolution Segmentation for object-based classification and accuracy assessment of land use/land cover classification using remotely sensed data. J Indian Soc Remote Sens 36:189–201

Rahman MR, Rahman A (2023) Urban green and blue spaces dynamics—a geospatial analysis using remote sensing, machine learning and landscape metrics in Rajshahi Metropolitan City, Bangladesh. In: Rahman A, Sen Roy S, Talukdar S, Shahfahad (eds) Advancements in urban environmental studies. GIScience and geo-environmental modelling. Springer, Cham. https://doi.org/10.1007/978-3-031-21587-2_10

Rahman MR, Lateh H, Islam MN (2018) Climate of Bangladesh: temperature and rainfall changes, and impact on agriculture and groundwater—a GIS-based analysis. In: Islam M, van Amstel A (eds) Bangladesh I: climate change impacts, mitigation and adaptation in developing countries. Springer climate. Springer, Cham. https://doi.org/10.1007/978-3-319-26357-1_2

RMDP (2004) Rajshahi Metropolitan Development Plan (2004–2024), Volume 1, Structural and Master Plan, Rajshahi Development Authority (RDA), RDA Bhaban, Bonolata Commercial Area, Airport Road, Rajshahi-6203, Bangladesh

Roostaei S, Alavi SA, Nikjoo MR, Kamran VK (2012) Evaluation of object oriented and pixel based classification methods for extracting changes in urban area. Iran Int J Geomat Geosci 2(3). ISSN 0976-4380

Santoro M, Wegmüller U, Lamarche C, Bontemps S, Defourny P, Arino O (2015) Strengths and weaknesses of multi-year Envisat ASAR backscatter measurements to map permanent open water bodies at global scale. Remote Sens Environ 171:185–201

Strayer DL, Dudgeon D (2010) Freshwater biodiversity conservation: recent progress and future challenges. J N Am Benthol Soc 29(1):344–358. https://doi.org/10.1899/08-171.1

Sun XX, Li LW, Zhang B, Chen DM, Gao LR (2015) Soft urban water cover extraction using mixed training samples and Support Vector Machines. Int J Remote Sens 36:3331–3344

Talukdar S, Singha P, Mahato S, Shahfahad PS, Liou Y-A, Rahman A (2020) Land-use land-cover classification by machine learning classifiers for satellite observations—a review. Remote Sens 12:1135. https://doi.org/10.3390/rs12071135

TerrSet (2020) TerrSet-geospatial monitoring and modelling software, Clark Labs, Clark University, 950 Main Street, Worcester MA 01610-1477, USA

Varis O, Vakkilainen P (2001) China's 8 challenges to water resources management in the first quarter of the 21st century. Geomorphology 41:93–104

Viala E (2008) Water for food, water for life a comprehensive assessment of water management in agriculture. In: Molden D et al (ed) EarthScan London and International Water Management Institute, 2007 Colombo ISBN-13: 978-1844073962

Wang Z, Gao X, Zhang Y (2021) HA-Net: a lake water body extraction network based on hybrid-scale attention and transfer learning. Remote Sens 13(20):4121

Wang Z, Liu Q, Liu Y (2020) Mapping landslide susceptibility using machine learning algorithms and GIS: a case study in Shexian County, Anhui Province, China. Symmetry. https://doi.org/10.3390/sym12121954

Wik M, Varner RK, Anthony KW, MacIntyre S, Bastviken D (2016) Climate-sensitive northern lakes and ponds are critical components of methane release. Nat Geosci 9(2):99–105. https://doi.org/10.1038/ngeo2578

Wu W, Li Q, Zhang Y, Du X, Wang H (2018) Two-step urban water index (TSUWI): a new technique for high-resolution mapping of urban surface water. Remote Sens 10:1704

Xie C, Huang X, Zeng W, Fang X (2016) A novel water index for urban high-resolution eight-band WorldView-2 imagery. Int J Digital Earth 9(10):925–941. https://doi.org/10.1080/17538947.2016.1170215

Xu H (2006) Modification of normalized difference water index (NDWI) to enhance open water features in remotely sensed imagery. Int J Remote Sens 27(14):3025–3033

Yang XC, Zhao SS, Qin XB, Zhao N, Liang LG (2017) Mapping of urban surface water bodies from Sentinel-2 MSI imagery at 10 m resolution via NDWI-based image sharpening. Remote Sens 9:596

Yin YQ, Li GJ, Yu T, Yang HY, Zhang YH (2015) The study of object-oriented water body extraction method based on high resolution RS image. Surv Mapping Bull 1:81–85

8 Investigating Factors Influencing Groundwater in the Upper Yamuna Basin, India: A Geospatial Analysis

Ratnadeep Dutta, Milap Punia, and Suvamoy Pramanik

Abstract

Groundwater is one of the most precious resources, as it is a significant component in securing multiple sustainable development goals (i.e. food security and livelihood sustainability). However, the continuous anthropogenic activities put pressure on the groundwater, alarming those precious resources, especially in the upper Yamuna basin, India. This article aims to investigate spatial variation of determinant factors of the groundwater in the upper Yamuna basin using statistical and geospatial techniques. The study utilised the open-source databases, including precipitation, evapotranspiration, drainage density, vegetation cover areas, agriculture areas, groundwater extraction, and population density. The result indicates that the groundwater depth is alarmingly high in Rajasthan, Uttar Pradesh, and Delhi (> 50 m). The spatial and statistical analysis revealed that groundwater abstraction is a significant determinant of lowering groundwater table depth in the intensive agricultural regions of Rajasthan, Haryana, and Uttar Pradesh. However, vegetation coverage depicts negative effects on groundwater depletion, for instance, in Tehri Garhwal, where vegetation cover is 97.24%, associated with lower groundwater depth and a strong negative correlation coefficient ($b = \sim -0.40$). The results of this study can provide better insight into the causes behind groundwater depletion, thus supporting better groundwater management.

Keywords

Groundwater table · Open-source data · Water table · Geospatial techniques · Upper Yamuna basin

8.1 Introduction

Groundwater, an indispensable resource crucial for sustaining agriculture, industry, and domestic needs, plays a pivotal role in the socio-economic development of India (Dangar et al. 2021; Saha and Ray 2019). Understanding the changing hydrological regimes, especially the sub-surface hydrology, is crucial in building the pathway for a water sustainable future. Therefore, basin and watershed management has been considered for many years by hydrologists and geographers.

R. Dutta · M. Punia · S. Pramanik (✉)
Centre for the Study of Regional Development, Jawaharlal Nehru University, New Delhi 110067, India
e-mail: suvamoy.vu@gmail.com; suvamoy.pramanik@sbse.du.ac.in

S. Pramanik
Assistant Professor, Department of Geography, Shaheed Bhagat Singh Evening College, University of Delhi, New Delhi 110017, India

S. Talukdar et al. (eds.), *Water Resource Management in Climate Change Scenario*, GIScience and Geo-environmental Modelling, https://doi.org/10.1007/978-3-031-61121-6_8

The arid and semi-arid regions of North and Northwest India have long relied on groundwater to maintain agricultural productivity and support the livelihoods of millions (Everard 2015; Shekhar et al. 2020). Nevertheless, this invaluable resource confronts an increasingly severe threat in the form of rapidly declining groundwater levels, thus heightening the vulnerability of both the aquifers and the dependent communities. The upper Yamuna sub-basin is one of the important regions in northern India as it encompasses multiple states, including Delhi, Haryana, Rajasthan, and parts of Uttar Pradesh and a substantial portion of the population. The region has significant ecological, economic, and cultural importance. A prominent feature of the upper Yamuna sub-basin is the existence of substantial urban clusters, with the National Capital Region (NCR) at its core, centred around Delhi. This region includes rapidly expanding urban centres like Jaipur, New Delhi, Gurgaon, Faridabad, Ghaziabad, Saharanpur, and Meerut, among others. This rapid urban growth has led to a substantial rise in the demand for water resources, resulting in an excessive reliance on groundwater (Foster 2020; Foster et al. 2022). Recent studies have raised alarming concerns about the declining groundwater levels in areas such as Ghaziabad, illustrating the pressing issue of groundwater depletion in the region. Furthermore, the impacts of urbanisation and industrialisation have introduced complex dimensions to the groundwater crisis (Haque et al. 2013; Sarker et al. 2021). These regions, historically recognised as the "breadbasket" of the nation, have emerged as an agriculturally prominent region primarily after the success of high yielding variety of seeds developed during the Green Revolution in the mid-twentieth century (Sudha and Jansirani 2019). The increased water demand for growing crops across this vast region has led to an irrigation infrastructure backed by canals and groundwater reserves (Das et al. 2021; Malakar et al. 2022; Tiwari et al. 2009).

Groundwater extraction emerged as a cornerstone of this agricultural transformation, providing a dependable water source to supplement erratic monsoon rains (Garg et al. 2020). However, the over-reliance on groundwater for irrigation and inefficient water management practices have led to precipitous groundwater depletion (Dangar et al. 2021; Das et al. 2021). Over the past few decades, the region has witnessed an alarming decline in groundwater levels, jeopardising the very foundation of its agricultural and economic prosperity (Panda et al. 2021; Rodell et al. 2009). This decline is primarily attributed to excessive irrigation groundwater pumping, inefficient water management practices, and inadequate recharge mechanisms (Reddy 1989; Schwartz and Ibaraki 2011). Understanding the gravity of this situation is crucial. The factors contributing to groundwater vulnerability and rapid depletion in these regions are intricate and interrelated. They encompass hydrogeological characteristics, agricultural practices, policy frameworks, and dynamics. Addressing this pressing issue, a comprehensive interdisciplinary approach is engaging hydrologists, agronomists, policymakers, and local communities.

The consequences of this groundwater crisis are multifaceted and far-reaching. First and foremost, it poses a significant risk to food security, given the region's role as a major producer of wheat, rice, and other essential crops for the nation (Panda et al. 2021). For instance, in the northern states of Punjab and Haryana, often dubbed the "Granary of India" for their substantial wheat and rice production, severe groundwater depletion due to excessive irrigation reliance poses a significant threat to staple crop sustainability (Bhattarai et al. 2021; Jain et al. 2021). Similarly, in the southern state of Tamil Nadu, a major rice producer, groundwater depletion has led to water scarcity during critical growing seasons, impacting both rice production and farmer livelihoods (Chinnasamy and Agoramoorthy 2015; Varghese et al. 2013). Meanwhile, in Maharashtra, known for its fruit orchards, reliance on groundwater for irrigation has raised challenges in maintaining orchard productivity due to declining water levels (Amarasinghe et al. 2021). These examples underscore the direct impact of the groundwater crisis on agricultural output, affecting the livelihoods of millions and raising broader concerns about national food security. Effective, sustainable groundwater management is imperative to safeguard India's agrarian water resources and, consequently, its food security. Moreover, declining groundwater levels exacerbate the vulnerability of marginalised rural communities, as their access to water for drinking and sanitation becomes increasingly precarious (Sekhri 2014). Additionally, the environmental repercussions are profound, encompassing the depletion of aquifers leading to land subsidence, degradation of water quality, and the desiccation of once-thriving ecosystems (Danielopol et al. 2003).

The research gap that underscores the importance of this study lies in the absence of comprehensive, district-level analyses of groundwater dynamics in water-stressed regions, particularly in 2020. Prior research has predominantly focused on broader regional or national trends, overlooking the need for granular insights for effective local-level groundwater management (Ashraf et al. 2017). Additionally, understanding the specific underlying controls shaping groundwater levels and quality within each district has often needed to be more sufficiently explored, hindering the design of targeted interventions (Patle et al. 2016). Furthermore, limited assessments of the effectiveness of existing mitigation measures pose a challenge, as many studies tend to identify issues without providing actionable, evidence-based recommendations for sustainable groundwater management. Given the potential shifts in environmental conditions over time, including climate patterns and land use, conducting this study is crucial to capturing the contemporary situation and informing adaptive strategies that account for changing dynamics. Therefore, this study aims to bridge these research gaps by conducting a district-wise analysis, unravelling underlying controls, assessing mitigation measures, and formulating evidence-based recommendations to enhance groundwater management practices in water-stressed regions. Based on the identified research gaps, our study is driven by two primary objectives—(i) to evaluate the spatial pattern of groundwater status in the upper Yamuna sub-basin for the year 2020 and (ii) to explore and analyse the factors influencing groundwater dynamics in the study region.

8.2 Database and Methodology

8.2.1 Study Area

The present study focused on the upper Yamuna River basin is one of the extensive river basins in northern India, which is a part of the larger Ganga basin with an area of 41,489 km^2 from traversing the states of Uttarakhand and Himachal Pradesh in the north to the part of Rajasthan state in the south. Administratively, the study area comprises 45 districts in six states (Fig. 8.1). Physiographically, it can be divided into two regions—the upper Himalayan highlands to the north and the lower plains along the Yamuna River. The southern part of the region consists mainly of the Yamuna River's floodplain, known as Bhabar, the extended plain called Bangar, the south ridge of the Aravalli mountains, and the plains of Rajasthan. The study area exhibits diverse climatic characteristics according to Köppen's climate classification. In Uttarakhand's mountainous terrain, the prevailing climate falls into the Et or Tundra Type, marked by average temperatures ranging from 0 to 10 °C, with annual rainfall showing variability from year to year. Meanwhile, in the higher elevations of Himachal Pradesh, the E or Polar-Type climate prevails, characterised by the hottest month's temperature within the range of 0–10 °C. Across much of the study area, the dominant climate type is Cwg or Monsoon Type with Dry Winters, featuring an annual rainfall range of 100–200 cm, further contributing to the region's climatic diversity. The study area is home to approximately 800 million people, with the majority residing in urban clusters. Tier one cities like New Delhi drive rapid economic and urban development, leading to increased migration. Other urban centres experiencing rapid growth include Ghaziabad, Jaipur, Dehradun, Gurugram,

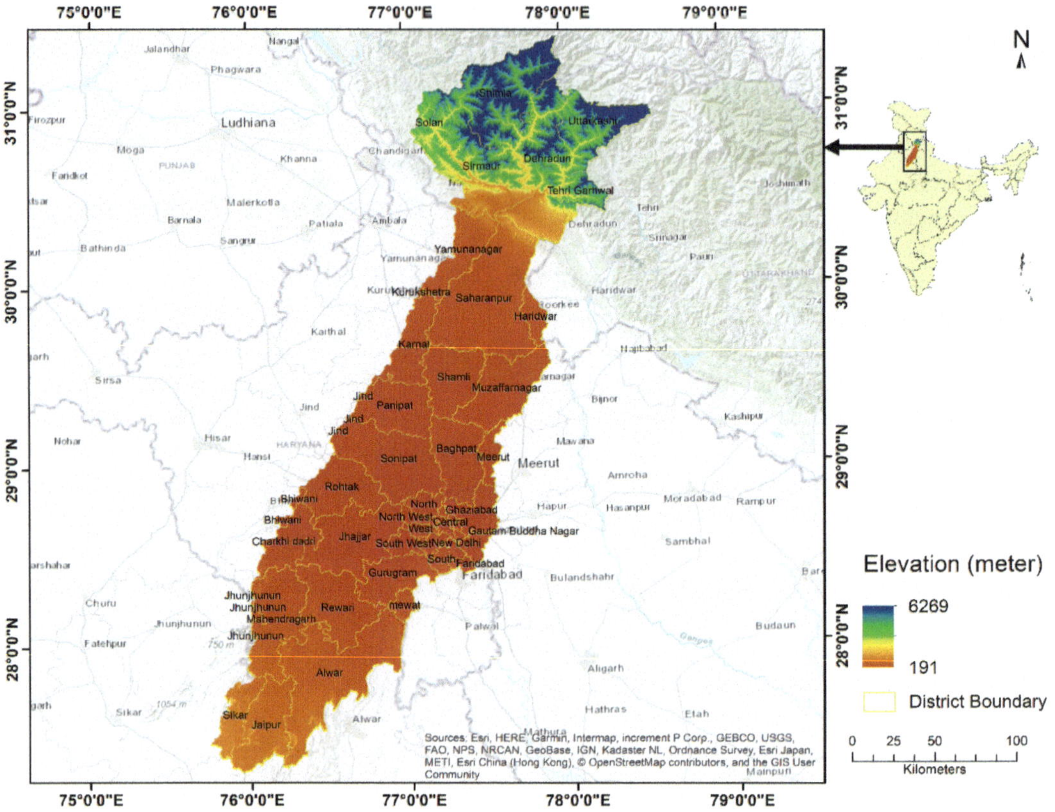

Fig. 8.1 Location of upper Yamuna sub-basin

Faridabad, Meerut, and Muzaffarnagar. These cities have substantial populations and are witnessing significant industrialisation, which, in turn, has led to skyrocketing groundwater extraction. States like Haryana, Rajasthan, and parts of Uttar Pradesh have high levels of irrigation, accounting for 90.56%, 24.26%, and 82.33% of the total irrigated area, respectively. The primary method of irrigation in this region is groundwater extraction. This relentless exploitation of groundwater resources exacerbates the already critical situation in the study area.

8.2.2 Database

Based on the literature survey, potential determinants are selected, which is represented in Table 8.1. We have incorporated various statistical and geospatial information to satisfy our objectives. For example, ground water level and extractions data are statistical information provided by Central Ground Water Board through in dedicated well-depth measurement ground survey. Similarly, the precipitation data provided by the Indian Meteorological Department is also statistical information generated by the station-based precipitation measurement (Pai et al. 2014). Besides, several geospatial information are collected/prepared—like potential evapotranspiration collected from MODIS satellite imageries (Running et al. 2017), drainage density prepared from Cartosat DEM (National Remote Sensing Centre 2014) by following proper drainage network assessment techniques (Kale et al. 2021), vegetation cover area and built-up cover area from ESA 100 m global land cover data (Zanaga et al. 2021). The population density data was collected from NASA's Socio-Economic Data and Application Centre

Table 8.1 Data source for multiple regression analysis

Data used	Data specification	Source	Purpose
Groundwater level	Station/monitoring well Annual mean (m)	Central Groundwater Board (CGWB)	To assess the depth to groundwater level in the study area
Rainfall	$0.25° \times 0.25°$ Annual mean (mm)	Indian Meteorological Department (IMD)	Assessment of average annual rainfall in the study area. As rainfall is the critical driver of groundwater recharge, it is taken as a variable
Potential evapotranspiration	1 km × 1 km (kg/m^2)	MOD16A2: MODIS Global Terrestrial Evapotranspiration	This dataset offers valuable insights into the amount of water lost to evaporation and transpiration from the land surface, a key component in the groundwater balance
Drainage density	Mean density (km/km^2)	Cartosat-1 CartoDEM Version-2 R1	The dataset was employed to calculate drainage density, a crucial indicator of the land's ability to retain water. This dataset was selected for its high-resolution terrain representation, allowing for precise calculations of drainage density
Vegetation cover area	Percentage per unit area	European Space Agency (ESA) WorldCover 10 m v100	Vegetation cover can be crucial in positively influencing groundwater recharge
Built-up cover area	Percentage per unit area	European Space Agency (ESA) WorldCover 10 m v100	Built-up cover acts as an obstacle to infiltration and, therefore, hinders the recharge process. This data has been taken to understand the relationship between urbanisation and groundwater dynamics
Groundwater abstraction	Billion cubic metres (bcm)	India Water Resource Information System (WRIS)	To quantify the amount of groundwater withdrawn for various purposes. This information is essential for assessing the impact of groundwater usage on aquifer levels
Population density	Persons per km^2 per pixel	Gridded Population of the World v4.11—NASA Socio-economic Data and Applications Centre (SEDAC)	It was selected to analyse the relationship between population density and groundwater demand

(Doxsey-Whitfield et al. 2015). Here, we primarily selected open-source data so the reproducibility of the research would retain.

8.2.3 Methods

8.2.3.1 Data Preprocessing

The data collected for this study consists mostly of raster images, except for groundwater depth and groundwater abstraction, which are collected differently. For groundwater level data, we used monitoring wells across the study area. We interpolated these values using the universal kriging method in ArcGIS to generate a raster map representing groundwater levels. Since we focus on districts, groundwater abstraction data at the district level required no additional processing; we simply integrated it into the analysis. To assess vegetation and built-up cover, we utilised the ESA WorldCover classified map. Extracting information from this map, we calculated the proportion of each cover class within district boundaries. All the collected data were then combined into a unified district-level map. To align parameter values with the district boundaries, we employed zonal statistics. As the variables used in the study have different units of scales, we normalised the data using minimum-maximum normalisation to ensure that different features contribute equally to the analysis (Eq. 8.1).

$$\text{Normalized } x_i = \frac{x_i - x_{\min}}{x_{\max} - x_{\min}}, \tag{8.1}$$

where x_i is each value of the variable to be normalised, $x_{\min}$ is the minimum value for the variable, and $x_{\max}$ is the maximum value of the variable X.

8.2.3.2 Multicollinearity Assessment

Multicollinearity, the high correlation between independent variables in regression analysis, was assessed to enhance the reliability of our model. It is a crucial issue to address as it can inflate standard errors, render coefficient estimates unstable, obscure statistical significance, and make it challenging to identify the true drivers of the dependent variable (Daoud 2017; York 2012). By conducting a Variance Inflation Factor (VIF) test, we aimed to mitigate these problems. Addressing multicollinearity ensures that our regression results provide more accurate and interpretable insights into the complex relationships between independent variables and groundwater levels in the study area, thus enhancing the overall reliability of our analysis (Schreiber-Gregory 2018; Senaviratna and Cooray 2019). Calculation of VIF was done using Eq. 8.2.

$$\text{VIF} = \frac{1}{\left(1 - R^2\right)}, \tag{8.2}$$

where R^2 is the coefficient of determination of the regression equation. A VIF score of 1 indicates that there is no multicollinearity among the independent variables. Each variable is completely unrelated to the others, which is rare in real-world data. VIF values below 5 generally indicate low-to-moderate multicollinearity. This suggests that the independent variables are not highly correlated and should not cause significant issues in your regression model. VIF values in the range 5–10 indicate moderate multicollinearity. VIF values above 10 indicate high multicollinearity and are a cause for concern. High multicollinearity can make it challenging to interpret the regression coefficients accurately. In such cases, one should strongly consider removing one or more of the correlated independent variables from the model or using techniques like Principal Component Analysis (PCA) to reduce multicollinearity. In cases where variables displayed high levels of correlation with each other that is variables with high VIF scores, we carefully removed them from the model. This step was taken to enhance the model's accuracy and stability.

8.2.3.3 Multiple Regression Analysis

Multiple Regression Analysis serves as a powerful tool for unravelling the intricate relationships between a dependent or criterion variable and multiple independent or predictor variables (Almonroeder 2022). In our study, this analysis was employed to discern the intricate dynamics impacting groundwater levels, represented in Eq. 8.3.

$$y_i = \beta_0 + \beta_1 x_1 + \beta_2 x_2 + \cdots + \beta_n x_n + \varepsilon \quad (8.3)$$

where y_i signifies the dependent variable that is groundwater level below the ground, while β_0 represents the intercept, and β_1 through β_n denote the regression coefficients. The independent variables, x_1 through x_n, encompass a range of dynamic environmental and anthropogenic properties. ε is the error term.

8.2.3.4 Dominance Analysis

Dominance analysis (DA) serves as a technique for assessing the relative significance of predictors in a multiple regression setting. This approach gauges the superiority of one predictor in comparison to another by examining their supplementary R^2 contributions across various subset models (Azen and Budescu 2003). The objective of predictor importance identification, within the scope of DA, is not cantered on model selection but rather on revealing the distinct impacts of predictors relative to one another within a designated model. In dominance analysis, there are primarily two types of dominances: general dominance and conditional dominance. These dominances help assess the relative importance of predictors in a multiple regression model. General dominance measures the overall contribution of each predictor to the R-squared value (the proportion of variance explained by the model) in the multiple regression model, represented in Eq. 8.4.

$$R_X^2 - \frac{1}{p}\sum_{i=1}^{p} R_i^2, \quad (8.4)$$

where R_X^2 is the R-squared value for predictor "X" in the full model. p is the total number of predictors in the model. R_i^2 is the R-squared value for predictor "i" in the full model.

Conditional dominance assesses how a predictor's contribution to the R-squared value changes as you include more predictors in the model, represented in Eq. 8.5.

$$R_X^2(k) - R_X^2(k-1), \quad (8.5)$$

where $R_X^2(k)$ is the R-squared value for predictor "X" in the model with k predictors. $R_X^2(k-1)$ is the R-squared value for predictor "X" in the model with $(k-1)$ predictors.

8.2.3.5 Geographically Weighted Regression (GWR)

The Geographically Weighted Regression (GWR) analysis was selected as our modelling approach due to its unique advantages over traditional regression methods in this context. GWR is particularly suited to our study because it accounts for spatial heterogeneity by allowing regression coefficients to vary across different geographical areas within the study region. This spatially explicit approach is essential for capturing localised variations in the relationships between independent variables and groundwater levels, which can be critical in hydrological studies (Koh et al. 2020; Yeh et al. 2022; Yu et al. 2020). Unlike traditional regression, which assumes uniform relationships across the entire study area, GWR provides insights into how these relationships change geographically. This is especially relevant in regions like North and Northwest India, where groundwater dynamics can vary significantly over short distances due to factors such as hydrogeological characteristics and land use. GWR's ability to capture these spatial nuances makes it a powerful tool for our analysis, enhancing the precision and interpretability of our results. Therefore, the final set of variables, devoid of multicollinearity concerns, was employed in the Geographically Weighted Regression (GWR) analysis using ArcGIS (Eq. 8.6).

$$y_i(u_i, v_i) = \beta_0(u_i, v_i) + \beta_1(u_i, v_i)x_1 + \beta_2(u_i, v_i)x_2 + \cdots + \beta_n(u_i, v_i)x_n + \varepsilon_i, \quad (8.6)$$

where (u_i, v_i) is the location of a phenomena and β_i and x_i are the regression coefficients and predictors at that location. Through this comprehensive analytical approach, we aimed to unveil the complex dynamics impacting groundwater resources in our study area, thereby contributing to a more profound understanding of groundwater vulnerability and the factors driving its decline.

8.3 Results and Discussion

8.3.1 Spatial Pattern of Groundwater Depth and Its Determinants

The spatial status of groundwater level and its drivers, namely, rainfall, drainage density, vegetation cover, built-up cover, groundwater abstraction, population density, and potential evapotranspiration, for the year is represented in Fig. 8.2. The variation in depth to groundwater within the study area, ranging from 4.17 to 57.03 m with an average of 23.53 m, highlights significant regional hydrogeological disparities. The clustering of maximum depth to the water level in the southern part of the basin, encompassing districts like Sikar, Alwar, Gurugram, South and Southeast Delhi, and Faridabad, indicates a critical situation. This result is similar to previous researches (Gupta and Sarma 2016; Patle et al. 2016; Gopal et al. 2015).

The lower geographical elevation of the southern region, combined with extensive groundwater extraction driven by urbanisation and industrialisation, leads to the deepening of the water table. This underscores the urgent need for stringent groundwater management practices, including regulating extraction and promoting recharge methods. Conversely, the presence of minimum depth to groundwater (less than 6.38 m) in Rohtak and Bhiwani districts of Haryana highlights a different set of challenges. The proximity to the water table can lead to issues such as waterlogging and salinity, affecting crop productivity and soil health (Houk et al. 2006).

Proper drainage and salinity management measures are essential in such areas to ensure sustainable agriculture. The cluster of depth to groundwater ranges between 6 and 14 m in the northernmost part of the basin, including districts like Solan, Shimla, Uttarkashi, Sirmaur, Dehradun, Tehri Garhwal, Yamunanagar, Kurukshetra, Saharanpur, and Haridwar, presents a more balanced groundwater scenario. However, monitoring and managing groundwater resources here are crucial to prevent over-extraction and maintain a sustainable balance between water availability and demand.

The observed mean annual rainfall for 2020 varies significantly across the study area, ranging from 0.88 to 3.81 mm. Notably, the Tehri Garhwal region stands out with the highest recorded rainfall, while the Bhiwani district in Haryana experiences the lowest. This variation is primarily attributable to the geographical diversity of the region, with the Himalayan mountainous region receiving substantially more rainfall compared to the lower parts of the basin, which include portions of Rajasthan, Haryana, and Delhi (Fig. 8.2f). Areas receiving higher rainfall tend to experience more significant groundwater recharge due to a surplus of water infiltrating the ground. This contributes to maintaining healthy groundwater levels and can offset extraction rates. Conversely, regions with lower annual rainfall, like parts of Rajasthan, Haryana, and Delhi, are susceptible to groundwater depletion. Limited recharge potential can lead to a significant reliance on groundwater extraction resulting in falling water tables.

In the study area, we observe variations in drainage density, with some districts exhibiting notably higher values. Specifically, districts like Jhajjar, Northwest Delhi, Northeast Delhi, Central Delhi, and Gautam Budh Nagar stand out with a maximum drainage density of 0.14 km per square km, which suggests a more efficient natural drainage system. Higher drainage density generally implies a better natural capacity to channel excess water away, potentially enhancing groundwater recharge (Barron et al. 2013; Luijendijk 2021). In areas with higher drainage density, rainwater is more likely to infiltrate the ground rather than cause surface runoff, which can replenish groundwater aquifers. Areas with higher drainage density may experience more stable groundwater levels as excess water is efficiently managed. In contrast, regions with lower drainage density may be prone to waterlogging, which can hinder groundwater recharge and potentially lead to fluctuations in the water table (Fig. 8.2c).

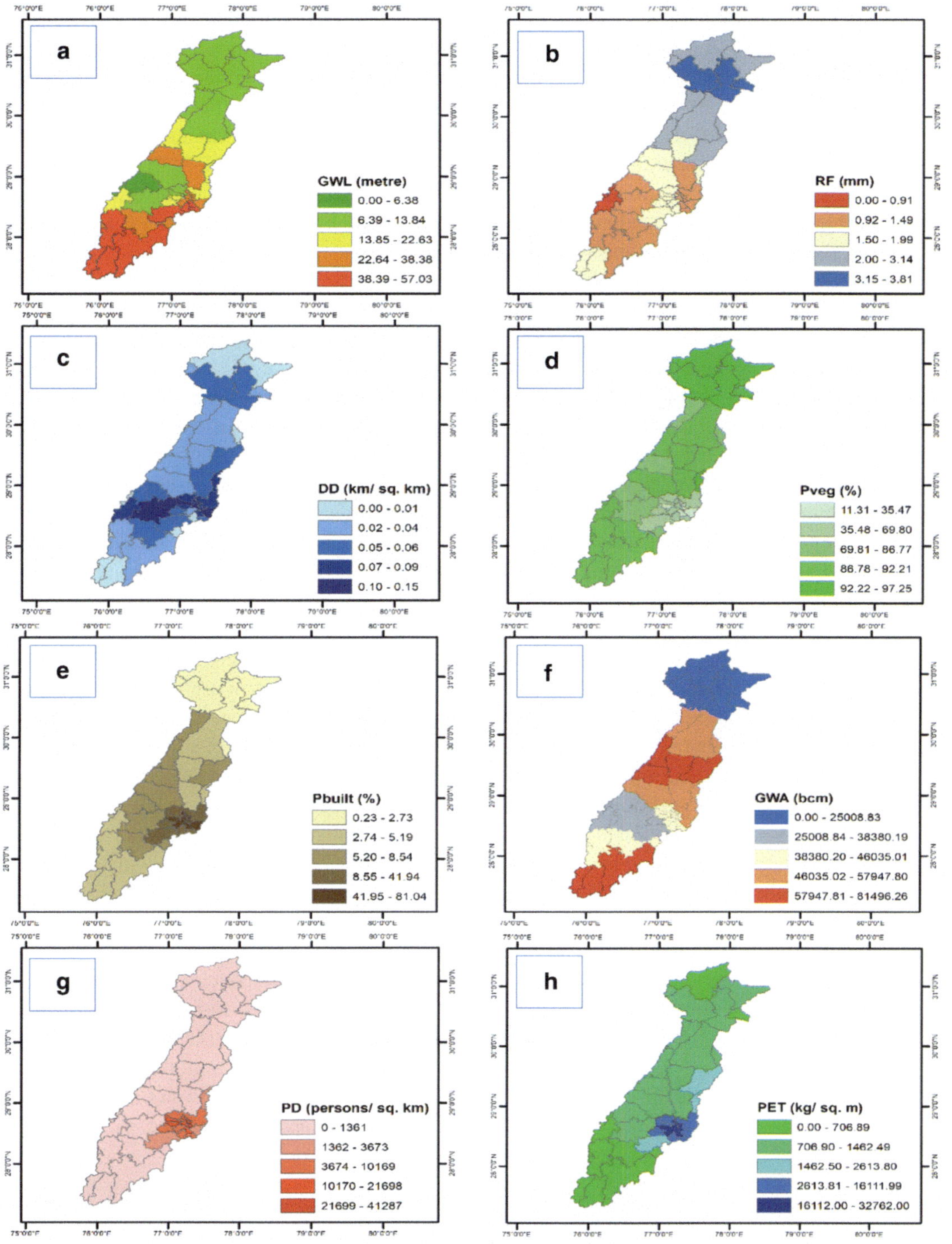

Fig. 8.2 Spatial pattern of **a** groundwater depth, **b** rainfall, **c** drainage density, **d** percentage vegetation cover, **e** percentage built-up cover, **f** groundwater abstraction, **g** population density, and **h** potential evapotranspiration

The highest proportion of vegetated land is found in the district of Tehri Garhwal, where it is approximately 97.24%. On the other hand, East Delhi has the lowest proportion of vegetation cover (~ 11.31%). This can be attributed to the fact that as vegetation cover and built-up development are intimately interrelated, and highly urbanised locations influence the distribution of tree cover. In the Himalayan mountains, where there is very little population density and built-up land, a higher proportion of tree cover can be found. Similarly, the percentage of built-up land in a district shows very similar pattern. High values of Pbuilt presence can be found in a cluster in and around the capital region of Delhi including district of Northeast, Central, West, New Delhi, East, Southeast, Gautam Budh Nagar. Here, the Pbuilt varies between 41 and 81%. This is surrounded by a cluster of rapidly developing regions, namely Gurugram or Gurgaon, Faridabad, and Ghaziabad. Lowest Pbuilt can be found in the upper part of the basing specifically in the hilly districts of Shimla, Solan, Sirmaur, Dehradun, Uttarkashi, and Tehri Garhwal (Fig. 8.2d and e).

The pattern of distribution of groundwater abstraction, which is an anthropogenic control of groundwater dynamics, is very unique, forming clusters of very high to bearable amount of groundwater withdrawal. High values of GWA can be found in the central and lower parts of the basin. These regions are intensive agricultural regions of Rajasthan, Haryana, and Uttar Pradesh. Maximum groundwater withdrawal for the year 2020 can be observed in the district of Karnal in Haryana (~ 81,000 bcm) followed by Jaipur in Rajasthan (~ 75,000 bcm). These regions show continuous decline in groundwater table historically as well (Panda et al. 2021). The clustered pattern of groundwater withdrawal can be distinctly observed in Fig. 8.2f.

Population density shows a very unique conglomeration as most of the study area experiences PD below 1361 p/km^2, while areas surrounding the national capital of Delhi experience very high PD (more than 3500 p/km^2.). In districts of Sahadra, Northeast and East, the population density is more than 30,000 p/km^2 (Fig. 8.2g). The pattern of potential evapotranspiration across the study area is not random; high PET is found in the Delhi–Ghaziabad–Gurgaon region, while districts of Rajasthan, Himachal Pradesh, and Uttarakhand show relatively lower values of PET. The PET ranges between 291.35 kg/m^2 (in Alwar, Rajasthan) and 32,762 kg/m^2 in Shahdara (Delhi) (Fig. 8.2h).

8.3.2 Bivariate Relationship

The scatter plots represented in Fig. 8.3 depict the relationships between the five determinants and the dependent variable, groundwater level. There is a negative relationship between rainfall and groundwater depth. In simpler terms, an increase in rainfall leads to a decrease in groundwater depth. Drainage density shows a negative correlation with groundwater depth. Higher drainage density is associated with shallower groundwater levels. This observation is particularly important in regions with fewer rivers and streams. The data suggests that regions with a higher percentage of vegetation cover tend to have shallower groundwater. In essence, an increase in vegetation cover results in a decrease in groundwater depth. Groundwater abstraction exhibits a positive relationship with groundwater depth. In other words, when groundwater is extracted at higher rates, groundwater levels tend to increase. This is an important finding for understanding the impact of water demand on groundwater resources. Surprisingly, population density is negatively correlated with groundwater depth. This implies that in regions with lower population density, groundwater levels tend to decline more rapidly compared to areas with higher population density.

8.3.3 Multiple Linear Regression Analysis

The results of our regression analysis aimed at understanding the factors influencing groundwater level (GWL) in the study area. We performed

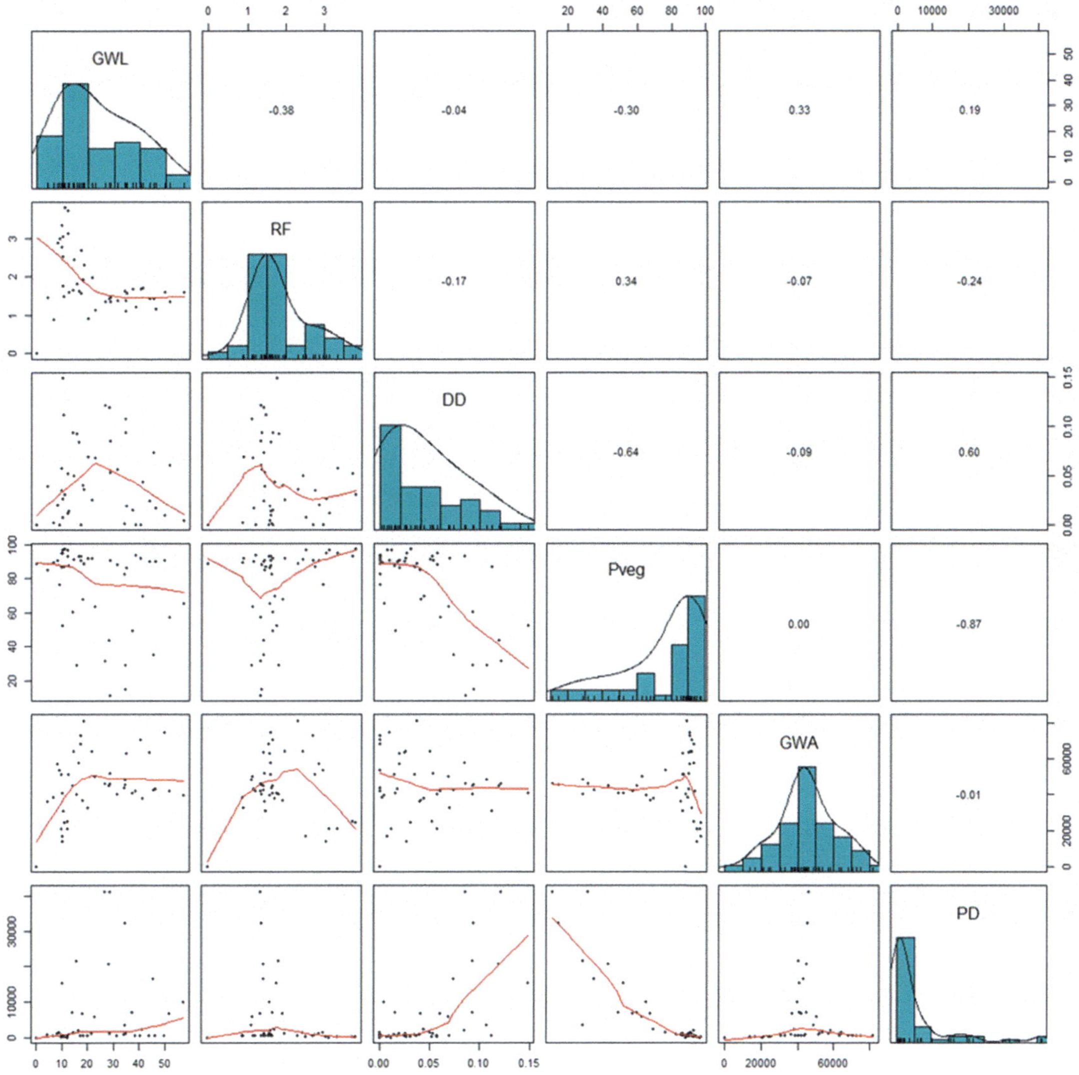

Fig. 8.3 Scatter plot showing the correlation between groundwater depth and rainfall, drainage density, vegetation cover, groundwater abstraction, and population density

five regression models to explore the relationships between GWL and various predictor variables. The models were sequentially built, each one removing a specific predictor variable to assess its impact on the model's performance and multicollinearity. Table 8.2 presents the VIF values at each step of the model modification.

After removing both Pveg and PET, VIF values for the remaining variables are generally low, indicating minimal multicollinearity. A multiple linear regression was performed on these five independent variables (Model 5), the result of which is shown in Table 8.3.

Table 8.3 shows the estimates, standard errors, *t* values, and *p*-values for the five variables included in the linear regression model. The intercept has an estimated coefficient of 0.8736, which means that when all other variables are held constant, the expected decline in the groundwater level is 0.8736 m. Thus, in practical terms, if the intercept variable represents some characteristic or condition in the study area that

Table 8.2 Variance inflation factors on step-wise regression analysis

Model No.	Variance inflation factor (VIF) score							
	RF	DD	Pveg	Pbuilt	GWA	PD	PET	Remarks
Model 1	1.199	1.745	33.859	52.529	1.035	6.531	14.562	Discarded
Model 2	1.199	1.745	**Removed**	12.511	1.025	6.582	14.123	Discarded
Model 3	1.173	1.745	8.064	**Removed**	1.031	6.530	10.809	Discarded
Model 4	1.154	1.735	32.839	38.991	1.032	5.158	**Removed**	Discarded
Model 5	1.150	1.732	4.751	**Removed**	1.018	4.171	**Removed**	Final model

Table 8.3 Result of multiple linear regression

Variables	Estimate	Std. error	*t*-value	*p*-value
Intercept	0.8736	0.2558	3.415	0.0015**
RF	– 0.3207	0.1711	– 1.874	0.0684
DD	– 0.3108	0.1623	– 1.915	0.0629
PVEG	– 0.5063	0.2554	– 1.982	0.0545
GWA	0.3412	0.1607	2.123	0.0401*
PD	– 0.1600	0.2703	– 0.592	0.5572

*Significant at the 0.05 level ($p<0.05$)
**Significant at the 0.01 level ($p<0.01$)

can vary (such as land use, geology, or demand patterns), then for each unit increase in that characteristic or condition, we would expect the groundwater level to decrease by approximately 0.8736 m, provided that all other factors like climate, extraction rates, etc., remain constant

RF has an estimated coefficient of – 0.3207. The negative sign denotes that increase in rainfall will decrease the depth to groundwater. Specifically, all else being equal, a unit increase in RF is associated with a decrease of 0.3 m of groundwater depth. However, its *p*-value (0.0684) is greater than 0.05, which suggests that this result might be due to chance. Our observation that increased rainfall is associated with decreased groundwater depth aligns with earlier studies (Malakar et al. 2022; Tiwari et al. 2009). It underscores the significance of precipitation as a primary driver of groundwater recharge in regions with variable rainfall patterns.

DD has a similar estimated coefficient of – 0.3108, but its *p*-value is slightly lower (0.0629). This negative connection highlights the importance of considering drainage systems in groundwater management, especially in areas with low drainage density.

Pveg has an estimated coefficient of – 0.5063. In practical terms, if Pveg represents the percentage of land covered by vegetation such as forests, grasslands, or green spaces, a higher percentage of vegetation cover is linked to a slower rate of groundwater decline. This could be due to several factors: vegetation helps reduce surface runoff and enhances groundwater recharge by allowing rainwater to infiltrate the soil, plant roots can stabilise the soil and prevent erosion, which can protect aquifers from siltation, and vegetation can also contribute to a more balanced local hydrological cycle, helping to maintain groundwater levels.

GWA has a positive estimated coefficient of 0.3412 and a significant *p*-value of 0.0401. This implies that higher values of GWA correspond to higher values of the response variable. In other words, as groundwater abstraction increases (meaning more groundwater is being withdrawn), it is associated with an increase in sub-surface water depth. This suggests that increased levels of groundwater extraction may be contributing to certain trends or changes in the study area, as indicated by the response variable. The low p-value reinforces the statistical significance of this relationship, indicating a substantial association between groundwater abstraction and the response variable.

Finally, contrary, our study reveals a surprising negative relationship between population density and groundwater depth. While this finding may appear counterintuitive, it echoes the observations made by Kalhor and Emaminejad (2019), who emphasise the water demand challenges posed by densely populated areas. PD has an estimated coefficient of – 0.1600. However, plotting the groundwater depth as a function of population density gives an obvious positive trend indicating the association of higher population densities with deeper groundwater levels.

In brief, the model's adjusted *R*-squared is 0.34, explaining only 34% of groundwater dynamics variability. Results are mixed, with some variables showing a connection to the response variable, but others don't. Most *p*-values are above the conventional significance level of 0.05, indicating uncertainty.

RF, DD, and GWA seem influential in groundwater fluctuations, but further investigation is needed, especially for RF and DD, which have borderline significant relationships. Additional data or a deeper analysis is required to fully understand these associations.

8.3.4 Dominance Analysis

A dominance analysis was applied on the final model to understand the relative contribution of the different predictor variables. The result of the analysis is shown in Table 8.4. The general dominance values for each predictor in the regression model represent the contribution of each predictor to the R^2 statistic, indicating how much variance in the dependent variable it explains. This is used to rank the predictor variables based on their contribution to explaining the variability of the dependent variable. A lower rank indicates a higher level of impact on the dependent variable. In our model, RF has the highest general dominance value of approximately 0.0978 and the lowest standardised rank of 1, indicating that it is the most important predictor in the model. Groundwater abstraction has the second-highest general dominance value of approximately 0.0922 and a standardised rank of 2. Pveg has the third-highest general dominance value of approximately 0.0852 and a standardised rank of 3. DD has a general dominance value of approximately 0.0431 and a standardised rank of 4. PD has the lowest general dominance value of about 0.0248 and the highest standardised rank of 5, indicating that it has the most negligible impact on the dependent variable.

Table 8.4 General dominance of predictor variables: rainfall (RF), drainage density (DD), percentage of vegetation cover (Pveg), groundwater abstraction (GWA), and population density (PD)

Variables	General dominance	Standardised dominance	Ranks
RF	0.09778901	0.28500526	1
DD	0.04311702	0.12566420	4
Pveg	0.08522147	0.24837726	3
GWA	0.09223331	0.26881322	2
PD	0.02475219	0.07214006	5

The conditional dominance values indicate how each predictor's dominance changes as you include more predictors in the model. Table 8.5 is organised by subset size, ranging from 1 predictor up to 5 predictors. For example, at subset size 2, RF has a conditional dominance value of approximately 0.1185, suggesting its contribution to R^2 when combined with another predictor. When coupled with other subsets, all the predictor variables except drainage density show decreasing conditional dominance. This implies that the effectiveness of these predictors in explaining variance decreases when they are combined with other predictors, compared to when they are used individually.

8.3.5 Result of Geographically Weighted Regression (GWR) Analysis

The Geographically Weighted Regression (GWR) analysis allows us to examine the spatial variations in the relationships between predictor variables and groundwater depth (Fig. 8.4).

The GWR analysis reveals a noticeable variation in the R^2 values across the study area. The upper region, including the districts of Shimla, Uttarkashi, Dehradun, Tehri Garhwal, Solan, and Sirmaur, exhibits the most pronounced variability, with R^2 values ranging from 0.45 to 0.52.

Table 8.5 Conditional dominance of predictor variables: rainfall (RF), drainage density (DD), percentage of vegetation cover (Pveg), groundwater abstraction (GWA), and population density (PD)

Variables	Subset size: 1	Subset size: 2	Subset size: 3	Subset size: 4	Subset size: 5
RF	0.140714943	0.11854782	0.09519807	0.07533020	0.059154038
DD	0.001914168	0.03668810	0.05396115	0.06128100	0.061740685
Pveg	0.092236700	0.09717600	0.09078289	0.07971929	0.066192457
GWA	0.106610720	0.10234079	0.09274218	0.08352564	0.075947238
PD	0.036923698	0.03622441	0.02764454	0.01706326	0.005905057

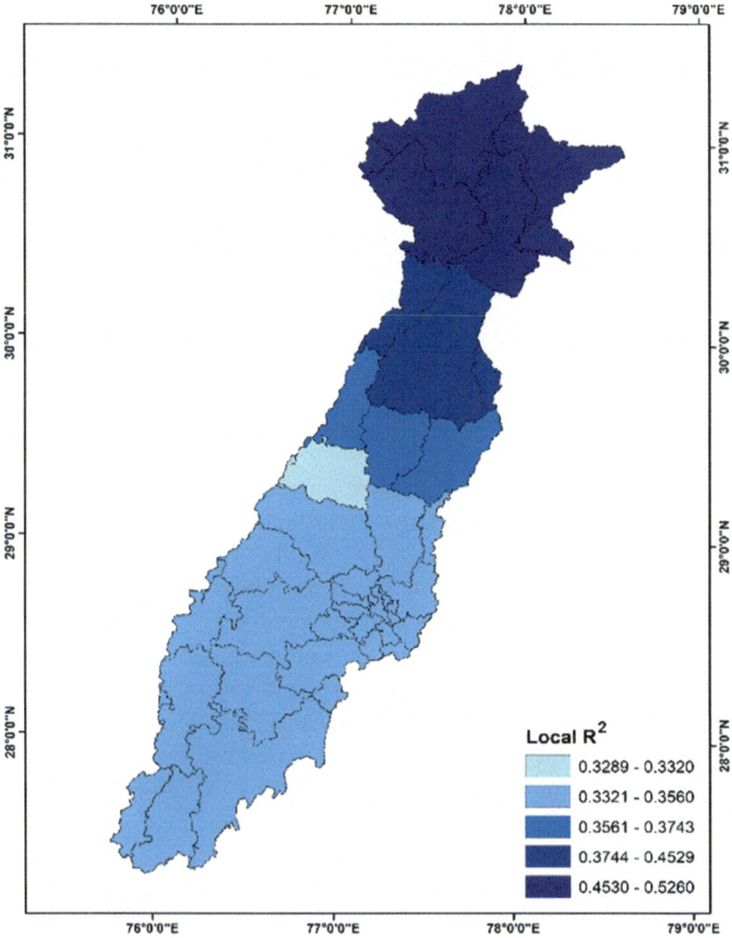

Fig. 8.4 Local *R*-squared of GWR

In contrast, the lower part of the region, which includes the districts Baghpat, Sonipat, Meerut, Rohtak, Ghaziabad, Gautam Buddha Nagar, Faridabad, Gurugram, Jhajjar, Bhiwani, Jind, Charkhi Dadri, Mewat, Rewari, Jhunjhunun, Alwar, Jaipur, Sikar, and the entire Delhi state, has R^2 values ranging from 0.32 to 0.35 signifying an explained variability of 32–35%. This suggests that, in these areas, additional factors such as geology, aquifer characteristics, land use patterns, and other variables exert a supplementary influence on sub-surface hydrology, contributing to the observed variation (Haaf et al. 2020).

The intercepts for all districts are positive, indicating that while keeping other factors constant, groundwater depth continues to increase.

The rate of increase varies across districts, with the northern part of the basin, including regions like Shimla, Solan, Dehradun, Uttarkashi, and Haridwar, experiencing the highest rates of decline. This is a significant concern as mountainous regions face groundwater decline more rapidly if not recharged.

Rainfall (RF): The coefficient for rainfall shows a similar negative relationship as observed in the scatter diagrams. However, the rate of increase in the groundwater level is much lower in the districts of Rajasthan falling in the study area, along with those of Haryana, including Gurgaon and Bhiwani. At the same time, the northern part shows a better rate of decrease in groundwater depth caused by rainfall recharge. Urban centres like Delhi, Ghaziabad, Faridabad are also in critical situations where even rainfall may not enhance the groundwater condition due to its highly concretised surface and groundwater extraction (Barron et al. 2013; Yazicigil et al. 2011).

Drainage Density (DD): The drainage density coefficient ranges from − 0.61 in the upper basin to − 0.15 in the western section. When drainage density is taken into account, regions in Rajasthan, Haryana, and Delhi are less affected by groundwater extraction. However, it is crucial to note that other factors like groundwater abstraction dominate in controlling groundwater dynamics in these areas (Dangar et al. 2021). In the basin's upper part, drainage density effects are more prominent due to increased stream presence and subsequent groundwater recharge (Fig. 8.5).

Vegetation Cover (Pveg): Pveg is negatively correlated with groundwater depth, meaning that areas with higher vegetation cover generally have shallower groundwater. For instance, a 1% increase in vegetation cover can lead to a 0.60-m rise in groundwater levels in regions like Delhi, Ghaziabad, Meerut, and Rohtak. This effect is less pronounced in upper Himalayan districts.

Groundwater Abstraction (GWA): GWA is positively related to groundwater depth, indicating that increased groundwater extraction leads to deeper groundwater levels. This effect is particularly evident in parts of Rajasthan, Haryana, Delhi, and Uttar Pradesh, including districts like Sikar, Jaipur, Jhunjhunun, Mewat, Rewari, Alwar, and Gurugram. Continued unsustainable groundwater exploitation in these regions could have severe consequences (Dangar et al. 2021; Danielopol et al. 2003).

Population Density (PD): Surprisingly, PD is negatively correlated with groundwater depth. This implies that groundwater levels tend to decline more rapidly in areas with lower population density than in densely populated regions. This unexpected finding warrants further investigation into the complex interplay between population growth, water demand, and groundwater dynamics (Ashoka et al. 2018).

8.3.6 Control of Geology

While groundwater availability is mainly controlled by the variables described above, its occurrence is influenced by the underlying rock structure. The study area falls under a unique set of geological formations. As the topography and rainfall virtually control surface runoff and groundwater recharge, the variability in the rock formations and variations of landforms ranging from the rugged terrain of Himalayan mountains through the alluvial plains of northern India to the aeolian parts of Rajasthan plays an essential role in the occurrence of groundwater. Although quantifying the influence of these static factors, namely geologic structure and topography, is outside the scope of this research, their influence is no less critical. The study area falls within three groundwater provinces (Fig. 8.6) as described in "India: A Regional Geography" (Singh 1971).

These provinces are (i) the Himalayan province, which is not very significant concerning groundwater resources; (ii) the

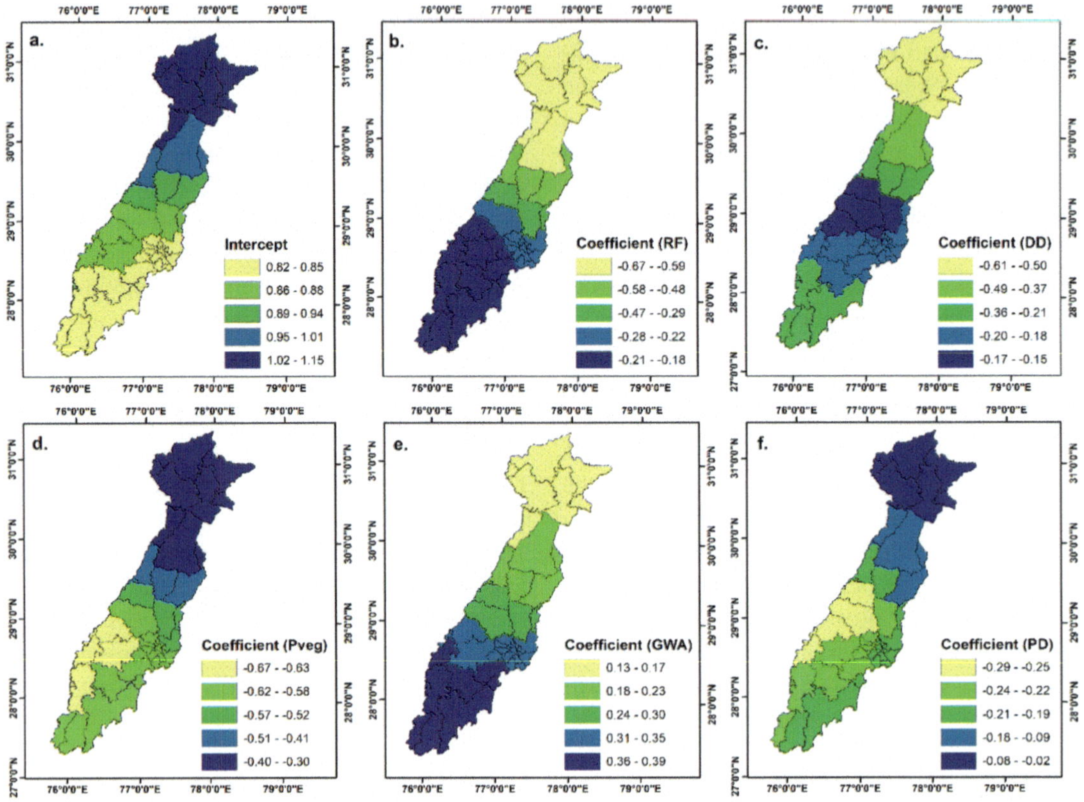

Fig. 8.5 Map showing results of GWR: **a** intercept, **b** coefficient of rainfall, **c** coefficient of drainage density, **d** coefficient of vegetation cover, **e** coefficient of groundwater abstraction, and **f** coefficient of population density

Ganga–Brahmaputra Alluvial province (including the districts of Haryana, Uttar Pradesh, Delhi), which constitutes one of the largest and most potential groundwater reservoirs in the world. The aquifer systems are extensive, thick, hydraulically interconnected, and moderate-to-high yielding. To the north of this tract, all along the Himalayan foothills occur the linear belt of Bhabar piedmont deposits and the Tarai belt down the slope with characteristic auto-flowing conditions; and (iii) Pre-Cambrian Sedimentary province (comprising parts of Rajasthan), which is not much suitable for groundwater development and contains the inadequate amount of groundwater. The groundwater availability quantified in our study by means of depth to groundwater table depicts the same. Furthermore, it has been seen from the study that in these provinces, the combined actions of anthropogenic and natural drivers will eventually cause groundwater depletion. Still, the rate of decline will vary depending on these underlying geological and pedological characteristics.

8.4 Conclusion

Spatially, the groundwater levels vary broadly throughout the study area. While in the districts of Himachal Pradesh, Uttarakhand, and parts of Uttar Pradesh and Haryana, the depth to groundwater is relatively lower (ranging from 6 to 14 m), in parts of Rajasthan, Uttar Pradesh and Delhi, the depth of groundwater is alarmingly high (as much as 50 m below ground level). Although the water futures for selected areas in Rajasthan and Delhi are virtually identical, illustrating a distinct water issue, the underlying drivers are considerably different, necessitating a region-specific groundwater and drinking water

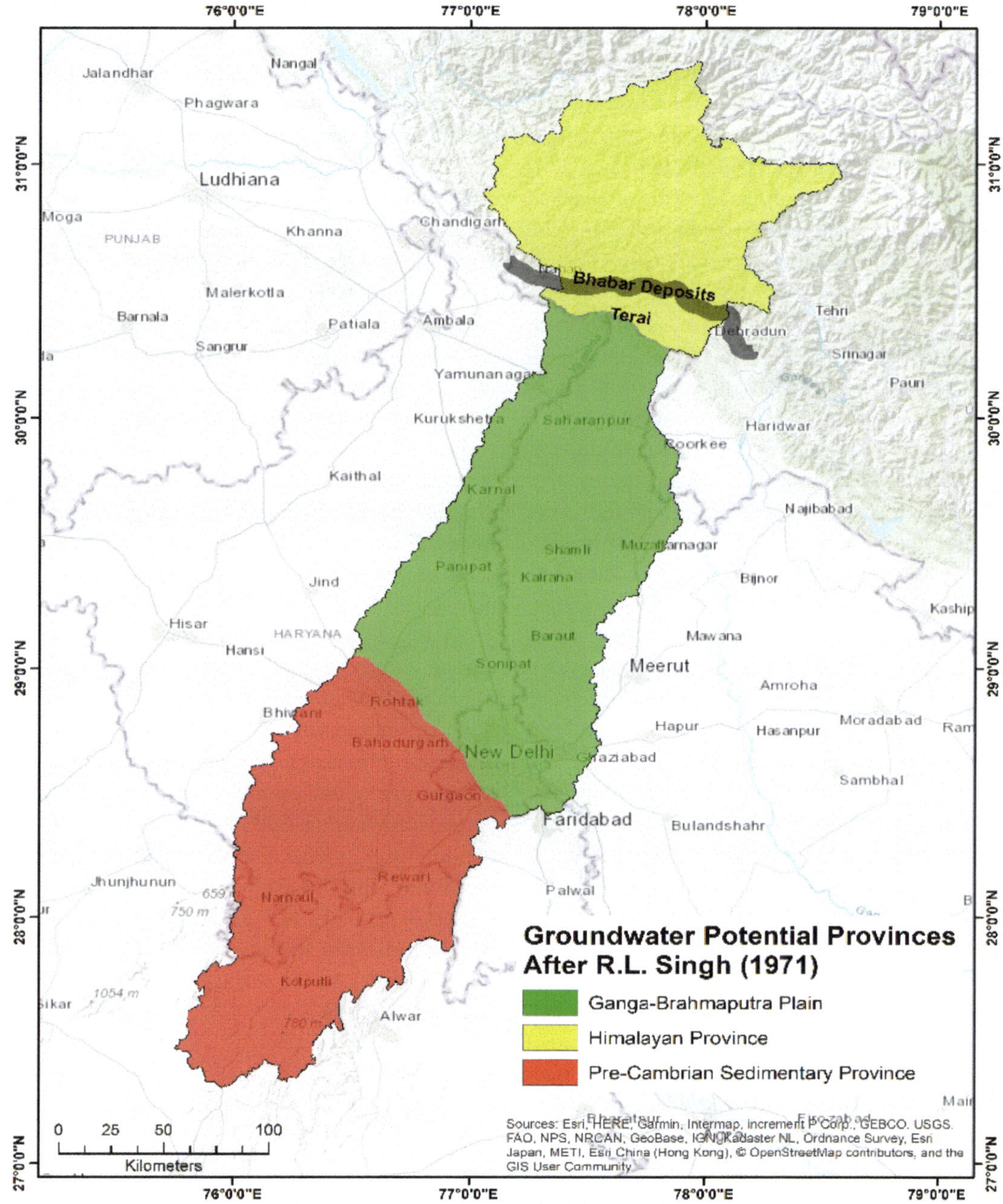

Fig. 8.6 Groundwater potential provinces of upper Yamuna sub-basin (After Singh 1971)

infrastructure development strategy. Rajasthan, a less potential groundwater province and a lower precipitation region, has naturally lower groundwater levels, which are exacerbated by unprecedented abstraction for irrigation. In contrast, in Delhi, the groundwater issue has become acute due to unrestricted extraction for drinking and industrial purposes, outstanding development, construction, and de-watering, reducing the recharge rate (Bhaduri and Rawat 2016).

Our study tried to understand this spatial variability using several dynamic and major anthropogenic drivers. After statistical cleaning, five independent variables were selected, namely rainfall, drainage density, vegetation coverage, groundwater abstraction, and population density, of which groundwater abstraction significantly influences groundwater availability throughout the study area. This is backed by our extensive literature review reflecting the negative impact of the ever-increasing rate of groundwater withdrawal on the groundwater depth at places within the sub-basin. However, interesting results came to our notice, such as the positive influence of vegetation cover and drainage networks on the availability of groundwater resources as they help groundwater recharge. The impact of these factors varies across the region (depending on the other related factors such as geology and soil properties). In places where groundwater depletion is a significant issue and the rate of it is increasing daily, afforestation will reduce this trend, such as in places like Rajasthan and Haryana. These areas encompass agriculturally intensive zones in Rajasthan and Haryana, highlighting the significance of our findings, given that more than 60% of the nation's irrigated agriculture relies on groundwater (Dangar et al. 2021). This is particularly concerning as certain parts of India already grapple with severe groundwater depletion. This poses not only an immediate threat to the local population but also raises long-term concerns about the sustainability of groundwater utilisation. This presents an immediate and tangible threat to the well-being of the local population, as it directly impacts their access to a vital source of freshwater. However, the implications extend far beyond the present day. While rainfall positively impacts groundwater recharge, the influence weakens due to urbanisation, deforestation, and climate change, leading to slower recharge rates than groundwater abstraction. Our study highlights exceptionally high abstraction rates in parts of Rajasthan, Haryana, and Uttar Pradesh, with an annual agricultural, domestic, and industrial extraction of over 55,000 billion cubic metres. With a burgeoning industrial economy, intensive agriculture, and a growing population, groundwater extraction rates will likely continue to escalate shortly.

Since groundwater supplies are limited and are being exhaustively used, the water table has declined, resulting in poor discharge of tube wells, depth of tube wells, and higher energy costs. This may have an impact on the socioeconomic situation of small farmers. In this situation, regular groundwater monitoring provides clues to understanding the trend of water level so that proper management practises such as reducing groundwater withdrawal and increasing groundwater recharge through rainwater harvesting, rivers and lakes, tube wells, dug wells, and agricultural fields can be adopted for sustainable groundwater use in this agriculture-dominated region. Prospective paths indicate us towards integrating cutting-edge technologies such as remote sensing and machine learning, coupled with real-time data monitoring, to refine groundwater assessments further. Furthermore, climate change projections and sustainability modelling constitute paramount considerations in anticipating the evolving impacts of a changing climate on groundwater resources. As we navigate this complex terrain, collaborative endeavours among scientists, policymakers, and communities emerge as pivotal avenues essential for formulating robust and adaptive strategies to preserve groundwater availability and quality, thereby ensuring the resilience and sustainability of this invaluable resource for the benefit of future generations.

References

Almonroeder TG (2022) Advanced statistics for physical and occupational therapy. Routledge

Amarasinghe UA, Sikka A, Mandave V, Panda RK, Gorantiwar S, Ambast SK (2021) Improving economic water productivity to enhance resilience in canal irrigation systems: a pilot study of the Sina irrigation system in Maharashtra, India. Water Policy 23(2):447–465. https://doi.org/10.2166/wp.2021.231

Ashoka A, Wada Y, Fishman R, Mishra V (2018) Strong linkage between precipitation intensity and monsoon season groundwater recharge in India. Geophys Res Lett 45:5536–5544. https://doi.org/10.1029/2018GL078466

Ashraf B, AghaKouchak A, Alizadeh A, Mousavi Baygi M, Moftakhari HR, Mirchi A, Anjileli H, Madani K (2017) Quantifying anthropogenic stress on groundwater resources. Sci Rep 7(1):12910. https://doi.org/10.1038/s41598-017-12877-4

Azen R, Budescu DV (2003) The dominance analysis approach for comparing predictors in multiple regression. Psychol Methods 8(2):129–148. https://doi.org/10.1037/1082-989X.8.2.129

Barron OV, Donn MJ, Barr AD (2013) Urbanisation and shallow groundwater: predicting changes in catchment hydrological responses. Water Resour Manage 27(1):95–115. https://doi.org/10.1007/s11269-012-0168-0

Bhaduri A, Rawat BS (2016) Alarm bells ring for Delhi's groundwater. India Waterportal. Retrieved from: https://www.indiawaterportal.org/articles/alarm-bells-ring-delhis-groundwater

Bhattarai N, Pollack A, Lobell DB, Fishman R, Singh B, Dar A, Jain M (2021) The impact of groundwater depletion on agricultural production in India. Environ Res Lett 16(8):085003. https://doi.org/10.1088/1748-9326/ac10de

Chinnasamy P, Agoramoorthy G (2015) Groundwater storage and depletion trends in Tamil Nadu State, India. Water Resour Manage 29(7):2139–2152. https://doi.org/10.1007/s11269-015-0932-z

Dangar S, Asoka A, Mishra V (2021) Causes and implications of groundwater depletion in India: a review. J Hydrol 596:126103. https://doi.org/10.1016/j.jhydrol.2021.126103

Danielopol DL, Griebler C, Gunatilaka A, Notenboom J (2003) Present state and future prospects for groundwater ecosystems. Environ Conserv 30(2):104–130. https://doi.org/10.1017/S0376892903000109

Daoud JI (2017) Multicollinearity and regression analysis. J Phys: Conf Ser 949:012009. https://doi.org/10.1088/1742-6596/949/1/012009

Das J, Rahman ATMS, Mandal T, Saha P (2021) Exploring driving forces of large-scale unsustainable groundwater development for irrigation in lower Ganga River basin in India. Environ Dev Sustain 23(5):7289–7309. https://doi.org/10.1007/s10668-020-00917-5

Doxsey-Whitfield E, MacManus K, Adamo SB, Pistolesi L, Squires J, Borkovska O, Baptista SR (2015) Taking advantage of the improved availability of census data: a first look at the gridded population of the world, Version 4. Papers Appl Geogr 1(3):226–234. https://doi.org/10.1080/23754931.2015.1014272

Everard M (2015) Community-based groundwater and ecosystem restoration in semi-arid north Rajasthan (1): socio-economic progress and lessons for groundwater-dependent areas. Ecosyst Serv 16:125–135. https://doi.org/10.1016/j.ecoser.2015.10.011

Foster S (2020) Global policy overview of groundwater in urban development—a tale of 10 cities! Water 12(2):456. https://doi.org/10.3390/w12020456

Foster S, Hirata R, Eichholz M, Alam M-F (2022) Urban self-supply from groundwater—an analysis of management aspects and policy needs. Water 14(4):575. https://doi.org/10.3390/w14040575

Garg KK, Singh R, Anantha KH, Singh AK, Akuraju VR, Barron J, Dev I, Tewari RK, Wani SP, Dhyani SK, Dixit S (2020) Building climate resilience in degraded agricultural landscapes through water management: a case study of Bundelkhand region, Central India. J Hydrol 591:125592. https://doi.org/10.1016/j.jhydrol.2020.125592

Gopal K, Singh RP, Tashi KS (2015) Water level fluctuation as the sum of environmental and anthropogenic activities in Southeast, Punjab (India). J Environ Analyt Toxicol 05(05). https://doi.org/10.4172/2161-0525.1000298

Gupta P, Sarma K (2016) Spatial distribution of various parameters in groundwater of Delhi, India. Cogent Eng 3(1):1138596. https://doi.org/10.1080/23311916.2016.1138596

Haaf E, Giese M, Heudorfer B, Stahl K, Barthel R (2020) Physiographic and climatic controls on regional groundwater dynamics. Water Resour Res 56(10). https://doi.org/10.1029/2019WR026545

Haque SJ, Onodera S, Shimizu Y (2013) An overview of the effects of urbanization on the quantity and quality of groundwater in South Asian megacities. Limnology 14(2):135–145. https://doi.org/10.1007/s10201-012-0392-6

Houk E, Frasier M, Schuck E (2006) The agricultural impacts of irrigation induced waterlogging and soil salinity in the Arkansas Basin. Agric Water Manag 85(1–2):175–183. https://doi.org/10.1016/j.agwat.2006.04.007

Jain M, Fishman R, Mondal P, Galford GL, Bhattarai N, Naeem S, Lall U, Balwinder-Singh, DeFries RS (2021) Groundwater depletion will reduce cropping intensity in India. Sci Adv 7(9):eabd2849. https://doi.org/10.1126/sciadv.abd2849

Kale R, Jose PG, Taloor AK, Kumar R (2021) Assessment of digital elevation models based on the drainage morphometric parameters for the Tawi River Basin. In: Rao CM, Patra KC, Jhajharia D, Kumari S (eds) Advanced modelling and innovations in water resources engineering. Lecture notes in civil engineering, vol 176. Springer, Singapore, pp 119–140. https://doi.org/10.1007/978-981-16-4629-4_10

Kalhor K, Emaminejad N (2019) Sustainable development in cities: studying the relationship between groundwater level and urbanization using remote sensing data. Groundw Sustain Dev 9:100243. https://doi.org/10.1016/j.gsd.2019.100243

Koh E-H, Lee E, Lee K-K (2020) Application of geographically weighted regression models to predict spatial characteristics of nitrate contamination: implications for an effective groundwater management strategy. J Environ Manage 268:110646. https://doi.org/10.1016/j.jenvman.2020.110646

Luijendijk E (2021) Transmissivity and groundwater flow exert a strong influence on drainage density [Preprint]. Physical: Landscape Evolution: modelling and field studies. https://doi.org/10.5194/esurf-2021-33

Malakar P, Bhanja SN, Dash AA, Saha D, Ray RK, Sarkar S, Zahid A, Mukherjee A (2022) Delineating variabilities of groundwater level prediction across the agriculturally intensive transboundary aquifers of South Asia. ACS EST Water 2:1–14

National Remote Sensing Centre (2014) Evaluation of Indian National DEM (Version-2) from Cartosat-1 Data. Indian Space Research Organisation, Department of Space, Government of India

Pai DS, Sridhar L, Rajeevan M, Sreejith OP, Satbhai NS, Mukhopadyay B (2014) Development of a new high spatial resolution (0.25° × 0.25°) long period (1901–2010) daily gridded rainfall data set over India and its comparison with existing data sets over the region. MAUSAM 65(1 (January 2014)):1–1

Panda DK, Ambast SK, Shamsudduha M (2021) Groundwater depletion in northern India: impacts of the sub-regional anthropogenic land-use, socio-politics and changing climate. Hydrol Process 35(2):e14003. https://doi.org/10.1002/hyp.14003

Patle GT, Singh DK, Sarangi A (2016) Modelling of declining groundwater depth in Kurukshetra District, Haryana, India. Curr Sci 111(4):717. https://doi.org/10.18520/cs/v111/i4/717-723

Reddy STS (1989) Declining groundwater levels in India. Int J Water Resour Dev 5(3):183–190. https://doi.org/10.1080/07900628908722432

Rodell M, Velicogna I, Famiglietti JS (2009) Satellite-based estimates of groundwater depletion in India. Nature 460(7258):999–1002. https://doi.org/10.1038/nature08238

Running S, Mu Q, Zhao M (2017) MOD16A2 MODIS/Terra Net Evapotranspiration 8-Day L4 Global 500m SIN Grid V006 [dataset]. NASA EOSDIS Land Processes Distributed Active Archive Center. https://doi.org/10.5067/MODIS/MOD16A2.006

Saha D, Ray RK (2019) Groundwater resources of India: potential, challenges and management. In Sikdar PK (ed) Groundwater development and management. Springer International Publishing, pp 19–42. https://doi.org/10.1007/978-3-319-75115-3_2

Sarker B, Keya KN, Mahir FI, Nahiun KM, Shahida S, Khan RA (2021) Surface and ground water pollution: causes and effects of urbanization and industrialization in South Asia. Sci Rev 73:32–41. https://doi.org/10.32861/sr.73.32.41

Schreiber-Gregory DN (2018) Ridge regression and multicollinearity: an in-depth review. Model Assist Stat Appl 13(4):359–365. https://doi.org/10.3233/MAS-180446

Schwartz FW, Ibaraki M (2011) Groundwater: a resource in decline. Elements 7(3):175–179. https://doi.org/10.2113/gselements.7.3.175

Sekhri S (2014) Wells, water, and welfare: the impact of access to groundwater on rural poverty and conflict. Am Econ J Appl Econ 6(3):76–102. https://doi.org/10.1257/app.6.3.76

Senaviratna NAMR, Cooray TMJA (2019) Diagnosing multicollinearity of logistic regression model. Asian J Prob Stat, 1–9. https://doi.org/10.9734/ajpas/2019/v5i230132

Shekhar S, Kumar S, Densmore AL, Van Dijk WM, Sinha R, Kumar M, Joshi SK, Rai SP, Kumar D (2020) Modelling water levels of northwestern India in response to improved irrigation use efficiency. Sci Rep 10(1):13452. https://doi.org/10.1038/s41598-020-70416-0

Singh RL (1971) India: a regional geography. National Geographical Society of India. https://books.google.co.in/books?id=nqItAAAAMAAJ

Sudha S, Jansirani S (2019) Green revolution and its impact on Indian agriculture. Indian J Public Health Res Dev 10(5):74. https://doi.org/10.5958/0976-5506.2019.00972.0

Tiwari VM, Wahr J, Swenson S (2009) Dwindling groundwater resources in northern India, from satellite gravity observations. Geophys Res Lett 36(18):L18401. https://doi.org/10.1029/2009GL039401

Varghese SK, Veettil PC, Speelman S, Buysse J, Van Huylenbroeck G (2013) Estimating the causal effect of water scarcity on the groundwater use efficiency of rice farming in South India. Ecol Econ 86:55–64. https://doi.org/10.1016/j.ecolecon.2012.10.005

Yazicigil H, Yilmaz KK, Erdemli BU, Yagbasan O (2011) Impacts of decreasing recharge rates on sustainable groundwater management. In: Baba A, Tayfur G, Gündüz O, Howard KWF, Friedel MJ, Chambel A (eds) Climate change and its effects on water resources, vol 3. Springer Netherlands, pp 43–50. https://doi.org/10.1007/978-94-007-1143-3_6

Yeh H-F, Chang J-C, Huang C-C, Chen H-Y (2022) Spatial correlation of groundwater level with natural factors using geographically weighted regression model in the Choushui River Alluvial Fan, Taiwan. Front Earth Sci 10:977611. https://doi.org/10.3389/feart.2022.977611

York R (2012) Residualization is not the answer: rethinking how to address multicollinearity. Soc Sci Res 41(6):1379–1386. https://doi.org/10.1016/j.ssresearch.2012.05.014

Yu H, Gong H, Chen B, Liu K, Gao M (2020) Analysis of the influence of groundwater on land subsidence in Beijing based on the geographical weighted regression (GWR) model. Sci Total Environ 738:139405. https://doi.org/10.1016/j.scitotenv.2020.139405

Zanaga D, Van De Kerchove R, De Keersmaecker W, Souverijns N, Brockmann C, Quast R, Wevers J, Grosu A, Paccini A, Vergnaud S, Cartus O, Santoro M, Fritz S, Georgieva I, Lesiv M, Carter S, Herold M, Li L, Tsendbazar NE, Fabrizio R, Arino O (2021) ESA WorldCover 10 m 2020 v100. https://doi.org/10.5281/zenodo.5571936

Effect of Settlement and Agriculture Expansion on Carbon Balance in Kanwar Lake, a Ramsar Site

9

Akthab Roshan, Swapan Talukdar, Mohd Rihan, Tanmoy Das, Md. Sarfaraz Asgher, and Shahfahad

Abstract

Kanwar Lake faces ecological degradation due to expanding agriculture and built-up areas, threatening biodiversity and altering the wetland landscape. Therefore, this study aims to understand the complex interaction between human activities and the ecosystem's carbon dynamics. Using satellite imagery, the changes in LULC were mapped using a Random Forest algorithm that reveals fluctuations in six classes. The modeling of carbon storage using the InVEST tool and the economic valuation of carbon storage used a pixel-based approach that assesses the net change in carbon sequestered or lost and converts it into an economic value. Results showed a significant increase in built-up areas from 2012 to 2022, fluctuating cropland, and potential agricultural expansion indicated by adjacent fallow land. Conflicts between agriculture and wetland conservation were noted, impacting regional food security and carbon sequestration. While some carbon storage recovery was observed, the trend indicated a landscape dominated by low-carbon storage zones, with a maximum observed storage of 12.24 t/ha. Fluctuations were attributed to agricultural intensification, urban development, conservation efforts, and climate change. The economic valuation of carbon sequestration revealed spatial variations in value over time. From 1992 to 2022, there was a decrease in both the total economic values of carbon sequestered and emitted, indicating a diminishing capacity of the ecosystem to provide carbon sequestration services. Carbon emission modeling across the lake indicated medium-level emissions in wetland and marshy areas in 1992, with a general reduction in emission intensity by 2002. By 2012, increased emissions were noted in croplands, continuing into 2022, despite a stabilization in the carbon storage capability of marshy areas. The findings highlight the importance of sustained land management and conservation strategies to enhance carbon sequestration and mitigate the economic and ecological impacts of carbon emissions.

A. Roshan · S. Talukdar · M. Rihan · T. Das · Shahfahad (✉)
Department of Geography, Faculty of Sciences, Jamia Millia Islamia, New Delhi 110025, India
e-mail: fahadshah921@gmail.com

M. S. Asgher
Department of Geography, University of Jammu, Jammu Tawi, Jammu & Kashmir 180006, India

S. Talukdar
Department of Geography, Asutosh College, Kolkata, West bengal 700026, India

S. Talukdar et al. (eds.), *Water Resource Management in Climate Change Scenario*, GIScience and Geo-environmental Modelling, https://doi.org/10.1007/978-3-031-61121-6_9

Keywords

Land use and land cover (LULC) · Carbon sequestration · Dynamics of carbon storage · Wetland degradation · Remote sensing · Ramsar site conservation

9.1 Introduction

The advent of the industrial age has led to a significant increase in atmospheric carbon dioxide (CO_2) levels (Bradu et al. 2022), with growing concerns about the implications for global climate change (Jones et al. 2023). Since the 1970s, this issue has captured the attention of the international community, prompting a surge in research aimed at understanding the circulation and distribution of carbon on a global scale. Scientists and policymakers have focused their efforts on identifying effective methods to capture atmospheric CO_2—primarily resulting from fossil fuel consumption—and sequester it in various natural sinks, such as soils and vegetation.

Wetlands, despite occupying only a small fraction of the Earth's surface, have emerged as critical players in the global carbon cycle (LaFond-Hudson and Sulman 2023; Maltby 2022). Their ability to store carbon at a rate that far surpasses that of other terrestrial ecosystems has placed them at the forefront of climate change mitigation research. These ecosystems are not just carbon sinks but also hotspots for biodiversity, playing a pivotal role in maintaining ecological balance (Li et al. 2023; Berde et al. 2022). They are involved in key biogeochemical processes, serving as natural filters for water and nutrients while sequestering CO_2 through high rates of primary productivity (Howard et al. 2017). The concept of 'coastal blue carbon' is a testament to the role of coastal ecosystems, such as salt marshes, mangroves, and seagrasses, in capturing organic carbon (Yin et al. 2023; Lovelock et al. 2023). These habitats have garnered attention for their ability to sequester carbon at densities an order of magnitude greater than terrestrial forests, thus acting as powerful counterbalances to atmospheric CO_2 levels (Solomon et al. 2023). This has made the conservation of wetlands not only an ecological priority but also a climate imperative.

However, wetlands are facing increasing threats from both climate change and human activities (Xiong et al. 2023). Changes in these ecosystems can lead to the release of their stored carbon, undermining their role as carbon sinks (Zheng et al. 2023). This is particularly concerning in regions where the expansion of agriculture and urban development encroaches upon these sensitive areas (Hu et al. 2023; Grecchi et al. 2014). In countries like India, where rapid development is a priority, the importance of wetlands and the services they provide are often underrepresented in environmental policy and land-use planning (Jacob and Dupras 2021; Sinclair et al. 2021). Research on wetland carbon sequestration has been extensive in regions such as the Arctic, the tropics, and central North America (Watts et al. 2023; Wang et al. 2022). However, there remains a significant gap in studies focusing on the Indian subcontinent, where the dynamics of wetland carbon storage are less understood. This gap is not just scientific but also extends to the incorporation of wetland ecosystem services into policymaking and economic planning. To bridge these gaps, a variety of methods have been employed, ranging from direct carbon flux measurements to manipulative field experiments and modeling approaches. Among these, the use of remote sensing technology offers a unique and valuable perspective, providing extensive, time-efficient, and cost-effective means of assessing long-term LULC changes. Remote sensing has the added advantage of enabling the analysis of remote and inaccessible areas, which is often the case with wetland ecosystems. Complementing remote sensing, the InVEST carbon storage and sequestration model stands out as a versatile tool, offering a global application with the flexibility of scale and providing biophysical and economic outputs. By integrating this model with remote sensing data, this study aims to deliver a comprehensive assessment of the LULC changes at Kanwar Lake and their implications for carbon storage and sequestration. The research endeavors to fill the existing knowledge void, providing a nuanced understanding of wetland carbon dynamics and offering a foundation for informed policy decisions to safeguard these vital ecosystems in the face of ongoing environmental changes.

Previous research on land-use and land cover (LULC) changes in wetland ecosystems often lacked a holistic approach, focusing either on ecological or anthropogenic aspects without fully integrating the two (Obubu et al. 2022; Nucci et al. 2022). Additionally, many studies did not account for the detailed spatiotemporal dynamics of carbon storage and sequestration, leaving a critical gap in understanding how LULC changes directly impact carbon dynamics in wetland ecosystems (Roy et al. 2022; Ranjan et al. 2023; Wu et al. 2023). The economic implications of these changes were also not thoroughly quantified, particularly in the context of developing regions where data accessibility can be a challenge (Cheng et al. 2023; Song et al. 2023). These gaps highlighted the need for an integrated methodological framework that could simultaneously address the ecological, carbon, and economic aspects of LULC changes.

The current study addressed these gaps by employing advanced remote sensing techniques coupled with machine learning algorithms, such as the Random Forest, to provide a more nuanced mapping of LULC changes over a period of thirty years. By integrating the InVEST model for carbon storage and sequestration, the study quantitatively analyzed the impact of LULC dynamics on carbon budgets, a novel approach specifically tailored to the unique wetland context of Kanwar Lake. Furthermore, it innovatively combined these analyses with an economic valuation of carbon, offering a more comprehensive understanding of the ecosystem services provided by the wetland. This multidimensional analysis is novel in its application to Kanwar Lake, providing critical insights that can inform targeted conservation strategies. The study's methodology and findings contribute a significant step forward in wetland management research, offering a replicable model for other wetlands experiencing similar pressures and serving as a reference point for policymakers and conservationists.

9.2 Materials and Methodology

9.2.1 Study Area

The Kabartal Wetland, also known as Kaabar Tal or Kanwar Jheel, is a significant ecological area located in the Begusarai district of northern Bihar State, India (Fig. 9.1). Covering an area of 2620 ha, it is part of the Indo-Gangetic plains and lies within the Kosi-Gandak interfan area of the East Ganga Plains, which is characterized by a dense distribution of wetlands of various sizes and configurations. This wetland was designated as a Ramsar site on July 21, 2020, highlighting its international importance, particularly in terms of wetland conservation and sustainable use. The wetland experiences a tropical monsoonal climate with three distinct seasons: a cool winter (8–5 °C) during December and January, a hot summer (25–38 °C) from April to June, and a monsoon season from July to September. The monsoon is the dominant season and significantly influences the region's LULC pattern and wetland dynamics. Kabartal is integral to the floodplain complex of the area, absorbing monsoon floodwaters to a depth of 1.5 m, thus providing a vital service in a state where 70% of the land is prone to inundation. It stands out as a perennial wetland in an area where most are seasonal, typically drying up during the non-monsoon months.

Kabartal Wetland boasts significant biodiversity, with records of 165 plant species and 394 animal species, including 221 bird species. This diversity is particularly notable given the pressures of agricultural use during the dry season when marshlands recede. The wetland serves as a critical stopover in the Central Asian Flyway, offering respite to 58 migratory water bird species. Its ecological significance extends to aquatic biodiversity, with more than 50 fish species identified in the area. The presence of five critically endangered species, including three vulture species (the red-headed vulture, white-rumped vulture, and Indian vulture)

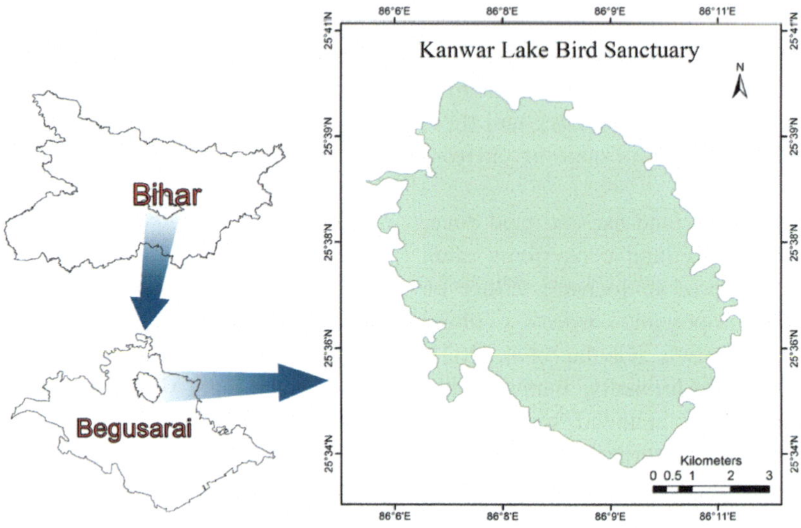

Fig. 9.1 Location of the study area, Kanwar Lake in Bihar

and two waterbirds (the sociable lapwing and Baer's pochard), underscores its conservation importance. However, the wetland faces several threats, mainly due to water management practices such as drainage, water abstraction, damming, and canalization, which alter its natural hydrological regime and potentially impact its ecological integrity. These challenges highlight the need for continued and effective conservation measures to preserve this vital ecosystem.

9.2.2 Materials

In this research, we utilized satellite imagery from both the Landsat 5 Thematic Mapper (TM) and Landsat 8 Operational Land Imager/Thermal Infrared Sensor (OLI/TIRS) to perform LULC classification using the Random Forest method. These images were obtained from the US Geological Survey (USGS) Earth Explorer website. The Landsat 5 TM images cover the years 1992, 2002, and 2012, while the Landsat 8 OLI images were used for the year 2022, with an acquisition date of April 29, 2022. This temporal range allows for a comprehensive analysis of LULC changes over three decades.

The Landsat 5 TM imagery comprises seven spectral bands with a spatial resolution of 30 m for bands 1–7. Notably, the thermal infrared band (band 6) originally has a resolution of 120 m, but it is resampled to 30 m to maintain consistency across all bands. Each scene from this satellite covers approximately 170 km north–south by 183 km east–west. In contrast, the Landsat 8 satellite includes both the OLI and TIRS. The OLI captures data in the visible, near-infrared, and shortwave infrared portions of the spectrum, while the TIRS measures land surface temperature in two thermal bands, employing advanced quantum physics principles for heat detection. Landsat 8 provides images with a spatial resolution of 15 m in the panchromatic band and 30 m in multispectral bands, covering a swath of 185 km. This detailed and varied imagery is instrumental in accurately classifying the seven identified LULC classes within the study area, facilitating a nuanced understanding of land cover dynamics over time.

9.2.3 Method for Land-Use Land Cover Mapping of Wetland Using Random Forest

The Random Forest (RF) algorithm, a nonparametric ensemble machine learning technique, was developed by Breiman (2001). It has gained prominence in solving various environmental challenges, including water resource management and natural hazard mitigation. RF is particularly adept at processing diverse data types, encompassing both satellite imagery and numerical data (Abdullah et al. 2019). This flexibility is crucial for applications in remote sensing, including LULC classification. The RF algorithm employs a multitude of decision trees in an ensemble learning framework, combining both regression and classification trees. This approach enhances the predictive accuracy and robustness of the model. The effectiveness of RF in LULC classification is attributed to its two fundamental parameters: the number of trees ('n-tree') and the number of features considered at each node ('m-try'). Each tree within the ensemble contributes a vote toward the final classification of a new dataset, with the aggregate of these votes determining the classification outcome.

In the realm of remote sensing, especially for LULC classification, the RF algorithm has demonstrated considerable efficacy. The use of a large number of decision trees enhances the model's accuracy in image classification and land-use modeling. Although increasing the number of trees beyond a certain threshold does not significantly impact the overall model performance, it can make the process unnecessarily lengthy without yielding proportional benefits (Breiman 2001). Feng and Timmermans (2015) identified the use of 200 decision trees in an RF model and found it to be highly accurate for their purposes. In this study, the RF algorithm is employed to classify the LULC into six distinct categories: wetland, built-up area, vegetation, marshy area, and fallow land. This detailed classification enables a nuanced understanding of the landscape, crucial for effective environmental management and planning in the context of wetlands.

9.2.4 Method for Carbon Storage Modeling Using InVest

The method for modeling carbon storage using the Integrated Valuation of Ecosystem Services and Tradeoffs (InVEST) carbon storage and sequestration (CCS) model starts with an understanding of carbon dynamics in the terrestrial biosphere (Butnar et al. 2020). Carbon is stored in various forms, both in living and dead organic matter above and below ground. Plants, through photosynthesis, play a crucial role in the uptake of atmospheric carbon and its transformation into different forms, such as short-lived pools (leaves, humus) and long-lived pools (soils). Understanding the distribution and dynamics of carbon storage and sequestration is critical for effective landscape management, particularly with regard to carbon mitigation strategies (Keenan and Williams 2018). This knowledge is critical for decision-making in conservation, restoration, and other strategies that impact ecosystem services.

The InVEST-CCS model, an open-source tool, is widely used for ecosystem services assessment (Nelson and Daily 2010). It assesses carbon storage in landscapes through the accumulation of carbon in aboveground and belowground biomass, dead organic matter, and soil. Aboveground biomass includes all living plant materials above the soil surface (bark, stems, branches, leaves), while belowground biomass includes root systems. The organic matter in the soil, which represents the largest organic carbon pool, is the organic component of the soil. Dead organic matter includes both litter and dead wood. The model estimates carbon storage in a landscape over time and uses LULC maps to determine how much carbon is stored in each carbon pool. It maps carbon storage densities over LULC maps and produces raster results showing storage, sequestration and total carbon expressed in Mg per pixel. When fitted with current and future LULC maps, the model can calculate net changes in carbon storage over time. It integrates carbon storage densities into LULC grids representing different land types such as forests, water bodies, and agricultural land and organizes the results into total aggregates, storage, sequestration, and value grid outputs. To operationalize the model, estimates of the amounts of carbon in at least one of the four basic pools (aboveground, belowground, soil, and dead organic matter) are required, expressed in metric tons per hectare (t/ha) for each LULC type. More comprehensive results are obtained with data for multiple pools. The model correlates these estimates with the LULC map to produce a map of carbon storage for each carbon pool. Carbon storage is measured at a spatial resolution of 30 m × 30 m per cell. Total carbon storage is calculated using the following equations:

$$C = C_{\mathrm{a}} + C_{\mathrm{b}} + C_{\mathrm{s}} + C_{\mathrm{d}}, \tag{9.1}$$

$$C_{\mathrm{T}} = \sum_{k=1}^{n} A_k * C_k, \quad (k = 1, 2, \ldots n), \tag{9.2}$$

where C represents the total carbon storage of each LULC type per unit area, C_{a} is aboveground carbon density, C_{b} is belowground carbon density, C_{s} is soil carbon density, C_{d} is dead organic matter density, A_k is the area of each LULC type, and C_{T} is the total carbon density of a cell.

9.2.5 Method for Economic Evaluation of Carbon Storage

The economic evaluation of carbon storage is a method used to estimate the financial value of carbon sequestered in ecosystems (Brouwer et al. 2022). It is particularly relevant for assessing the worth of carbon storage services provided by LULC classes over time. This evaluation is crucial for policymakers and environmental economists who aim to quantify the benefits of carbon sequestration in monetary terms. To conduct this evaluation, one must first ascertain the amount of carbon sequestered in a given land parcel. This is achieved by analyzing LULC maps that indicate the type of land cover (e.g., forests, wetlands) and correlating these with established carbon storage values for each LULC class. The model for economic evaluation calculates the market value of the carbon sequestered in the residual stock and the net amount of carbon stored in a land parcel over time. The valuation method involves the following formula (Eq. 9.3), which computes the present value of future carbon sequestration:

$$\text{value\_seq}_x = V\left(\frac{s_x}{q-p}\right)\sum_{t=0}^{q-p-1}\frac{1}{\left(1+\frac{r}{100}\right)^t\left(1+\frac{c}{100}\right)^t}, \tag{9.3}$$

where V is the price per metric ton of carbon, representing the current market or social value of carbon, s_x is the amount of carbon, in metric tons, sequestered on parcel 'x', q is the future year until which the valuation is projected, p is the current year or the baseline year for the valuation, r is the yearly market discount rate for the carbon price, reflecting the time value of money and the preference for immediate benefits over future benefits, and c is the yearly rate of change in the price of carbon, accounting for changes in market conditions, regulatory environments, and technological advancements that may affect the future value of sequestered carbon.

To calculate the value of carbon sequestration over time, the model requires both a current and a future LULC map. This allows for the assessment of net change in carbon storage over time, considering both sequestration and loss. The output of the model can be expressed as metric tons of carbon per pixel and, if necessary, translated into the value of sequestration in currency units per pixel. Discount rates play a pivotal role in this valuation method. The market discount rate r adjusts the future value of carbon to its present value, factoring in economic inflation and the uncertainty of future benefits. Meanwhile, the rate of change in the price of carbon c adjusts the social value of carbon sequestration over time. This adjustment reflects expectations about the impact of carbon emissions on climate change-related damages. If the anticipated impact of carbon sequestered today is greater than that of carbon sequestered in the future, the discount rate c should be positive. Conversely, if carbon sequestered today is expected to have a lesser impact than in the future, c should be negative. In practice, the model allows for the comparison of different future scenarios by applying the valuation to the current landscape and each projected future scenario. It then calculates the variation in carbon storage pixel by pixel, enabling the comparison of various land management and policy decisions. This method is essential for translating the ecological service of carbon storage into economic terms that can inform and justify environmental policy and management decisions.

9.2.6 Carbon Emission Modeling

For Carbon Emission Modeling, particularly in scenarios where wetlands are degraded due to human activities, the stored carbon—both above the surface, below the surface, and within the soil—may be released into the atmosphere as CO_2 (Chen et al. 2022). The extent of CO_2 emissions is contingent on the amount of carbon originally stored and the level of disturbance the area has undergone. The type and depth of disturbance are critical determinants of the amount of carbon released. Shallow disturbances may only affect the topsoil, leaving deeper soil layers and their stored carbon relatively undisturbed. In contrast, more severe disturbances can alter several meters of the soil profile, exposing more soil carbon to oxidation and subsequent emission as CO_2. To quantify emissions, disturbances are categorized into three levels of impact: high, medium, and low. This classification reflects the depth and extent of disturbance and its potential impact on carbon release. The emissions are estimated using Eq. 9.4.

$$\mathrm{LU}_{\mathrm{CE}} = \sum_{i=1}^{n} a_i * \varepsilon_i. \tag{9.4}$$

Here

LU_{CE} is the carbon emission from LULC
a_i—denotes the area of ith LULC type.
ε_i—indicates Carbon emission coefficient of i th LULC type.

Here, $\mathrm{LU}_{\mathrm{CE}}$ is the total carbon emission from LULC changes, a_i denotes the area of the ith LULC type, and ε_i is the carbon emission coefficient for the ith LULC type. The carbon emission coefficient reflects the amount of carbon expected to be released per unit area of the LULC type, given the level of disturbance it has experienced.

The total carbon emissions represent the volume of carbon released over the time period in question, determined by the magnitude and nature of changes or disturbances. By correlating LULC change data with carbon pool data, the model can calculate the potential carbon emissions that would have occurred during a specific time interval. When a land cover classification transitions to a state of emission, the estimated emissions are derived from the source land cover class's disturbance level. Emissions are categorized as low, medium, or high, based on the magnitude of the disturbance. Finally, the map values are interpreted as follows: negative values indicate a net carbon storage (sequestration), zero values denote no net change in carbon storage, and positive values indicate carbon emissions for the specified time interval. This method allows for a spatially explicit assessment of carbon dynamics and can inform management and policy decisions aimed at mitigating carbon emissions from LULC change.

9.3 Results and Discussion

9.3.1 Spatiotemporal Dynamics of LULC of Kanwar Lake

The spatiotemporal dynamics of LULC of Kanwar Lake, as demonstrated in the LULC maps from 1992, 2002, 2012, and 2022 (Fig. 9.2), and quantified in Table 9.1, reflect significant changes in the landscape over a 30-year period. These maps and figures depict a total of six LULC classes: wetland, marshy area, vegetation, built-up, cropland, and fallow land. The data show a complex interplay between natural ecosystem conservation and anthropogenic land-use pressures. In 1992, wetlands covered a mere 1.45% of the total area, which increased significantly by 2012 but showed a subsequent decrease by 2022 (Table 9.1). The marshy area, which represents a large share of the total area, displayed a peak in 2002, indicating a period of expansion, which then decreased before following an upward trend post-2022. This recovery after 2022 could be

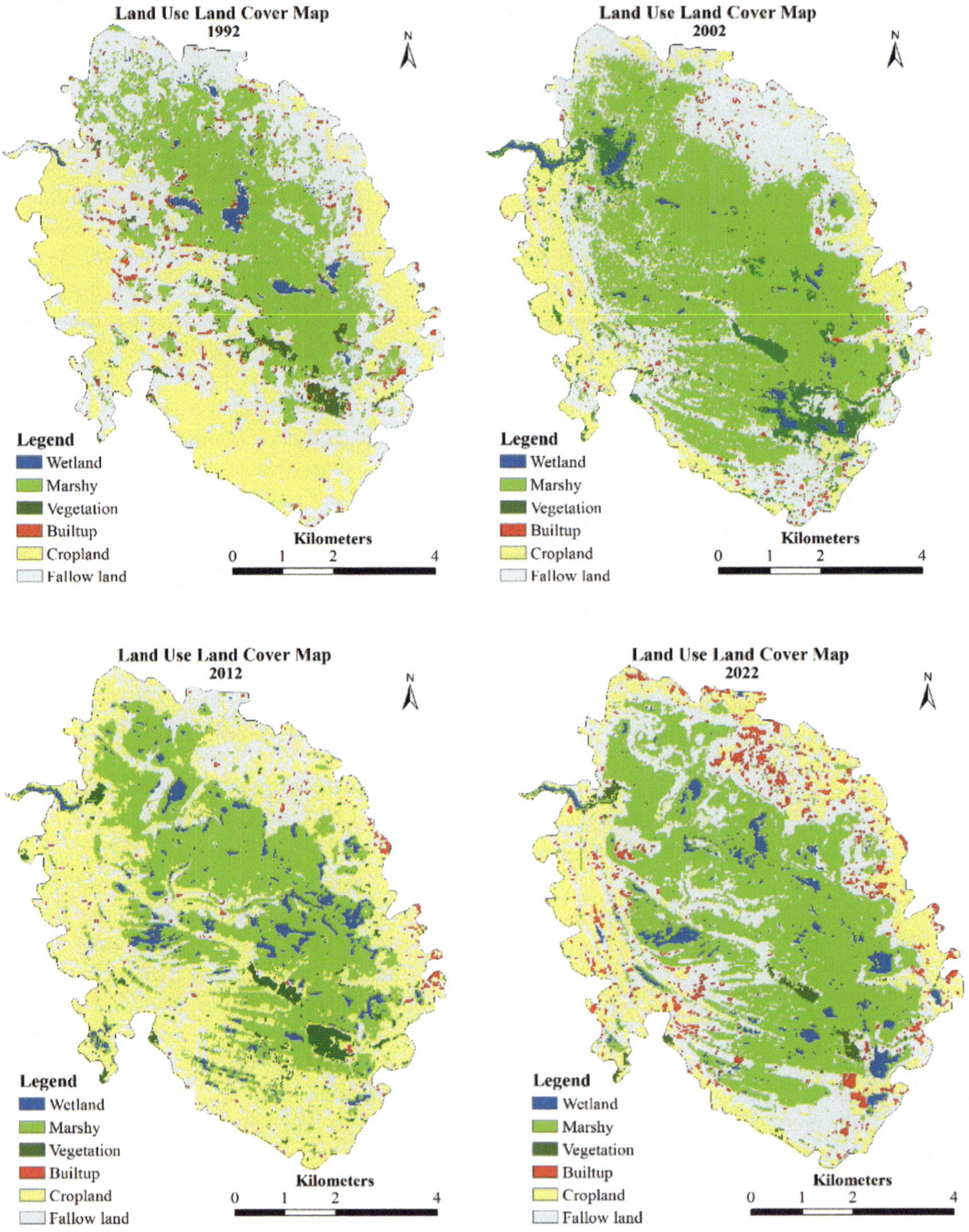

Fig. 9.2 Spatiotemporal LULC dynamics of Kanwar Lake from 1992 to 2022

attributed to the designation of Kabar Tal as a Ramsar site in 2020 and the associated conservation efforts. Vegetation coverage increased by 2002 but has since shown a downward trend. The built-up areas, which were minimal initially, have significantly increased from 2012 to 2022, suggesting urban or infrastructural development. The cropland, an essential LULC class

Table 9.1 Computation of area coverage under different LULC classes for 1992–2022

LULC classes	Area 1992 (km^2)	Area 2002 (km^2)	Area 2012 (km^2)	Area 2022 (km^2)
Wetland	0.761115	0.833572	2.198761	1.870163
Marshy	14.135251	24.432605	17.377791	21.17375
Vegetation	0.97437	2.670148	1.11942	0.55958
Built-up	0.862414	0.455932	0.33879	1.981427
Crop land	15.282894	7.14522	22.183577	9.583465
Fallow land	20.181839	16.6629	8.98801	17.038988
Total area	52.197883	52.200377	52.206349	52.207373

for the local economy, has undergone irregular changes over the three decades. After an initial decrease from 1992, there was a notable increase by 2012, followed by a sharp decrease by 2022. These fluctuations may be influenced by varying agricultural practices and policy changes over the years. Fallow land, which has alternated with cropland in various parts of the study area, has shown a decreasing trend from 1992 to 2012 but rose again after that. It is also notable that settlements have been developing on the northeastern side of the study area, which was previously characterized as fallow land, indicating a possible shift from agricultural to residential or other forms of land use.

This analysis suggests that while there is a dynamic interconversion between cropland and fallow land, the overall trend for cropland is decreasing, which could be a concern for food security in the region. However, the presence of fallow land adjacent to cropland suggests potential for agricultural expansion in the future. The perennial nature of the wetland, which is used for cultivation when dry during the summer season, reveals a conflict between agricultural use and the conservation of wetland carbon reserves. This practice is causing gradual damage to the wetland, which serves as a crucial carbon sink. The observed increase in built-up areas and fallow land over time signals a landscape in transition, with increasing anthropogenic influence that may further impact the region's ecological integrity and carbon sequestration capacity.

9.3.2 Carbon Storage Modeling and Its Dynamics

The carbon storage models and associated data (Table 9.1) indicate changes in carbon storage within the study area over three decades (1992–2022). These maps, which display total carbon storage in metric tons per hectare, reveal the spatial distribution and temporal dynamics of carbon storage across different zones classified as high, medium, and low storage (Fig. 9.3). In 1992, the high carbon storage areas comprised 15.33 km^2, medium storage areas covered 16.81 km^2, and low storage areas spanned 20.07 km^2. The map from 1992 shows that the high storage areas, represented in blue, were predominantly associated with cropland and fallow land, suggesting that these land uses were significant carbon sinks at the time. The medium and low storage areas, depicted in green and yellow, corresponded to mostly fallow land and marshy areas, respectively. This distribution highlights the importance of both agricultural lands and natural wetlands in carbon sequestration. Over the years, there has been a substantial fluctuation in the extent of these carbon storage zones. By 2002, there was a noticeable decrease in high carbon storage areas to 7.2018 km^2 and medium storage to 4.0266 km^2, while low storage areas expanded dramatically to 40.9923 km^2. This could indicate a change in land management practices or a shift in land cover types that are less effective at storing carbon. By 2012, the high storage areas rebounded

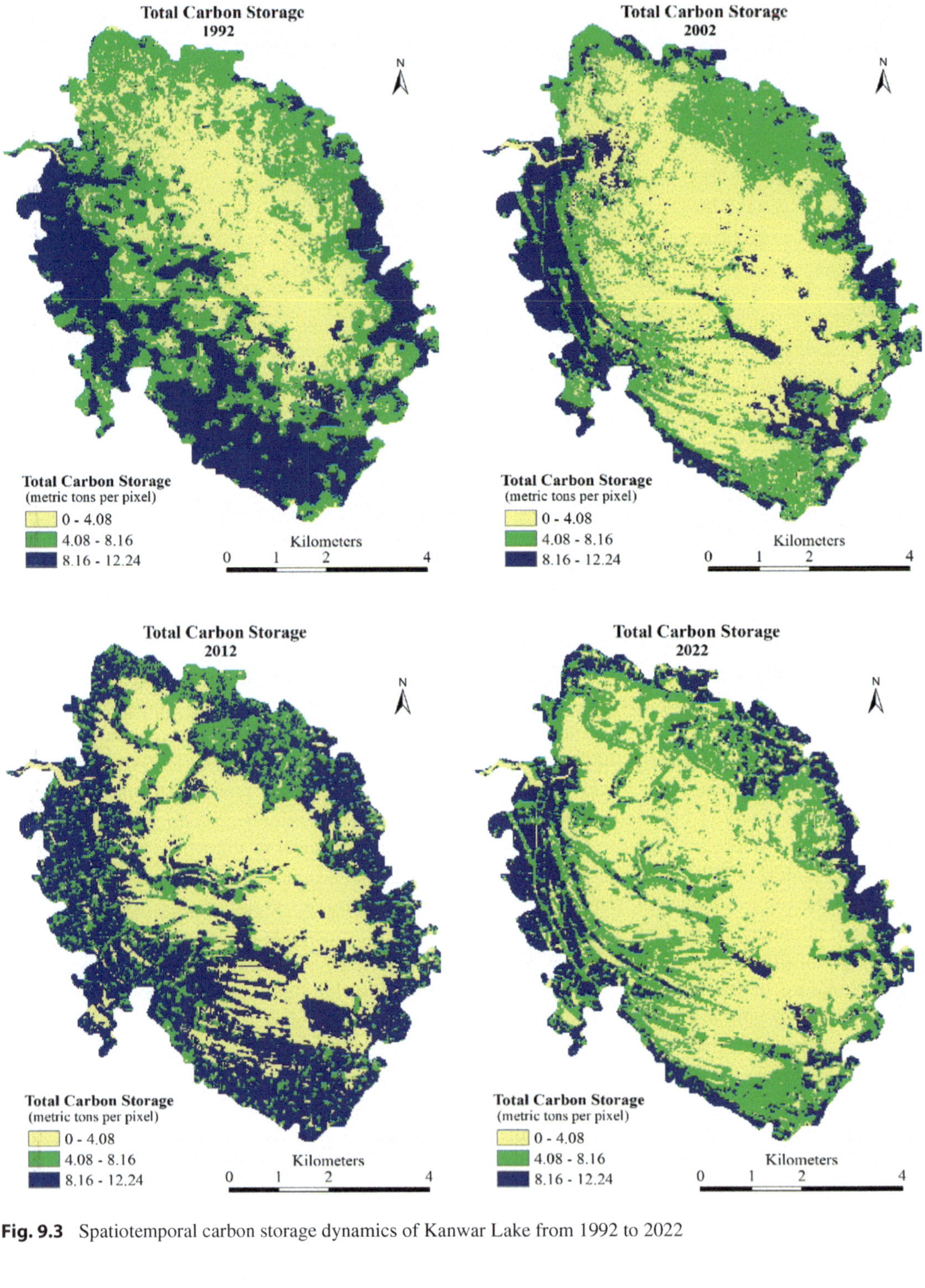

Fig. 9.3 Spatiotemporal carbon storage dynamics of Kanwar Lake from 1992 to 2022

to 21.9906 km^2, suggesting an improvement in carbon sequestration potential or restoration of carbon-rich land covers. However, medium storage areas diminished to just 0.3897 km^2, indicating a potential loss of moderately carbon-dense ecosystems. Low storage zones decreased to 29.8575 km^2 but still covered a significant portion of the study area. In 2022, the high storage areas again decreased to 11.58 km^2, and medium storage areas saw a slight increase to 1.88 km^2. The low storage areas continued to cover the largest portion of the study area at 38.74 km^2. This latest map suggests that while there has been some recovery in carbon storage capacity, overall, the trend is toward a landscape with a predominance of low-carbon storage zones. The maximum carbon storage observed is 12.24 t/ha, found in the high storage zones. The dynamic changes in these zones, particularly the decrease after an initial increase, could be influenced by various factors, including agricultural intensification, urban development, conservation efforts, and climate change impacts. The increasing blue areas in the 2012 map and then their reduction by 2022 could reflect the transient nature of carbon storage, potentially associated with temporary land management practices or environmental policies that were not sustained.

Overall, the data and models reveal a landscape experiencing significant changes in land use and management, with implications for carbon sequestration. The increase in areas of low-carbon storage over time could reflect ongoing pressures on the land, such as development or changes in agricultural practices that are less conducive to carbon storage. These findings are critical for informing land management and conservation strategies aimed at enhancing carbon sequestration to mitigate climate change (Table 9.2).

Table 9.2 Computation of area coverage of different carbon storage zones

Carbon storage zones	Areas in km^2			
	1992	2002	2012	2022
High storage	15.33	7.2018	21.9906	11.58
Medium storage	16.81	4.0266	0.3897	1.88
Low storage	20.07	40.9923	29.8575	38.74

9.3.3 Spatiotemporal Dynamics of Economic Valuation of Carbon Sequestration

The sequence of maps provided (Fig. 9.4) illustrates the spatiotemporal dynamics of the economic value of carbon sequestration around Kanwar Lake for three different decade-long periods: 1992–2002, 2002–2012, and 2012–2022. The maps are color-coded to indicate different economic valuation zones of carbon sequestration, with negative values representing economic losses due to carbon emissions and positive values representing gains from carbon sequestration. From the maps, it is evident that there are spatial variations in the economic value of carbon sequestration across the study area for each period. The economic values assigned to each pixel correspond to the net change in the carbon sequestered or lost, translated into a dollar value. To quantify the total economic impact, the average value of carbon sequestration or loss per pixel is multiplied by the pixel count in each class. Table 9.3 presents a summary of the economic valuation, categorized into five classes based on the dollar value per pixel. In the first period (1992–2002), the highest losses (categorized as − 805 to − 483 dollars) amounted to \$480,484, while the highest gains (categorized as 483 to 805 dollars) were significantly lower at \$336,812. This suggests that the overall economic impact of carbon emissions was greater than that of carbon sequestration during this decade. The subsequent period (2002–2012) also showed a significant economic impact from carbon dynamics, with the highest losses slightly increased to \$612,444 and the highest gains more substantially increased to \$902,424. This indicates an increased economic value associated with carbon sequestration, possibly due to improved carbon storage or higher valuation of carbon. However, in the most recent period (2012–2022), there is a noticeable reduction in both the highest losses and gains, with values of \$287,868 and \$99,176, respectively. This significant decrease in economic value from carbon emissions and sequestration could

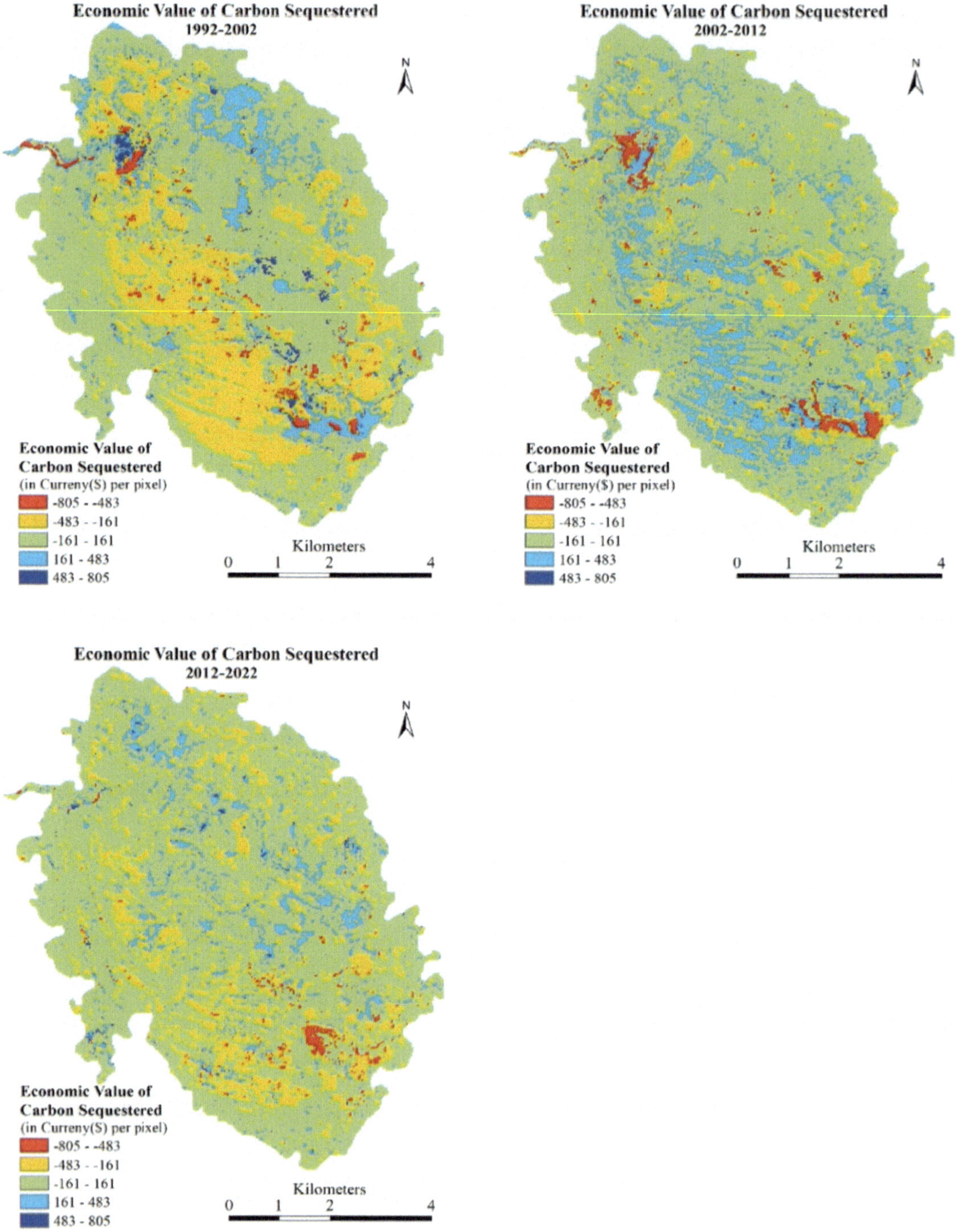

Fig. 9.4 Spatiotemporal dynamics of economic valuation carbon sequestration of Kanwar Lake for different periods

reflect a reduced rate of change in carbon storage or a lower valuation of carbon sequestration benefits over this period.

Overall, the analysis of the results indicates that there has been a decrease in both the total economic value of carbon sequestered and

Table 9.3 Computation of area coverage of different economic valuation zones under different carbon sequestration zones

Classes (economic value of sequestered carbon in $/pixel)	Total value in $ (1992–2002)	Total value in $ (2002–2012)	Total value in $ (2012–2022)
− 805 to − 483	480,484	612,444	287,868
− 483 to − 161	3,262,646	1,134,084	2,233,070
− 161 to 161	4,304,818	4,956,063	5,168,744
161 to 483	1,382,024	2,363,480	998,844
483 to 805	336,812	9024	99,176

emitted from 1992 to 2022. The sharp decline in economic value associated with carbon sequestration over the last decade is a concerning trend, suggesting that the ecosystem's ability to provide this valuable service may be diminishing. This decline could be due to various factors, including changes in land use, management practices, or ecological conditions that affect carbon storage capacity. The findings underscore the importance of implementing and maintaining effective land management and conservation strategies to enhance carbon sequestration and mitigate economic losses from carbon emissions.

9.3.4 Spatiotemporal Carbon Emission Modeling

The series of maps (Fig. 9.5) provide a visual representation of the carbon emission levels across Kanwar Lake for four distinct time periods: 1992, 2002, 2012, and 2022. The color-coded maps indicate areas of high, medium, and low-carbon emission, with the associated Table 9.4 quantifying the spatial coverage of these zones in square kilometers. In 1992, the map indicates that medium-level emissions (yellow zones) predominantly occurred in wetland and marshy areas, suggesting that these ecosystems were experiencing changes that led to carbon loss. High emission areas (red zones) were concentrated on cropland, particularly on the west, east, and southern borders of the study area. This is indicative of land-use activities such as intensive agriculture that typically result in higher carbon emissions. Low emission areas (green zones) were mostly associated with fallow land, which may reflect a lower disturbance regime in these areas, hence lower emissions. By 2002, there was a marked shift in emission patterns. The extent of medium and high emission areas decreased substantially, while low emission areas expanded across the study region, except on the western and eastern borders. This could be interpreted as a general reduction in carbon emission intensity, perhaps due to changes in land management or natural successional processes that increased carbon uptake or reduced carbon loss. In 2012, the landscape underwent further transformation, with wetland, marshy areas, vegetation, and fallow land largely falling into the low emission category. This transition may relate to either an increase in carbon sequestration due to conservation efforts, natural maturation of ecosystems, or decreased anthropogenic disturbance. Cropland, however, showed increased emissions, with some areas reaching the highest emissions of 12.22 metric tons, highlighting continued carbon loss associated with agricultural practices. The 2022 map reveals that wetlands and vegetation areas shifted to medium-level emissions, while marshy areas, which occupy a significant portion of the central study area, exhibited the lowest emissions. This may reflect a stabilization or improvement in the carbon storage capability of the marshy areas, possibly due to conservation measures or reduced disturbance. Linking these observations to the previous analysis of carbon storage and sequestration, it is evident that areas with

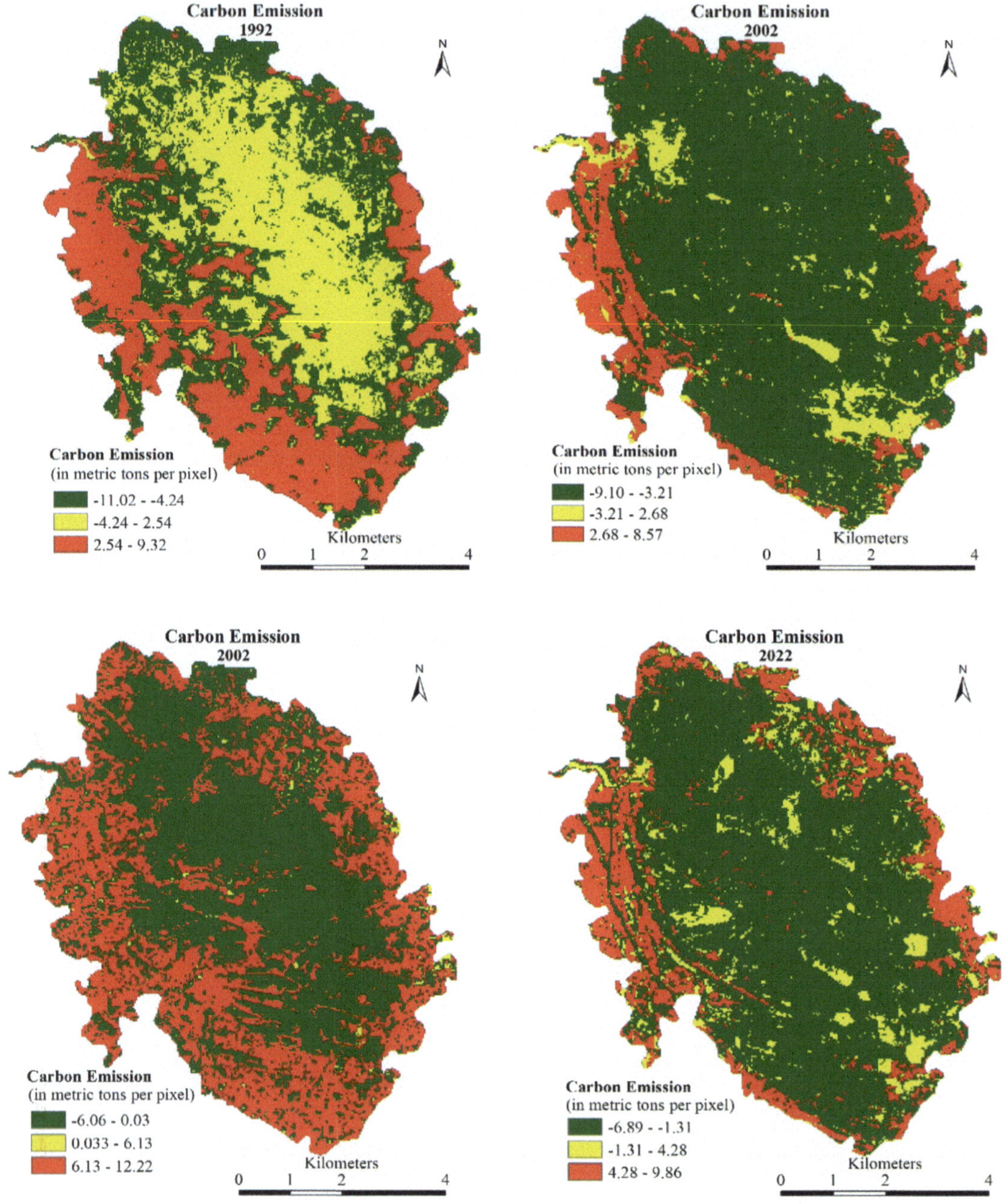

Fig. 9.5 Spatiotemporal dynamics of carbon emission from Kanwar Lake for different periods

lower emissions correspond to those with higher carbon storage and vice versa. This inverse relationship is expected as increased carbon storage is generally associated with lower emissions. The spatiotemporal analysis indicates that while certain areas have seen a reduction in emission levels, others, particularly croplands, remain sources of high emissions. It is essential to consider that these maps do not only represent carbon dynamics but are also influenced by socio-economic factors such as land-use policy, agricultural practices, and conservation efforts.

Table 9.4 Computation of area coverage of different carbon emission zones of Kanwar Lake

Carbon emission zones	Areas in km^2			
	1992	2002	2012	2022
− 11.02 to − 4.24	14.77	30.15	21.96	27.97
− 4.24 to 2.54	12.36	2.96	0.27	3.4
2.54 to 9.32	11.32	5.35	16.23	7.1

The observed changes in emission levels over time highlight the need for continued efforts in land management to maintain and enhance carbon sinks, thereby mitigating overall carbon emissions from the landscape.

9.4 Conclusion

This study provided a detailed analysis of the LULC change at Kanwar Lake, a critical Ramsar site and bird sanctuary, over a 30-year period. Employing the Random Forest algorithm for LULC mapping and the InVEST model for carbon storage quantification, the research revealed substantial shifts in land cover, marked by an increase in built-up areas from 1.45 to 5.55% and a decrease in cropland from 29.29 to 18.35% of the total area. Carbon storage capacity showed a worrying trend, with high storage areas reducing from 15.33 km^2 in 1992 to 11.58 km^2 in 2022 and low storage areas expanding to 38.74 km^2, indicating an ecosystem under stress. The economic valuation of carbon sequestration highlighted a decline in both sequestration and emissions' values, with the highest losses amounting to $480,484 in 1992–2002 and reducing to $287,868 in 2012–2022 and the highest gains decreasing from $336,812 to $99,176 in the same periods. This reflects a decrease in the ecosystem's ability to provide valuable carbon sequestration services. The significance of this work lies in its comprehensive approach to understanding the environmental impacts of LULC changes in a key wetland habitat. It underscores the urgent need for effective land management and conservation strategies to maintain and enhance the wetland's ecological integrity and carbon sequestration capacity.

One limitation of the study is the reliance on remote sensing data, which may not capture all on-the-ground nuances. Additionally, the economic valuation of carbon is based on current value estimates, which may not reflect future market and policy changes. Future research should focus on integrating ground-truthing to validate remote sensing data and include predictive modeling to anticipate future LULC changes. Long-term monitoring and implementation of sustainable land management practices are imperative to ensure the conservation of Kanwar Lake's biodiversity and its role as a carbon sink.

References

Abdullah H, Skidmore AK, Darvishzadeh R, Heurich M (2019) Sentinel-2 accurately maps green-attack stage of European spruce bark beetle (*Ips typographus*, L.) compared with Landsat-8. Remote Sens Ecol Conserv 5(1):87–106

Berde VB, Chari PVB, Berde CV (2022) Wetland and biodiversity hotspot conservation. In: Research anthology on ecosystem conservation and preserving biodiversity. IGI Global, pp 775–784

Bradu P, Biswas A, Nair C, Sreevalsakumar S, Patil M, Kannampuzha S, Mukherjee AG, Wanjari UR, Renu K, Vellingiri B, Gopalakrishnan AV (2022) Recent advances in green technology and Industrial Revolution 4.0 for a sustainable future. Environ Sci Pollut Res, 1–32

Breiman L (2001) Random forests. Mach Learn 45:5–32

Brouwer R, Pinto R, Dugstad A, Navrud S (2022) The economic value of the Brazilian Amazon rainforest ecosystem services: a meta-analysis of the Brazilian literature. PLoS ONE 17(5):e0268425

Butnar I, Li PH, Strachan N, Portugal Pereira J, Gambhir A, Smith P (2020) A deep dive into the modelling assumptions for biomass with carbon capture and storage (BECCS): a transparency exercise. Environ Res Lett 15(8):084008

Chen H, Ju P, Zhu Q, Xu X, Wu N, Gao Y, Feng X, Tian J, Niu S, Zhang Y, Peng C, Wang Y, Wang Y (2022) Carbon and nitrogen cycling on the Qinghai-Tibetan Plateau. Nat Rev Earth Environ 3(10):701–716

Cheng SH, Costedoat S, Sigouin A, Calistro GF, Chamberlain CJ, Lichtenthal P, Mills M, Nowakowski AJ, Sterling EJ, Tinsman J, Wiggins M, Brancalion PHS, Canty SWJ, Fritts-Penniman A, Jagadish A, Jones K, Mascia MB, Porzecanski A, Zganjar C, Brenes CLM, Brenes CLM (2023) Assessing evidence on the impacts of nature-based interventions for climate change mitigation: a systematic map of primary and secondary research from subtropical and tropical terrestrial regions. Environ Evidence 12(1):21

Feng T, Timmermans HJ (2015) Detecting activity type from GPS traces using spatial and temporal information. Eur J Transp Infrastruct Res 15(4):662–674

Grecchi RC, Gwyn QHJ, Bénié GB, Formaggio AR, Fahl FC (2014) Land use and land cover changes in the Brazilian Cerrado: a multidisciplinary approach to assess the impacts of agricultural expansion. Appl Geogr 55:300–312

Howard J, Sutton-Grier A, Herr D, Kleypas J, Landis E, Mcleod E, Pidgeon E, Simpson S (2017) Clarifying the role of coastal and marine systems in climate mitigation. Front Ecol Environ 15(1):42–50

Hu Y, Li Y, Li Y, Wu J, Zheng H, He H (2023) Balancing urban expansion with a focus on ecological security: a case study of Zhaotong City. China. Ecological Indicators 156:111105

Jacob C, Dupras J (2021) Institutional bricolage and the application of the No Net Loss policy in Quebec: can we really engender 'social fit' for more sustainable land use planning? J Environ Planning Policy Manage 23(1):114–129

Jones MW, Peters GP, Gasser T, Andrew RM, Schwingshackl C, Gütschow J, Houghton R, Friedlingstein P, Pongratz J, Le Quéré C (2023) National contributions to climate change due to historical emissions of carbon dioxide, methane, and nitrous oxide since 1850. Sci Data 10(1):155

Keenan TF, Williams CA (2018) The terrestrial carbon sink. Annu Rev Environ Resour 43:219–243

LaFond-Hudson S, Sulman B (2023) Modeling strategies and data needs for representing coastal wetland vegetation in land surface models. New Phytol 238(3):938–951

Li Y, Xiong L, Zeng K, Wei Y, Li H, Ji X (2023) Microbial-driven carbon fixation in natural wetland. J Basic Microbiol

Lovelock CE, Adame MF, Bradley J, Dittmann S, Hagger V, Hickey SM, Hutley LB, Jones A, Kelleway JJ, Lavery PS, Macreadie PI, Maher DT, McGinley S, McGlashan A, Perry S, Mosley L, Rogers K, Sippo JZ (2023) An Australian blue carbon method to estimate climate change mitigation benefits of coastal wetland restoration. Restor Ecol 31(7):e13739

Maltby E (2022) The wetlands paradigm shift in response to changing societal priorities: a reflective review. Land 11(9):1526

Nelson EJ, Daily GC (2010) Modelling ecosystem services in terrestrial systems. F1000 Biol Rep 2

Nucci A, Angiolini C, Manolaki P, Vogiatzakis IN (2022) An integrated approach to support a river ecological network: a case study from the Mediterranean. Land Use Policy 119:106217

Obubu JP, Odong R, Alamerew T, Fetahi T, Mengistou S (2022) Application of DPSIR model to identify the drivers and impacts of land use and land cover changes and climate change on land, water, and livelihoods in the L. Kyoga basin: implications for sustainable management. Environ Syst Res 11(1):1–21

Ranjan AK, Parida BR, Dash J, Gorai AK (2023) Evaluating impacts of opencast stone mining on vegetation primary production and transpiration over Rajmahal Hills. Sustainability 15(10):8005

Roy PS, Ramachandran RM, Paul O, Thakur PK, Ravan S, Behera MD, Sarangi C, Kanawade VP (2022) Anthropogenic land use and land cover changes—a review on its environmental consequences and climate change. J Indian Soc Remote Sens 50(8):1615–1640

Sinclair M, Sagar MV, Knudsen C, Sabu J, Ghermandi A (2021) Economic appraisal of ecosystem services and restoration scenarios in a tropical coastal Ramsar wetland in India. Ecosyst Serv 47:101236

Solomon T, Gupta V, Ncho CM (2023) Balancing livestock environmental footprints with forestry-based solutions: a review. Ecologies 4(4):714–730

Song S, Chen X, Hu Z, Zan C, Liu T, De Maeyer P, Sun Y (2023) Deciphering the impact of wind erosion on ecosystem services: an integrated framework for assessment and spatiotemporal analysis in arid regions. Ecol Ind 154:110693

Wang K, Bastos A, Ciais P, Wang X, Rödenbeck C, Gentine P, Chevallier F, Humphrey VW, Huntingford C, O'Sullivan M, Seneviratne SI, Sitch S, Piao S (2022) Regional and seasonal partitioning of water and temperature controls on global land carbon uptake variability. Nat Commun 13(1):3469

Watts JD, Farina M, Kimball JS, Schiferl LD, Liu Z, Arndt KA, Zona D, Ballantyne A, Euskirchen ES, Parmentier F-J, Helbig M, Sonnentag O, Tagesson T, Rinne J, Ikawa H, Ueyama M, Kobayashi H, Sachs T, Nadeau DF, Kochendorfer J, Jackowicz-Korczynski M, Virkkala A, Aurela M, Commane R, Byrne B, Birch L, Johnson MS, Madani N, Rogers B, Du J, Endsley A, Savage K, Poulter B, Zhang Z, Bruhwiler LM, Miller CE, Goetz S, Oechel WC (2023) Carbon uptake in Eurasian boreal forests dominates the high-latitude net ecosystem carbon budget. Glob Change Biol 29(7):1870–1889

Wu W, Zhang J, Bai Y, Zhang S, Yang S, Henchiri M, Seka AM, Nanzad L (2023) Aboveground biomass dynamics of a coastal wetland ecosystem driven by land use/land cover transformation. Remote Sensing 15(16):3966

Xiong Y, Mo S, Wu H, Qu X, Liu Y, Zhou L (2023) Influence of human activities and climate change on wetland landscape pattern—a review. Sci Total Environ 879:163112

Yin S, Wang J, Zeng H (2023) A bibliometric study on carbon cycling in vegetated blue carbon ecosystems. Environ Sci Pollut Res, 1–18

Zheng K, Shi M, Wu H, Gu H, Jiang P, He P, Dong T, Liu Z, Wang X, Zhang N, Wei R (2023) Estimation and simulation of carbon sequestration in typical dryland areas of China under future climate change scenarios. Front Ecol Evol 11:1250586

10 An Overview of the Integration of Water Features in the Planning of Chandigarh

Nisar Khan and Zeeshan Ibrar

Abstract

Water-Sensitive Urban Planning (WSUP) is a holistic approach that acknowledges the role of water in urban environment by seamlessly integrating the urban water cycle and natural streams with spatial planning. This paper delves into the less-explored facets of WSUP in the urban planning of Chandigarh, a city celebrated for its modern architecture and urban design. Often regarded as a masterpiece of brutalist architecture, grid-iron layout, and segregated land use, the city's remarkable integration of natural water streams tends to be ignored in discussions. Located on the foothills of the Siwalik range, the site of Chandigarh City was traversed by a network of streams draining from the hills. It is interesting to know that these meandering streams were carefully intertwined into the geometric planning of the city. This paper conducts an exploratory study to uncover the integration of streams with the masterplan of the city, resulting in the enhancement in its built environment. Furthermore, it examines the consequences of unregulated urbanisation on the natural water stream network. In the paper, authors employ comparative method between archival maps and GIS-based maps to study the natural water network with the layers of urbanisation. Finally, the paper discussed the need for WSUP in the Indian context, emphasising on the integration of natural water bodies with the city infrastructure and planning. These principles are critical for fostering sustainable urban development in the diverse landscape of Indian cities.

Keywords

Urban landscape · Urban morphology · Water-sensitive urban planning · Water resource · Water-centric planning · Water ecology · Chandigarh

10.1 Introduction

Water-Sensitive Urban Planning (WSUP) is a sustainable approach for urban development that aims to integrate natural water resources with the spatial planning. Also known as Water-Sensitive Urban Design, this approach offers several advantages, such as mitigating the risk of urban flooding, improving public amenities, and reducing the cost of installation of the drainage system. It is increasingly relevant in today's

N. Khan (✉) · Z. Ibrar
Department of Architecture, Jamia Millia Islamia, New Delhi 110025, India
e-mail: nkhan2@jmi.ac.in

Z. Ibrar
e-mail: gf.zibrar@jmi.ac.in

S. Talukdar et al. (eds.), *Water Resource Management in Climate Change Scenario*, GIScience and Geo-environmental Modelling, https://doi.org/10.1007/978-3-031-61121-6_10

cities experiencing unpredictable weather events. According to Kumar et al. (2023), the primary objective of WSUP is to seamlessly integrate water planning with urban planning. Rohilla et al. (2017) mention WSUP as an emerging strategy that enhances a city's resilience to water-related disasters. Wong and Brown (2009) advocate three important WSUP principles: considering cities as water supply catchments, using cities for ecological services, and cultivating water-sensitive communities inside cities. While the success of WSUPs hinges on its social and institutional resources, there are two vital spatial components for its implementation, viz. land-use zoning and hydro-geographical orientations (Roy et al. 2008). This involves delineating planning zones and designing development control regulations. Land-use zoning divides the city into zones based on land use, topography, and planning regulations. In WSUP, presence of water bodies is given consideration in the zoning decisions where city rivers may be designated as distinct planning zones. Hydro-geographical considerations encompass evaluating existing road layouts, determining the location, size, and distribution of green spaces like district parks, assessing existing open spaces such as urban woodlands, and drainage systems. The WSUP results in the preservation of water bodies, reconfiguration of road networks, optimisation of open spaces, and enhancement of drainage systems.

Dolman et al. (2013) underscore the significance of inclusive land-use decisions in urban planning for reducing environmental impacts. An empirical study on the interplay of water resources and land use underscores a persistent connection between water availability, water demand, and water contamination resulting from land use. This underscores the necessity of integrating them into comprehensive masterplans (McGrane 2016). These approaches are readily applicable to greenfield projects in Indian cities (Kumar et al. 2023). Water-Sensitive Urban Planning is in response to the environmental concerns, caused by urbanisation, and provides an inclusive urban planning strategy for our cities. Chandigarh is a distinct case of using the principles of WSUP in its masterplan for its futuristic approach because the understanding on the subject was developed in later decades only.

Chandigarh is famous as the first planned city in independent India. The city was planned as a replacement to Lahore, which went to Pakistan after partition in 1947. The first master plan of Chandigarh was prepared by American planner Albert Mayer and Mathew Nowicki who designed a curvilinear, fan-shaped layout which was based on the Garden City concept originally proposed by Ebenezer Howard in 1902 (Siddiqui et al. 2019). Tragically, the untimely demise of Mathew Nowicki led to the discontinuation of the project by Albert Mayer. Consequently, in 1951, the project was transferred to Charles Eduard Jeanneret, more famously known as Le Corbusier (Joshi 1999; Chandigarh Administration 2023). Le Corbusier also followed the principles of the Garden City in the preparation of his masterplan of the city; however, he deviated from the curvilinear layout earlier prepared by Albert Mayer. The masterplan of Corbusier was characterised by the grid-iron layout, neighbourhood sectors, land-use zoning, extensive road network, and modernist concrete architecture (Curtis 1982). These attributes of architecture and urban planning of Chandigarh are more pronounced and often overshadow discussions on other aspects of its planning. One particular aspect that is often ignored is the inclusion of water streams into the planning and design of the city. It is interesting to discover how the streams with meandering courses were integrated into the rectilinear layout of the city. On one hand, this integration of water streams with the city layout preserved these water bodies and also enhanced the urban environment of the city. What makes this more remarkable is that the implementation of WSUP is employed in the masterplan of Chandigarh, before it was developed as an urban planning approach.

This paper aims to look into the WSUP characteristics in Chandigarh's planning, providing insights that could be applied to urban development in other cities. The overarching research question that guides this study is: "How did Le Corbusier integrated the natural water features in the planning of Chandigarh, and what is the impact of unregulated urbanisation on the same afterwards?"

10.2 Materials and Methods

The paper analyses the attributes of WSUP in the urban planning of the first planned City of India, i.e. Chandigarh. The study employs elevation map prepared using GIS to study the physical features of the site the city, particularly the presence of mountain range towards the Northeast and the natural drainage basin. Further archival maps are used to determine the presence of pre-existing natural water streams at the site selected for the planning of the city. The plan of the city is overlaid on the archival maps to understand how the rectilinear layout interacted with the streams already existing on the site. The change in urban morphology is assessed through changes in land-cover area for two time periods, i.e. 2005 and 2023. The datasets and their sources for assessing stream network and land-cover classifications are summarised in Table 10.1.

The digital elevation model (DEM) is an essential source for deriving slope and stream networks in a region (Yadav and Ibrar 2022). The Shuttle Radar Topography Mission Digital Elevation Model (SRTM-DEM) is used to process elevation data to generate the area's slope angles. The data are obtained from the US Geological Survey (USGS) portal. These elevation data give open dissemination of this high-resolution global dataset and provide worldwide coverage of void-filled data at a resolution of 1 arc-second (30 m). The slope angle in degrees was calculated using the hydrologically corrected SRTM-based DEM in the GIS extension ArcGIS Spatial Analyst tool for the study area.

This study uses the "Strahler method" to number the stream order. The stream order is obtained through SRTM-DEM data in the working environment of hydrological tools in the software ArcGIS version 10.5. The spatial analyst and hydrological tools are applied to extract the streams, and then, the stream order tool is used to derive the ordering of the streams. The stream order is calculated in three stages using SRTM-DEM as base data. In the first stage, the Fill tool is used on DEM input data to get filled DEM data as the raster output. The second stage

Table 10.1 List of datasets and sources for stream network and land-cover classification

S. No.	Components	Datasets	Source
1	Stream order network	SRTM 1 Arc Second Global	Authors, USGS, 2023
2	Elevation	DEM	Authors, USGS, 2023
3	Land-cover classification (for the 2023 time period)	Landsat 8–9 Operational Land Imager (OLI) and Thermal Infrared (TIRS) Collection 2 Level-2	Authors, USGS, 2023
4	Land-cover classification (for the 2005 time period)	Landsat 4–5Thematic Mapper (TM) Collection 2 Level-1 30-m multispectral data	Authors, USGS, 2023
5	Archival maps (copyright free)	India and Pakistan 1:250,000 Series U502, U.S. Army Map Service, 1955; Sheet No. NH 43-8	University of Texas at Austin, PCL Map Collection 2021
6	Archival maps (copyright free)	Delhi Environs 1:100,000, 1914 Perry-Castañeda Library Map Collection Maps from Baedeker's Indien: Handbuch Für Reisende, 1914	University of Texas at Austin, PCL Map Collection 2021

Source Compiled by Authors

is the calculation of flow direction in which filled DEM is used as input data. This step gives the flow direction value for each pixel, which is the direction in which water flows over that pixel as it makes its way downstream. The third stage is a calculation of flow accumulation. In this step, the flow direction raster is used as input data. The accumulated flow is based on the total number or fraction of cells flowing into each cell in the output raster. The raster calculator and stream order tools give the study area's final stream order network. The stream order is then overlaid with the archival maps to analyse the change in network pattern. The land-cover classification uses an interactive supervised classification tool in Arc GIS. The following section discusses the site settings of Chandigarh.

10.3 Results and Discussion

10.3.1 Site Characteristics of Chandigarh

The Union Territory of Chandigarh covers a total area of 114 km^2 and as per the Census (2011), it has a population of 1,055,450. Chandigarh enjoys the unique status of being the capital of two Indian states, Punjab and Haryana. Chandigarh stands out as a well-planned city, offering high standard of civic amenities to its residents (Rodríguez-Lora et al. 2021). Originally designed to accommodate a population of 0.5 million, the city has seen significant population growth over the years. According to the Chandigarh Administration (2023), the city holds the distinction of ranking first in India in various key indicators, such as the Human Development Index, quality of life, and e-readiness.

The city of Chandigarh is situated on the foothills of the Siwalik hills in the North region of India, a part of the sensitive Himalayan ecosystem. On the eastern side, Chandigarh shares its boundary with the state of Haryana, while Punjab borders the remaining sides (Fig. 10.1). The location was carefully selected, reaping the benefits of both hills in the close proximity and plains in the surroundings. A number of villages surrounded by agricultural fields were present on the location, with Mani Majra being the only prominent settlement nearby. The predominant rural nature offered several opportunities for the design and planning of Chandigarh. Firstly, the site provided vast open land with a gradual slope towards southwest, free from the constraints of steep contours present at the foothills nearby. This almost flat topography acted as a blank canvas for the urban planners and architects in the creation of a well-defined city layout (Fig. 10.2).

Secondly, the location offered a picturesque backdrop of Siwalik range facilitating the creation of vistas and adding to the city's scenic value. A network of streams flowed down the terrain of Siwalik range providing water availability for the city. It is this network of streams that was employed in the Water-Sensitive Urban Planning of the city (Sarin and Walden 1974). The forthcoming sections delve into the relationship between the City's master plan as prepared by Le Corbusier and its integration with the pre-existing network of water streams. The subsequent section addresses the contemporary challenges due to unregulated urbanisation on this relationship of water streams and the city.

10.3.2 Macro-aspects of WSUP in the Master Plan

To study the water-sensitive planning of Chandigarh, a stream network map of the site was prepared from the US Army map surveyed prior to the construction of city. The presence of major and minor streams on the site can be seen in Fig. 10.3. It can be seen that the area is primarily drained by two seasonal rivulets: Sukhna Choa in the East and Patiali-Ki-Rao in the West along with a stream called Choa Nadi that can also be seen passing through the city (Sarin and Walden 1974). These rivulets are part of the larger watershed system that originates in the Himalayas and ultimately converges into Ghaggar River situated on the East of Chandigarh (Prakash 2002). This network of

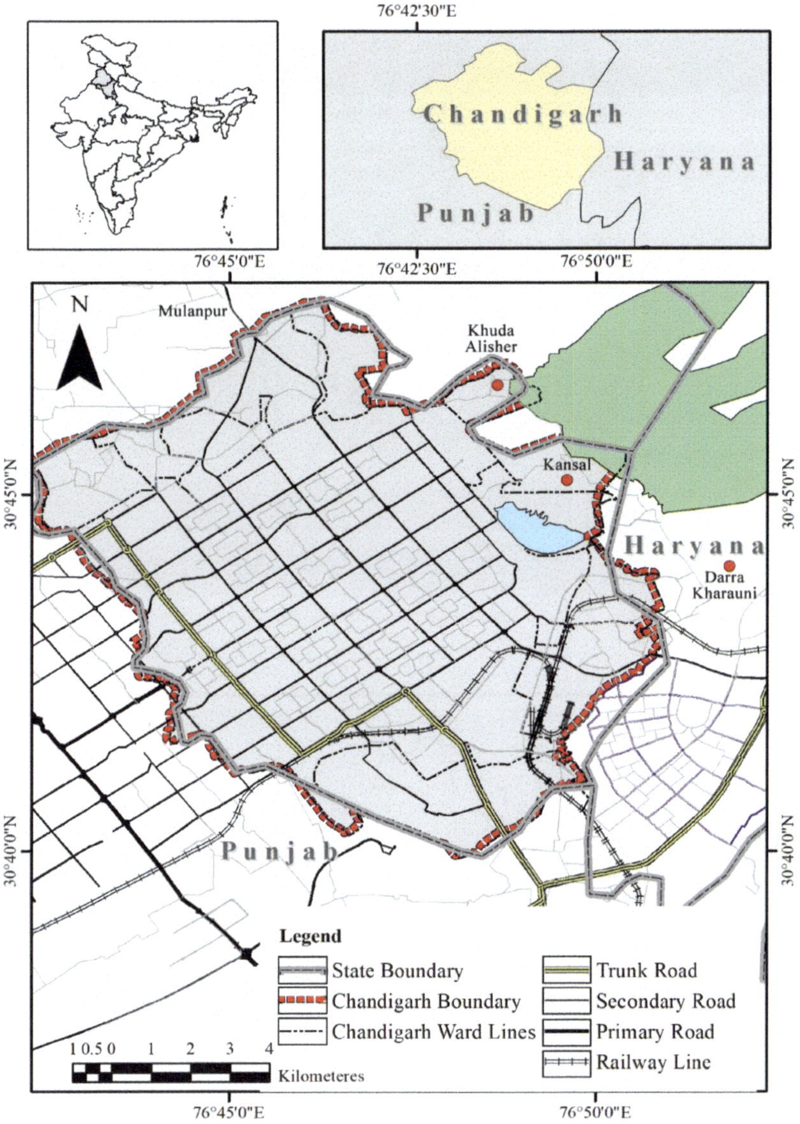

Fig. 10.1 Locational setting of Chandigarh City. *Source* Open Street Map (2023)

streams played a crucial role of serving as the city's primary source of water by maintaining a continuous recharge of the aquifers. The significance of this water network was understood by Corbusier, who integrated it in a seamless manner in the masterplan of the city. This is discussed in the following sections.

The first instance of integration of the natural stream network in the masterplan is seen in the sensible selection of the site itself. The selection of this site for planning of Chandigarh was a decision taken in recognition with the pre-existing stream network (Corbusier 1997). In the North, a catchment basin is located between the site and the Siwalik mountain that fed into the Patiali ki Rao, Choa Nadi, and Sukhna Choa. This area was earmarked in the master plan as green zone ensuring that the relationship between the city, water, and hills is not disturbed. At the site, the streams were well defined and were not closely spaced thus making developable land available in large parcels. While the streams of

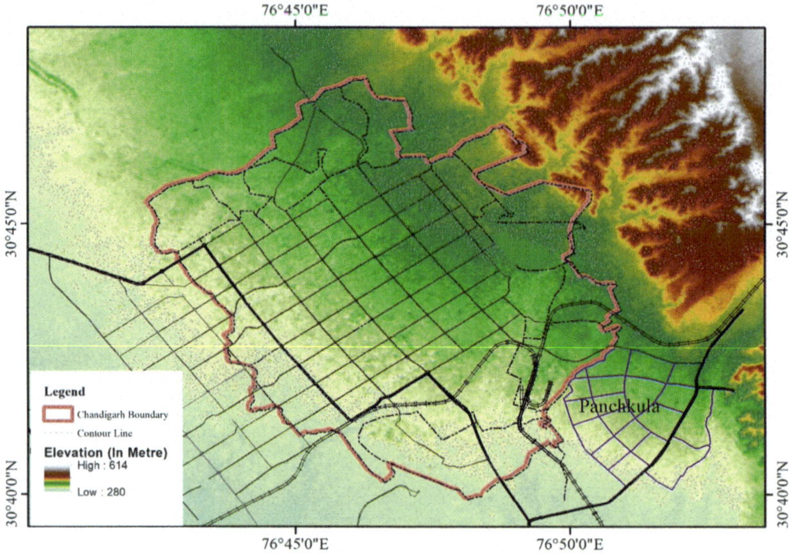

Fig. 10.2 Elevation map of Chandigarh City

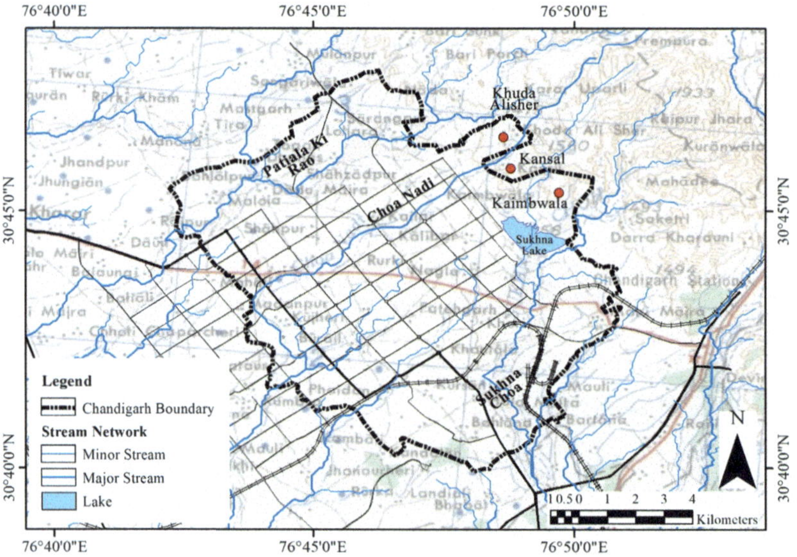

Fig. 10.3 Natural stream network in Chandigarh City. *Source* Base map—India and Pakistan 1:250,000 Series U502, U.S. Army Map Service, 1955; Sheet No. NH 43-8 Courtesy of the University of Texas Libraries, The University of Texas at Austin

Patiali ki Rao and Sukhna Choa were used to define the extent of the city in the West and East, respectively, the presence of fewer minor streams within the site allowed for the development without fragmenting or altering the natural course of the water streams. In the vicinity where the stream network is denser, it is obvious that the road network and other built components would not have resulted in this harmonious relationship between the master plan and the stream network.

The proximity of the Ghaggar River to the city is also a unique aspect of the planning of

Chandigarh. Throughout history, the cities have been located along the rivers, be it Indus Valley Civilisation, Nile Civilisation, or the eternal city of Varanasi located along Ganga. This is true even in the modern cities, such as Gandhinagar which is located along the Sabarmati River, Bhubaneswar is located along the Kuakhai River, and Jamshedpur is located along the Subarnarekha River. The latest example of Amaravati planned as the new capital of Andhra Pradesh, is also located on the banks of Krishna River. However, there exists a contradiction in the case of Chandigarh, where the city is situated approximately seven kilometres away from the Ghaggar River. This deliberate separation of city from river is likely due to the Ghaggar's susceptibility to flooding and its tendency to change course during the monsoon season (Bhattacharya and Kumar 2010; Chaba 2019). Thus, the choice of the location of the city provided safeguard against the river flooding during the monsoons (Saini and Kaushik 2012). At the same time, the presence of the streams like Patiali ki Rao, Choa Nadi, and Sukhna Choa in the city serves as drainage channels, preventing inundation within the city during the rainy season. This calculated planning ensures that while the city is not situated directly on the riverbanks, it remains integrated with the natural water streams that enhance its overall resilience to flooding. From this discussion, it can be said that a balanced approach was adopted in the selection of site for Chandigarh location imparting flood resilience and acting as the basis for the water-sensitive planning of the city.

The second aspect of the Water-Sensitive Urban Planning is the orientation of the city. The city is unusually oriented diagonally with the cardinal directions. Normally in order to achieve climatic comfort, orientation in cardinal directions is preferred in this part of the world. By making the orientation diagonal to the cardinal directions, the city was aligned with the Siwalik range, resulting in making the major road network parallel to the flow of streams. This decision offers several advantages. First and foremost, it preserves the natural course of the water bodies, ensuring that they flow unimpeded through the city, thus preventing any disruption to the natural water system. This alignment allows the city to harmoniously coexist with the water streams, resulting in the integration of nature into the urban environment. Here, it is worth mentioning that since this diagonal orientation isn't ideal from the solar radiation perspective, use of various types of sun-shades and louvres is employed to control the heating in the design of the buildings. It becomes obvious that Corbusier prioritised the water-sensitive approach for the master planning while employing shading devices to solve the heat gain problem.

10.3.3 Micro-aspects of WSUP in the Master Plan

This water-sensitive approach was also continued in detailed aspects of the master plan. Much like the selection of site and the orientation of the city, which were influenced by the presence of water features, the form of the city was also shaped by the course of the Sukhna Choa and Patiali ki Rao. These rivulets flow close to each other in the Northern part of the site; thus, it can be seen that the layout is kept narrow at the top. It is at this location, and Capitol Complex was located to suitably use the available site. The layout widens progressively as the two rivulets diverge towards the south of the city (Fig. 10.4). This form of the city can be appreciated because it overlays in tandem with these water bodies by following their natural flow instead of attempting to control or alter their course. This approach highlights Corbusier's attempt to create a water-sensitive masterplan by preserving the integrity of natural water elements.

Furthermore, the floodplains of the two rivulets were earmarked as greenbelts in the masterplan. Reserve forests and large botanical parks were planned along the banks of both the rivulets. These green belts along the rivulets served many purposes; firstly they acted as the extended catchment area; secondly, they provided recreational amenities to the people and also enriched the ecosystem of the water bodies. To further complement this objective,

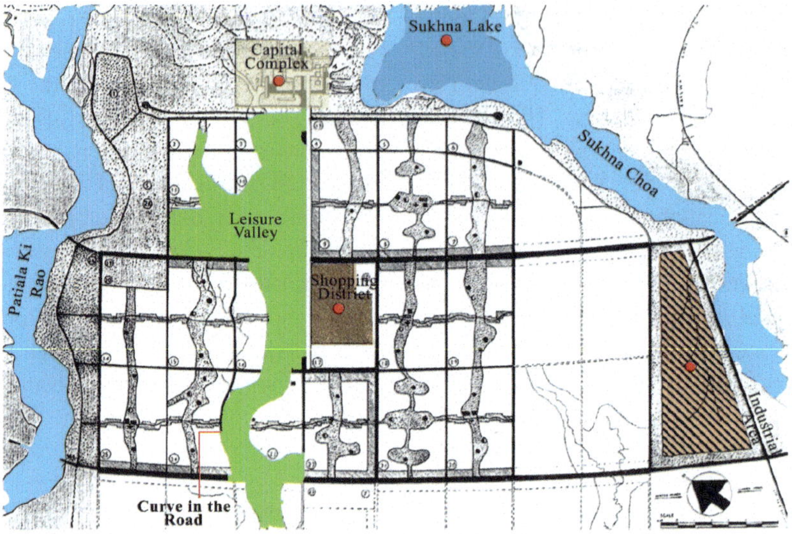

Fig. 10.4 Chandigarh master plan of the initial phase

universities and other institutions were strategically placed adjacent to the green belt of the Patiali ki Rao. These large open spaces in these institutions served as an extended buffer between the main city and the waterbodies. This arrangement not only ensured the preservation of the water bodies but also provided a tranquil and scenic setting for educational institutions (Corbusier 1961). This can be appreciated by comparing with the conventional urban planning approach where waterfronts are created along the edges of water bodies. In the conventional approach, it is often seen that the natural ecosystem of the water body gets disturbed and also the waste disposal due to the anthropogenic activities causes water pollution. In the planning of Chandigarh, sufficient buffer was provided along the natural rivulets and streams thus creating a thriving greenbelt which also acted as catchment area for water.

The city's planning also took into consideration the stream known as Choa Nadi (or N-Choa). It was a smaller seasonal stream in comparison to Sukhna Choa and Patiali-Ki-Rao and flows between them (Prakash 2002). It is interesting to know that Choa Nadi had dried up at the time of the planning of Chandigarh. However, Corbusier revived the stream by planning a citywide greenbelt along its course (Chalana 2015). This greenbelt was named Leisure Valley as it was designed as a recreational parkland, featuring different types of landscaped spaces including war memorial, rose gardens, orchards, lawns, and stadiums. Furthermore, important museums and Art College are located in the Leisure Valley which provides a serene environment to these amenities (Chandigarh Administration 2023). From water-sensitive approach, the planning of Leisure Valley along the Choa Nadi provided it with a catchment area which is crucial in the revival of the stream.

In the planning of the Leisure Valley, the original course of Choa Nadi was maintained. The flow of the natural course of the stream was prioritised over the built components such as buildings and roads. This is clearly visible in the road alignment of Udyan Path between Sectors 23 and 24, which is curved, to prevent disturbing the course of Choa Nadi which forms a bend at this location. This instance of curvilinear road is surprising, especially considering Le Corbusier's renowned emphasis on geometric precision in his designs. Since Choa Nadi is present within the city and at this location it required deviations from the rigidity of the gridiron structure in order to preserve the integrity of the water body, this is an important instance that reflects the water-sensitive planning preference in the masterplan. This approach resulted in the sustenance of overall ecological health of the stream because its flow and catchment remained undisturbed.

The concept of the Leisure Valley was extended across the city by providing green belts in every sector of the city, metaphorically mentioned as the lungs of the city (Corbusier 1961; Prakash 2002). The green belts follow the orientation of Leisure Valley and also aligns with the natural drainage of the site, as discussed in previous sections. This alignment allows the natural drainage of water to work effectively. Also, the alignment of the longitudinal roads follows the same orientation making them parallel to the streams. This overall arrangement enhances the effectiveness of the sewerage and drainage system installed along the roads.

The water-sensitive planning in Chandigarh is also visible in the preservation of the floodplains of the streams. Unlike many other cities, Chandigarh does not adopt the canalisation method for its water bodies. Canalisation can be understood from the recent project of Sabarmati Waterfront in Ahmedabad. This project is an example of large-scale use of concrete in creation of the waterfront promenades, thus altering the ecological characteristics of the river bank and its catchment (Mathur 2012). It is known that canalisation of natural water streams alters the flow and formation of waves, ripples, and pools; which impacts the process of oxygenation and self-cleaning ability of these streams. As a result the overall sustenance of marine flora and fauna is affected. In comparison, the approach of maintaining the natural characteristics of water bodies as seen in Chandigarh allowed the water bodies to flourish as habitats for rich biodiversity in the city.

From above discussion, it can be seen that the water streams were planned using nature-based approach in the masterplan of Chandigarh; however, there is one example of Sukhna Lake, which is an artificial reservoir located in the Northern part of the city. Despite Sukhna Lake being an artificial water body in Chandigarh, the approach of water-sensitive planning was carried forward in its overall planning. This again starts with the location of the Sukhna Lake, which is determined by the natural course of Sukhna Choa. The lake was planned to be fed from the Sukhna Choa, thus becoming part of the overall water network. From hydrological point of view, it was not possible to maintain the level of water in the lake without integrating it with the natural streams. Over the years, Sukhna Lake has evolved into the city's most popular recreational and ecological space. City residents utilise the promenade for everyday recreational activities, making it an integral part of their daily lives. It has also emerged as the preferred venue for significant touristic activities and cultural events. The lake is also a habitat for diverse range of flora and fauna, including various fish species, ducks, birds, and, on some occasions, even crocodiles have been sighted. The lake also serves as a temporary residence for many migratory birds during the winter months (Chandigarh Administration 2023).

The significance of Sukhna Lake can also be understood from that after the death of Pierre Jeanneret who was the partner of Le Corbusier and also served as the first Chief Architect of the city, his ashes were immersed in the lake (Moulis 2012; Prakash 2002). This act reflects the deep social, ecological, and spiritual connection that the lake developed with the city. In fact, it can be said that for Chandigarh, Sukhna Lake has acquired the status of the water monument, competing with the architectural monument of the Capitol Complex. This monumental status of an artificial reservoir was a result of the integration with the natural water network thus sustaining the water levels and thriving ecology in the lake.

The water-sensitive approach employed by Le Corbusier in integrating water bodies and greenbelts with the city's structure is notable, particularly when viewed in the context of that time, when extensive modification of sites to accommodate the desired city layout was considered as highly acceptable (McHarg 1969). For the sake of comparison, one prominent example is that of New Delhi, which was planned by Edwin Lutyens a few decades before Chandigarh. The two cities are similar for being

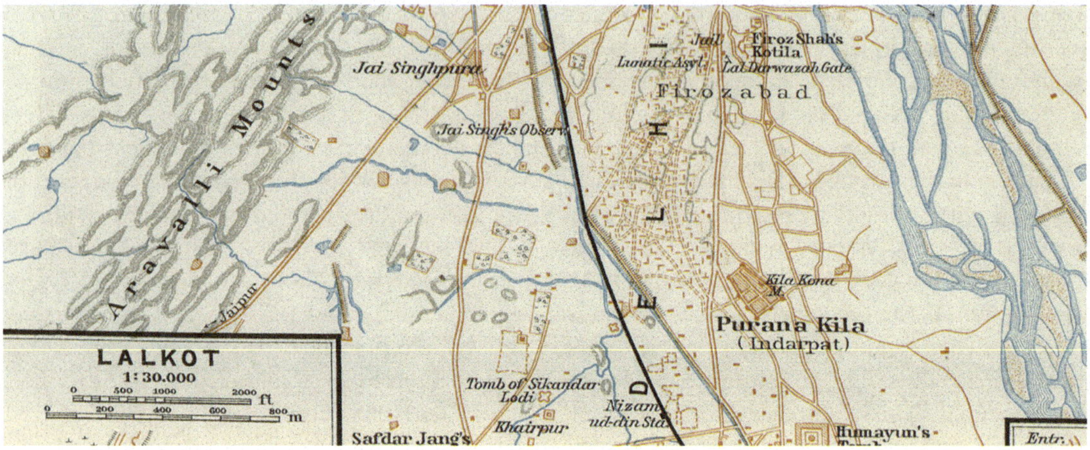

Fig. 10.5 Presence of water streams on the site of Lutyens Delhi. *Source* Delhi Environs 1914, Courtesy of the University of Texas Libraries, The University of Texas at Austin

planned as imposing new capitals by using the Garden City concept in a well-defined geometric pattern. However, the approach adopted by Lutyens in dealing with the water features on the site was fundamentally different, when compared to that of Chandigarh.

The site chosen for New Delhi was situated to the south of Shahjahanabad, the last Mughal capital city. The site is bordered on the East by river Yamuna and Aravali range on the west. The site had the presence of a network of natural streams which flowed from the hills to the river (Fig. 10.5). In the case of New Delhi, which was planned in a radial pattern, the natural water features were largely ignored (Wescoat 2016). Except for the historic Barapullah Nalah, other minor streams were filled and levelled in the construction of New Delhi. Even the Barapullah Nalah which was a rivulet to the river Yamuna was converted into sewerage channel for the city of New Delhi. While Chandigarh's planning integrated its pre-existing water streams into the masterplan, New Delhi on the other hand modified the site extensively to create a clean slate for the new city (Hasan and Patel 2014). In other instance, the Central Vista in New Delhi is the central green space and thus can be likened to Chandigarh's Leisure Valley. While, on the both sides of the Central Vista, large landscaped lawns and artificial pools are provided (Byron 1931), the Leisure Valley in Chandigarh is planned retaining the natural water and green features.

Le Corbusier visited New Delhi and considered it as a good example (Frampton 1992). He also compared the plan of Chandigarh with that of New Delhi. While there is a similarity in the wide roads in both the cities, the approach towards the water features was entirely different. While New Delhi's planning demonstrated a strong sense of discipline and control, reminiscent of the British rulers' authority over Indian subjects, Chandigarh adopted a more harmonious approach, that valued the integration of the city with its natural features, creating a sustainable and aesthetically pleasing urban built environment. While Chandigarh's water-sensitive planning can be termed as a visionary approach for its time, the unregulated urbanisation in the urban periphery that occurred in the following decades adversely impacted the relationship of the water streams with the city. This impact is discussed in the next section.

10.3.4 Provisions of WSUP in the Periphery of the City

The discussion in the previous sections has revealed that the planning of Chandigarh was sensitive to the water features by developing them as recreational greenbelts and ecological zones. Corbusier also understood that these water features are part of large Himalayan ecosystem and altering the surroundings of the city would impact the performance of the streams and greenbelts in his masterplan. In order to conserve the natural setting and catchment around Chandigarh, Le Corbusier proposed to prohibit any construction on the land between the city and the hills. He considered the hills in the backdrop as part of sacred landscape. He viewed any unregulated construction in the surroundings of the city as akin to profanity to his masterplan (Samuel 2007). His resolve in preserving the catchment area as an unbuilt zone can be understood from an instance when he even sought the direct intervention of Prime Minister Nehru to stall an Army Headquarters project that threatened the natural surroundings (Kalia 1987).

To safeguard the city's natural surroundings as a statutory provision, Le Corbusier proposed the creation of a greenbelt around Chandigarh, a concept that likely drew inspiration from the Garden City movement (Foundation Le Corbusier 2016). This provision was formalised by the state legislature as the Periphery Control Act in the year 1952 (Sarin and Walden 1974). The primary aim of the Periphery Control Act was to provide an ecological buffer to the city by prohibiting urbanisation in 8 km from the city's boundary on all sides (Chalana 2015). In the year 1962, the extent of this green zone was further increased to 16 km by an amendment to the Periphery Control Act (Fig. 10.6). The Periphery Control Act was an exceptional provision that was enacted in an Indian city. It can be appreciated for recognising the understanding of water-sensitive planning on a regional scale. However, it is tragic irony that the Periphery Control Act failed to be effective in the preservation of the buffer around Chandigarh. Both government and private entities carried on with the urbanisation activities by exploiting the inconsistency in the letter and spirit of the act.

10.3.5 Impact of Unregulated Urbanisation on the Water Features

After Chandigarh became populated, numerous planned and unplanned settlements began to emerge in its periphery. Cities like Mohali and Panchkula were planned to the South and East

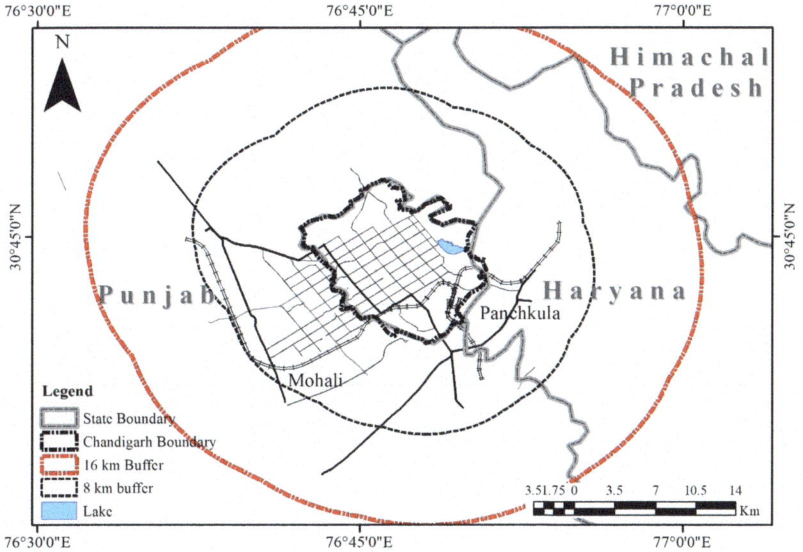

Fig. 10.6 Schematic layout of Chandigarh's periphery

of Chandigarh, resulting an urban agglomeration called as Tri-City. Unregulated ribbon development is also taking place along the highways approaching the city. Additionally, rural villages surrounding the city have undergone urbanisation. This urbanisation has disrupted the natural setting of the city and has an adverse impact on its environment. One of the most prominent cases is the unauthorised settlements in the Northern surroundings of the city. This area, originally occupied by a few villages, namely Kansal and Kaimbwala and their agricultural fields, served as the catchment area feeding Patiali ki Rao, Choa Nadi, and Sukhna Choa.

A comparative study of land-cover classification maps from 2005 and 2023 reveals

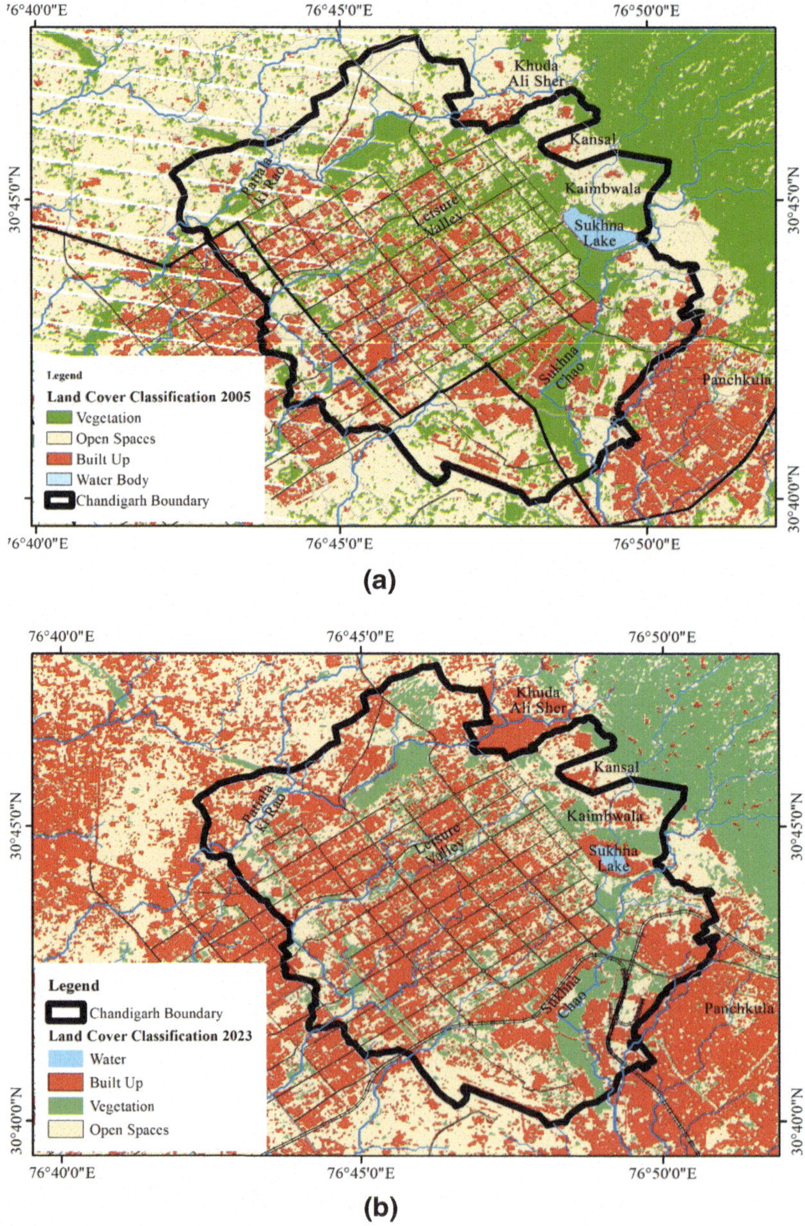

Fig. 10.7 **a** Land-cover classification in 2005. **b** Land-cover classification in 2023

this unregulated urbanisation as an increase in built-up area in the surroundings of the city (Fig. 10.7a and b). The expansion and densification of villages in the North of Chandigarh are prominently visible. As discussed earlier, this area is the catchment for the three rivulets which were integrated into the masterplan of the city. In the maps, it can be clearly seen that the unplanned settlements are growing in proximity to the rivulets, thus encroaching upon their floodplains and catchment area (Fig. 10.7b). Furthermore, the solid waste and sewage are disposed into these rivulets causing a decrease in the volume of water and also resulting in the water pollution (Kaur et al. 2017). On the regional scale, this catchment is fragmented due to this unplanned urbanisation.

In recent years, there is a significant reduction observed in the volume of water in the rivulets and also in the Sukhna Lake. The area of the lake has also decreased by half from its original of 228 ha due to the reduction in the flow of water in Sukhna Choa (Sareen 2017). Also, the volume of water in Choa Nadi, which flows in the Leisure Valley, has been reduced by the unplanned settlements around Kansal Village. The condition of Patiali ki Rao is worse, as it can be seen that most of the unplanned urbanisation has taken place on its bank. Rajendra Singh, famous as the Water Man of India, also called for the need to understand the behaviour of the catchment area for restoring the water levels in the Sukhna Lake (The Indian Express 2018). From this discussion, it can be understood that the territory of water-sensitive planning extends beyond the city boundary and that conservation of ecological zone around a city is crucial for its environment. Also, that in case of Chandigarh, Periphery Control Act was a visionary provision towards water-sensitive planning on regional scale.

10.4 Conclusion

There are three conclusions that can be derived from this discussion. Firstly, the water-sensitive planning in the celebrated case of Chandigarh by Le Corbusier, who remains known for Brutalist architecture and grid-iron pattern, is brought into the forefront of the discussion. Both Le Corbusier and Chandigarh are researched a lot; however, the water-sensitive aspects in the masterplan are ignored. The water-sensitive approach resulted in the protection of water bodies, creation of biodiversity, enhancement of built environmental, and creation of social spaces in Chandigarh. This approach undoubtedly was ahead of its time in the planning and was certainly not a result of coincidence. The writings and sketches of Corbusier indicate that he was aware about integrating the elements of nature with the city.

In his Final Testament, written before his passing, he articulated the concept of the "wedding" of the straight line (representing the planned city) with the axis of nature. In Corbusier's perspective, this concept called for a rediscovery of humanity and a connection with the fundamental laws governing life, nature, and the cosmos. In the planning of Chandigarh, Le Corbusier elevated elements of nature such as water and greenery to a sacred status. He also talks about "body-life balance" where he refers body as the built part of the city and equates life as the elements of nature.

Le Corbusier was also inspired by the biblical description of the Garden of Eden, and it seems that he aimed to replicate such a paradise in the landscape of Chandigarh. His integration of water streams and floodplains with the greenbelts became inhabited by rich flora and fauna resulting in the creation of delighting and surreal landscape in the city. Le Corbusier's preference for water bodies in the planning of Chandigarh is also evident in his drawings, where the water streams are prominently featured. In his conceptual sketches and master plan drawings, Le Corbusier depicted the presence of Patiali ki Rao, Sukhna Choa, and Choa Nadi as prominently as the city's road network and buildings, highlighting his deliberate intention to seamlessly integrate them into the city's form and structure.

Secondly, the need for restoring and conserving the ecological surroundings of Chandigarh is highlighted. Today, while Chandigarh is celebrated for its planning and architecture, the unregulated urbanisation in its surroundings is overlooked. Urban expansion beyond the city's edges has become so rampant that it has drastically altered the natural setting as envisioned in the masterplan. Urgent interventions are required to restore the natural setting of Chandigarh as much as possible, especially, in the zone to the North of the city, which should be declared as an ecologically sensitive zone in continuity with the Siwalik range, which is already designated as the reserved forest. In this manner, the relationship of the city with the nature may be sustained.

Finally, the learnings of Chandigarh can be used in other cities. Our cities now face multiple water-related problems. While on one hand many cities are experiencing water shortage, there are many instances of urban flooding due to poorly planned drainage system at the same time. These problems are expected to increase with the unpredictable climatic events. Chandigarh can be used as an example for how to carryout Water-Sensitive Urban Planning in our cities. So, in the planning of new urban settlements, streams and natural water bodies should be integrated, ensuring efficient water drainage and harvesting. In the case of existing cities, retrofitting using blue-green infrastructure can be carried out along with the revival of ponds, lakes, and streams to avail the benefits of Water-Sensitive Urban Planning.

References

Bhattacharya S, Kumar V (2010) Unprecedented Floods in Ghaggar Basin, South Asia Network on Dams. Rivers & People (SANDRP) 5(July):1–9

Byron R (1931) New Delhi. Archit Rev 69(410)

Census of India (2011) India—Census of India 2011—Chandigarh UT—Series 05—Part XII A—District Census Handbook, Chandigarh. Part XII A. Chandigarh: Directorate of Census Operations. Ministry of Home Affairs. Government of India. Available at: https://censusindia.gov.in/nada/index.php/catalog/324. Accessed: 6 Nov 2023

Chaba AA (2019) Explained: why Ghaggar overtops its banks often. Explained News—The Indian Express. Available at: https://indianexpress.com/article/explained/why-ghaggar-overtops-its-banks-often-5846301/. Accessed: 6 Nov 2023

Chalana M (2015) Chandigarh: city and periphery. J Planning Hist 14(1):62–84. Available at: https://doi.org/10.1177/1538513214543904

Chandigarh Administration (2023) Historical background | Chandigarh, The official website of the Chandigarh Administration, Official website of Chandigarh Administration. Available at: https://chandigarh.gov.in/know-chandigarh/planning-architecture/historical-background. Accessed: 5 Nov 2023

Corbusier L (1961) The modular. Second. Editorial Poseidon, S. R. L., Peru

Corbusier L (1997) The final testament of Père Corbu: a translation and interpretation of Mise au point by Ivan Zaknic. Yale University Press (Henry McBride Series in Modernism and Modernity), New Haven

Curtis W (1982) Modern architecture since 1900, 1st edn. Phaidon Press, New York

Dolman N, Savage A, Ogunyoye F (2013) Water-sensitive urban design: learning from experience. Proc Inst Civil Eng Municipal Eng 166(2):86–97. Available at: https://doi.org/10.1680/muen.12.00033

Foundation Le Corbusier (2016) Studies and research. Fondation Le Corbusier. Available at: https://www.fondationlecorbusier.fr/le-corbusier/oeuvres/architecture/etudes-et-recherches/. Accessed: 6 Nov 2023

Frampton K (1992) Modern architecture: a critical history. Thames & Hudson (World of Art), London

Hasan M, Patel D (2014) From Ghalib'sDilli to Lutyens' New Delhi: a documentary record. First. National Archives of India, Oxford Unversity Press

Joshi K (1999) Documenting Chandigarh: the Indian architecture of Pierre Jeanneret, Edwin Maxwell Fry, and Jane Beverly Drew. Mapin Publishing, Ahmedabad

Kalia R (1987) Chandigarh. In search of an identity. Southern Illinois University Press, Carbondale

Kaur H et al (2017) Geospatial modelling of flood susceptibility pattern in a subtropical area of West Bengal, India. Environ Earth Sci76(9). https://doi.org/10.1007/s12665-017-6667-9

Kumar A et al (2023) Water sensitive planning for the cities in the global south. Water 15(2):235. https://doi.org/10.3390/w15020235

Mathur N (2012) On the Sabarmati Riverfront: urban planning as totalitarian governance in Ahmedabad. Econ Pol Wkly 47(47/48):64–75

McGrane SJ (2016) Impacts of urbanisation on hydrological and water quality dynamics, and urban water management: a review. Hydrol Sci J 61(13):2295–2311. https://doi.org/10.1080/02626667.2015.1128084

McHarg IL (1969) Design with nature, 1st ed. Published for the American Museum of Natural History [by] the Natural History Press, Burlington, Canada

Moulis A (2012) An exemplar for modernism: Le Corbusier's Adelaide drawing, urbanism and the Chandigarh plan. J Archit 17(6):871–887

Open Street Map (2023) OpenStreetMap. Available at: https://www.openstreetmap.org/. Accessed: 5 Nov 2023

Prakash V (2002) Chandigarh's Le Corbusier: the struggle for modernity in postcolonial India. University of Washington Press, Washington DC

Rodríguez-Lora J-A, Rosado A, Navas-Carrillo D (2021) Le Corbusier's urban planning as a cultural legacy. An approach to the case of Chandigarh. Designs 5(3):44. https://doi.org/10.3390/designs5030044. Accessed: 6 Nov 2023

Rohilla SK et al (2017) Water sensitive urban design and planning: a practitioner's guide. Centre for Science and Environment (CSE), pp 1–94

Roy AH et al (2008) Impediments and solutions to sustainable, watershed-scale urban stormwater management: lessons from Australia and the United States. Environ Manage 42(2):344–359

Saini SS, Kaushik SP (2012) Risk and vulnerability assessment of flood hazard in part of Ghaggar Basin: a case study of Guhla block, Kaithal, Haryana, India. Int J Geomatics Geosci. Available at: https://www.semanticscholar.org/paper/Risk-and-vulnerability-assessment-of-flood-hazard-A-SainiKaushik/7e6630b278cadc3bcbb5a00583dfa44079ada527. Accessed: 6 Nov 2023

Samuel F (2007) Le Corbusier in detail. First. Routledge Book, Burlington. Elsevier Ltd. Available at: https://www.routledge.com/Le-Corbusier-in-Detail/Samuel/p/book/9780750663540. Accessed: 7 Nov 2023

Sareen J (2017) Chandigarh's once-beautiful Sukhna Lake goes dry, shrunk by 57%. The Quint. Available at: https://www.thequint.com/news/environment/chandigarh-sukhna-lake-dry. Accessed: 7 Nov 2023

Sarin M, Walden R (1974) Chandigarh as a place to live in. In the open hand. Available at https://mitp-arch.mitpress.mit.edu/pub/vgpgiezy/release/1. Accessed: 20 Nov 2023

Siddiqui A et al (2019) Smart Chandigarh tri-city region: spatial strategies of transformation. In: Vinod Kumar TM (ed) Smart metropolitan regional development. Advances in 21st century human settlements. Springer Singapore, Singapore, pp 403–450. Available at: https://doi.org/10.1007/978-981-10-8588-8_7

The Indian Express (2018) Sukhna Lake is dead, its capacity to discharge and recharge is over. Chandigarh News—The Indian Express. The Indian Express. Available at: https://indianexpress.com/article/cities/chandigarh/sukhna-lake-is-dead-its-capacity-to-discharge-and-recharge-is-over-rajendra-singh-5147918/. Accessed: 7 Nov 2023

University of Texas (2021a) Ambala NH-43-08 1955—India and Pakistan AMS Topographic Maps 1955—Perry-Castañeda Map Collection—UT Library Online. Available at: https://maps.lib.utexas.edu/maps/ams/india/. Accessed: 5 Nov 2023

University of Texas (2021b) Delhi Environs 1914—Perry-Castañeda Map Collection—UT Library Online. Available at: https://maps.lib.utexas.edu/maps/historical/baedeker_indien_1914/txu-pclmaps-delhi_environs_1914.jpg. Accessed: 5 Nov 2023

USGS (2023) EarthExplorer, U.S. Geological Survey Earth Explorer. Available at: https://earthexplorer.usgs.gov/. Accessed: 21 July 2023

Wescoat JL (2016) Barapula Nallah and its tributaries: watershed architecture in Sultanate and Mughal Delhi. Water Design: Environment and Histories. Marg Publications, Mumbai, pp 85–95

Wong THF, Brown RR (2009) The water sensitive city: principles for practice. Water Sci Technol 60(3):673–682. Available at: https://doi.org/10.2166/wst.2009.436

Yadav V, Ibrar Z (2022) Relationship of water stress and flood damage for sustainable development. Water Resour Manage 36(4):1323–1338. Available at: https://doi.org/10.1007/s11269-022-03083-7

Integration of Urban Water Management in the Town Proposed Under Industrial Corridor Development, India

11

Yogesh Bhardwaj, Asma Hamid, and Qamar Irshad

Abstract

Water is an important natural resource as it aids all lifeforms on the earth, directly or indirectly. As the urban areas have expanded and are set to expand at a faster pace globally, water resources in every form have been impacted. From overuse, misuse, and contamination to the scarcity of fresh water's availability, we have come to a point where management of water, in an urban area and beyond the urban area has become essential. In addition to human intervention, changing climate, haphazard, and organic planning in the cities also contribute to the complexity of the situation. Managing water in such a manner that complications at the baseline are addressed, various aspects of the urban water cycle are considered. To attain sustainable form of water resource usages new paradigms are adopted, which builds a holistic approach toward urban water system planning, provision, and management. Furthermore, they need to be integrated with the settlement planned for the area. In this chapter, we will study the importance of water, and the integrated urban water management that can be done after a detailed analysis in the town proposed under Industrial Corridor Development.

Keywords

Urbanization · Industrial towns · Water sources · Water scarcity · Integrated urban water management

11.1 Introduction

Cities are growing rapidly and house a larger population in cities than in rural areas (UNDESA, World Social Report 2020). This fast urbanization, along with inadequate public services and outdated urban planning models, has pushed many new arrivals into informal settlements and slums. This has also exacerbated inequality, urban poverty and has made it difficult to increase water security. Many cities are facing huge obstacles but one of the most crucial is water management which needs to be addressed urgently.

Nowadays there is a scarcity of potable water, and many water sources need expensive treatment (Salehi 2022). Volumes of wastewater are

Y. Bhardwaj · A. Hamid · Q. Irshad (✉)
Department of Architecture, Jamia Millia Islamia, New Delhi, India
e-mail: qirshad@jmi.ac.in

Y. Bhardwaj
e-mail: aryogeshbhardwaj@jmi.ac.in

A. Hamid
e-mail: archiasma123@jmi.ac.in

S. Talukdar et al. (eds.), *Water Resource Management in Climate Change Scenario*, GIScience and Geo-environmental Modelling, https://doi.org/10.1007/978-3-031-61121-6_11

also growing. People residing in cities often suffer from waterborne illnesses and lack access to safe drinking water in many parts of the world. The surrounding communities suffer as cities look upstream for new water sources and downstream for their effluent discharge. There is also disturbance to the aquatic systems and hydrological cycle, which includes essential ecosystem services (GWP 2013).

The quality, amount, and seasonality of water available to urban areas and their environs are predicted to change due to climate change as a result of more frequent and extreme weather events (Bahri et al. 2012). Low-lying coastal zones, home to 13% of the world's urban population, are particularly vulnerable. Due to climate change, there is an impact on water supply and sanitation, floods are increasing, water requires treatment and water availability is being reduced.

11.1.1 Water Scarcity

Nowadays, the biggest form of environmental threat is in the form of water scarcity. Cities are facing insufficient clean water supply. Because of severe financial and physical constraints on supply enhancement, water supply is not increasing at the same rate as demand due to rapid population growth and economic activity (Saleth 2011). According to recent studies, nine basins that account for more than four-fifths of India's total water use will physically run out of water by 2050 if the demand–supply gap widens (Amarasinghe et al. 2007).

According to UNDP, 2006 industrial and domestic water usage is expected to double by 2050 (UNDP Annual Report 2006), thus increasing the pressure on water resources. A region is considered water-stressed if its annual water resource per capita is less than 1700 m^3. The water scarcity criterion is 1000 m^3 of annual water resource per person, and the absolute threshold of water resource scarcity is 500 m^3 per person per year, according to the internationally recognized UNDP (indicator of water stress) (Wu 2002).

11.1.2 Wastewater Generation

Degraded urban water resources have an impact on ecosystems, human health, and livelihoods that depend on water. This is frequently the result of improper waste water treatment. Urbanization is leading to produce large amounts of wastewater, quality and quantity both are degraded. Gray/wastewater production is rising in line with the domestic water supply and city's rapid expansion. Approximately 70–80% of the water supplied for residential use is wasted, according to CPHEEO estimates. Class I and class II towns, which account for 72% of India's urban population, are estimated to generate 98 lpcd of wastewater per capita, whereas the National Capital Territory—Delhi alone—disposes of 3663 mld of wastewater per year, of which 61% is treated (Annual Report, CPCB 1999). Over the past several years, there has been a significant increase in the quantity and quality of wastewater dumped in rivers (Rashid et al. 2018).

In order to use water resources in a sustainable manner, the outdated paradigms for planning of water resources must be abandoned in favor of new ones, which call for infrastructure with updated drainage and landscape design techniques as well as better receiving water consideration (Capodaglio et al. 2016).

Urban water management is responsible for controlling waterborne diseases and epidemics, ensuring that people have access to services, and sanitation; managing runoff pollution, wastewater, rainwater, and stormwater drainage; and lowering the likelihood of water-related disasters (Bahri et al. 2012). In response to climate change, which will bring warmer temperatures, more frequent heat waves, droughts, floods, and an increase in sea level, urban water managers are creating strategies to make their water sectors more adaptive. Urban water management needs to be considered in conjunction with stormwater management, solid waste management, pollution control laws, and the built environment.

To reorient urban areas toward sustainability and resilience, it emphasizes more adaptive and integrated management of the total water cycle and efficient water resource use through diverse and flexible multi-scalar solutions (Leigh and Lee 2019).

11.1.3 Integrated Urban Water Management

Integrated Urban Water Management (IUWM) offers a collection of concepts in favor of resource management techniques that are more long-term, responsive, and coordinated (GWP 2013). An integrated approach to urban water systems has several advantages, one of which is that it may expand the opportunities for creating more sustainable systems (Mitchell 2006). In order to accomplish sustainable goals related to the economy, society, and environment, IUWM advocates for coordinating basin management with urban development (GWP 2013). The traditional methods were primarily top-down and hardly ever took stakeholder engagement into account. IUWM, on the other hand, is an adaptive and cyclical procedure that involves engagement with stakeholders and also encompasses environmental, economic, technical, social, and political aspects of water management (UNESCO and NIUA 2021).

The integrated urban water management is a process that should be planned and managed in a way that collective impact is minimized and collective system efficiency is maximized (Mitchell 2006). IUWM aims to enable multifunctionality of urban water services. Several international cities have embraced the IUWM concept to manage the urban water sector, including Singapore, Amsterdam, Melbourne, and Los Angeles (Rao 2022). A universal IUWM model does not exist. A solution that is effective in one city might not be in another. On the basis of the IUWM fundamental principles, cities must create their own models (UNESCO and NIUA 2021).

Integrated Urban Water Management employs a variety of techniques, including adaptive urban planning and design, which incorporates landscaping and urban layout; the use of unconventional water sources, such as roof runoff, storm water, gray water, and wastewater; the application of fit-for-purpose principles; pollution control and prevention from wastewater and storm water sources; storm water flow and quality management; hybrid technologies that combine ecological and structural elements; and non-structural strategies like pricing incentives, education, regulations, and restriction regimes (IRAP Toolkit Tool 2.1 2010).

11.2 Selecting the Case for Study

Indore stands as the largest and most populous city in Madhya Pradesh, India, positioned on the southern periphery of the Malwa Plateau at an average altitude of 553 m (1814 ft.) above sea level. Renowned as a prominent commercial and industrial hub, it also hosts numerous educational and cultural institutions. According to the 2011 census, Indore boasted a municipal corporation population of 1,994,397 and an urban agglomeration population of 3,570,295, ranking third in influence among the cities associated with the Delhi Mumbai Industrial Corridor, following Surat (4,501,610) and Jaipur (3,046,163). Spanning merely 530 km^2 (200 square miles), Indore emerges as the most densely populated major city in the central province. With a youthful population, the city holds a median age of 27 years. Impressively, Indore boasts a high literacy rate of 89% and a sex ratio of 935 females per 1000 males. Moreover, the city maintains a low unemployment rate, standing at a mere 3%.

Indore serves as a compelling case study within the context of the Delhi–Mumbai Industrial Corridor (DMIC) due to its strategic location and the multifaceted challenges it faces. As one of the key cities along the DMIC, it stands as a prime example of India's push for new industrial investments and economic growth. The DMIC project aims to transform Indore into a vibrant manufacturing and investment hub, showcasing the potential for economic development in the region (Scott Wilson India Pvt. Ltd. 2009). However, simultaneously, Indore's inclusion in the DMIC underscores the significant water stress it faces. The city's ongoing struggle with water scarcity, as highlighted in the proposal, presents a complex challenge that must be addressed in tandem with the ambitious industrial and infrastructure development plans. Ensuring a harmonious blend between driving industrial expansion and practicing sustainable, responsible management of water resources stands as an ideal case study of Indore's role as a focal point within the overarching DMIC initiative.

Balancing the pursuit of industrial growth with the need for sustainable and responsible water resource management is a central aspect of Indore's role as a case study in the broader DMIC initiative (Fig. 11.1).

Indore, a city experiencing rapid expansion, undergoes continual transformation in its urban layout, is segmented into two primary zones: the old city, characterized by its narrow lanes, traditional architecture, and vibrant marketplaces, and the newer sectors. Serving as a significant

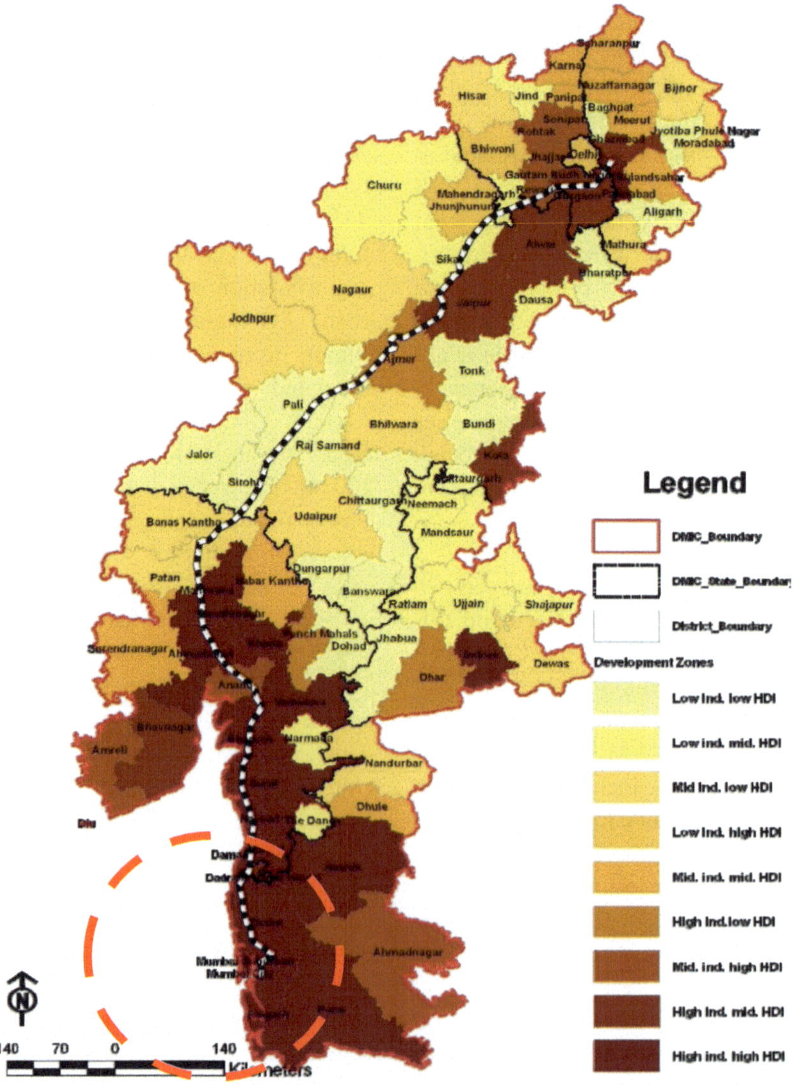

Fig. 11.1 Selection of Indore city as a case based on its distance from Delhi–Mumbai industrial corridor (DMIC) and socio-economic development of cities. *Image Source* Scott Wilson India (2009, p. 58)

hub for commerce and industry, Indore hosts several corporate giants like Reliance Industries, Tata Group, and Aditya Birla Group, fostering a robust economy in textile, pharmaceuticals, and engineering. Moreover, it stands tall as an eminent educational and cultural nucleus, housing prestigious institutions such as the Indian Institute of Technology, the Indian Institute of Management, and the Devi Ahilya Vishwavidyalaya. The city also boasts an array of museums, art galleries, and theaters. However, amid its rapid growth, Indore confronts multiple challenges in urban planning, contending with escalating urbanization, and surging demands for housing, infrastructure, and amenities. Additionally, it grapples with environmental concerns encompassing air and water pollution alongside burgeoning traffic congestion.

11.2.1 Urbanization and Water Scarcity: Indore

The fast expansion of urban areas and the development of peri-urban regions result in a mismatch between the demand for resources and their availability, particularly in environments with limited resources. As cities' water consumption increases, they are compelled to transport water from more distant sources, sometimes as far as 70 km from the Narmada River, incurring a cost of Rs. 45 per cubic meter, considering Unaccounted for Water (UFW). In this process, the conventional water sources within the city tend to be overlooked, as piped water supply systems conveniently deliver water directly to households and municipal bodies. Having served as a municipal area since 1870, Indore's early urban development stemmed from industrial expansion and military initiatives. Sir Patrick Geddes played a pivotal role by crafting the city's first-ever Development Plan in 1918, envisioning the prospects for a contemporary Indore. This visionary plan laid the groundwork for the city's subsequent growth trajectory and delineation of its planning zones (Table 11.1).

Indore is positioned within a semi-arid upper catchment zone, situated near the ridge line of the Ganga Basin, boasting an elevation of 553 m above sea level. While hydrologically designated as a recharge area, the city's hydro-geological features atop the Malwa Plateau paint a grim picture concerning groundwater accessibility. Consequently, Indore grapples with substantial water-related adversities due to these hydrogeological conditions and the recurring drought cycles it endures. Nestled in the Narmada sub-basin, the city rests along the banks of the Khan River, a tributary of the Narmada. Meandering through the urban landscape, the Khan River intricately divides Indore into distinct sections.

The northern part of the city is located in the Chambal sub-basin, while the southern part is located in the Narmada sub-basin. The following are the major sub-basins of the Narmada River in Indore: Khan sub-basin and Tawa sub-basin. The following are the major sub-basins of the Chambal River in Indore: Karoli sub-basin and Banas sub-basin. Both the Narmada and Chambal rivers are important sources of water for drinking, irrigation, and industrial use in Indore. The sub-basins are also home to a variety of ecosystems and biodiversity (Fig. 11.2).

Indore's water supply history illuminates the city's dependence on surface water reservoirs. Initially, Indore's water systems were sustained by rainwater harvested from Bilawal and Yashwant Sagar reservoirs, complemented by an array of traditional wells and step wells. Water scarcity in Indore has been a persistent issue throughout its history, primarily due to the factors mentioned earlier. The city's water

Table 11.1 Growth of Indore planning area (1975–2002), CDP under JNNURM, Mehta and associates

Indore planning area growth from 1975 to 2002			
S. No.	Year	Area (ha)	Growth (in %)
1	1975	2284	
2	1990	6115	167.73
3	1996	7747	26.68
4	2002	10,725	38.44

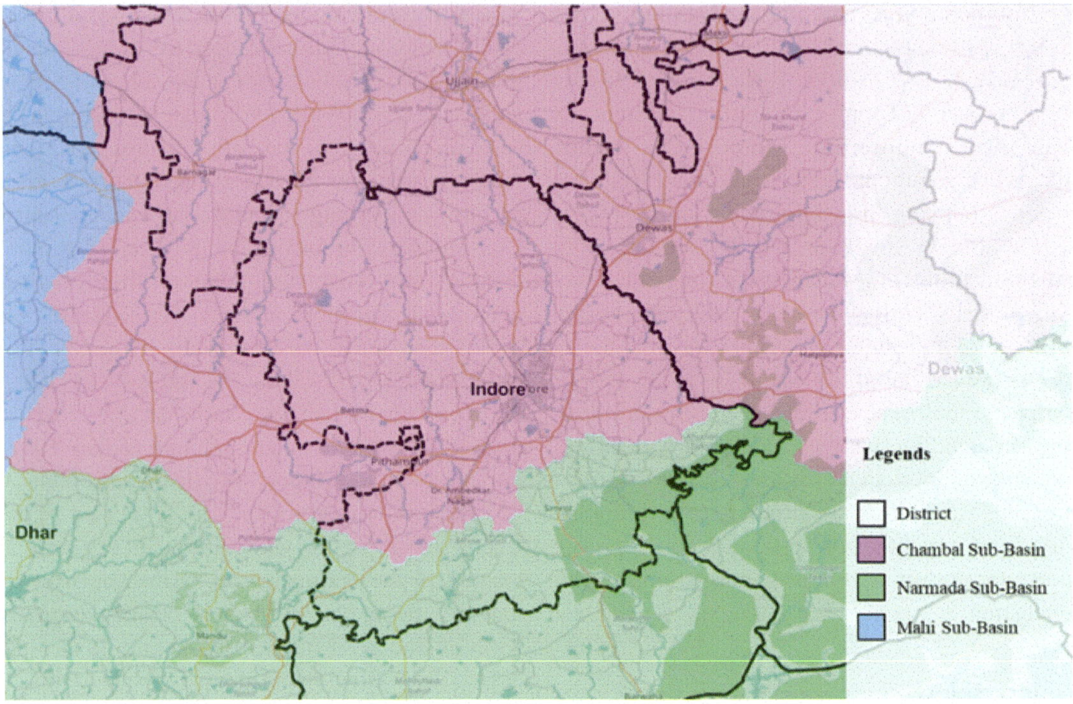

Fig. 11.2 Watershed/sub-basin of Indore city. *Source* Madhya Pradesh State Wetland Authority (2023)

supply, as depicted in Fig. 11.3, has continued to increase in response to growing demand. However, achieving equity and consistent availability of water has remained elusive. This issue elucidates the proportional surge in water supply against the backdrop of population growth in recent decades. With the city's populace and geographical expanse burgeoning, the water demand extended far beyond its confines, stretching into inter-basin territories. The Narmada River, located approximately 70 km away and at an altitude difference of over 500 m, became a crucial source to meet this growing water footprint (India Water Portal 2023).

11.2.2 Current Issues in the Urban Water System in the City

Indore, the most extensive and densely populated city in Madhya Pradesh, India, grapples with severe water scarcity. The city's prevailing water supply mechanism, drawing water from the Narmada River located 70 km distant, proves both economically unsustainable and costly. Moreover, the Indore Municipal Corporation (IMC) has yet to thoroughly investigate affordable and sustainable alternatives. These alternatives encompass the reuse of treated wastewater, revival of traditional local water sources, groundwater replenishment, and rainwater harvesting within the city's premises.

Indore heavily relies on the Narmada River as a primary source of drinking water. Nonetheless, the city possesses traditional resources that could potentially meet the demand for daily drinking water. As per Table 11.2, the Narmada River constitutes 75% of the city's overall drinking water supply, signifying a substantial reliance. The considerable distance of approximately 70 km between Indore and the Narmada River escalates the cost incurred in pumping water to the city, further highlighting the financial burden of this dependency.

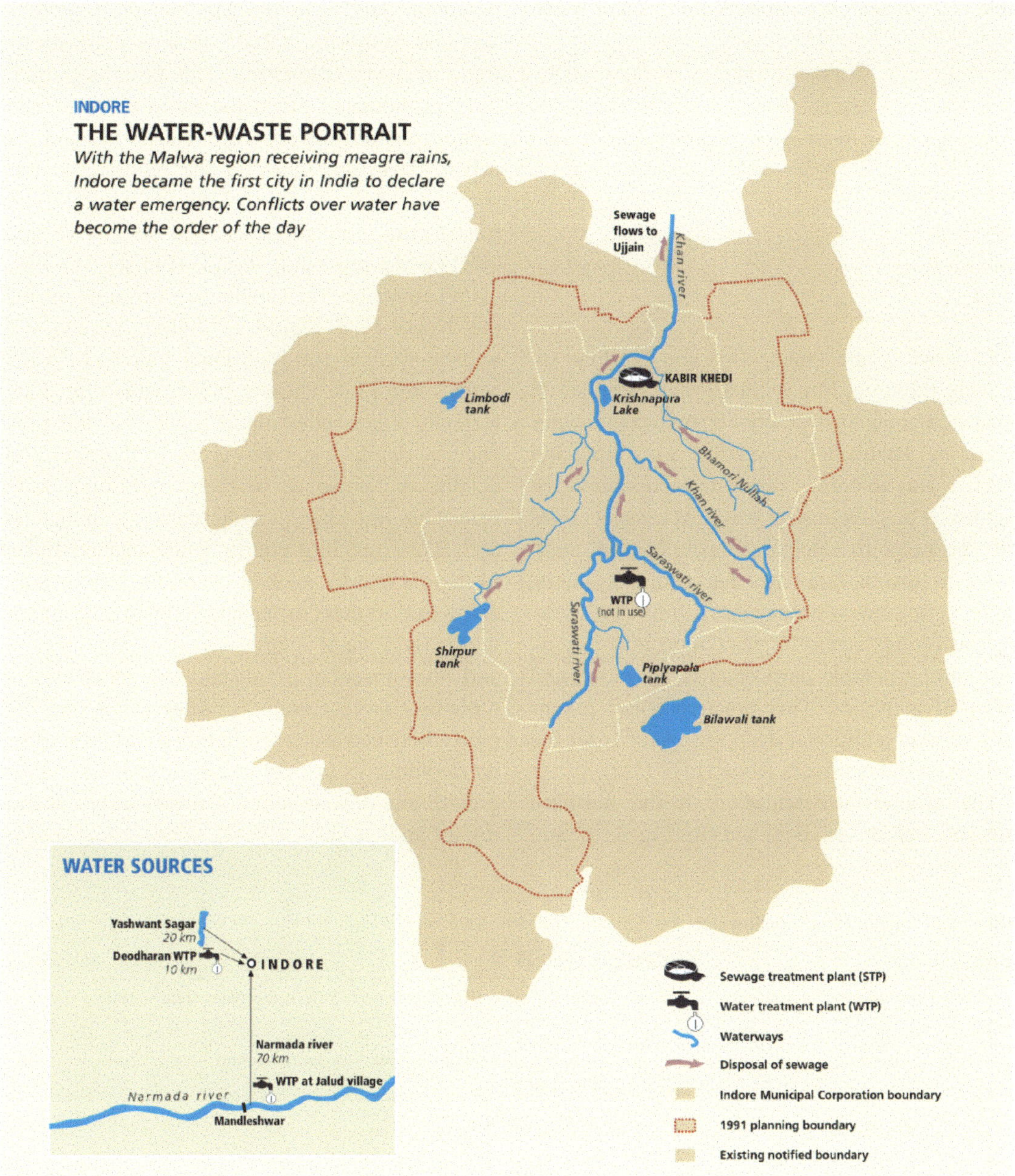

Fig. 11.3 Indore water resource and supply systems. *Source* Anon 2011, 71-City Water-Excreta Survey, 2005–06, Centre for Science and Environment, New Delhi

Table 11.3 shows that Indore has several traditional water resources, including wells, bawdis (step wells), and tanks. These resources have been neglected in recent years, but they have the potential to provide a significant amount of water for the city. The following are some of the measures, how Indore can utilize its traditional water resources to reduce its dependency on the Narmada River: Rejuvenate and conserve traditional water bodies: The government and local communities can work together to rejuvenate and conserve traditional water bodies such as

Table 11.2 Indore's water sources (Indore, The Water-Waste Portrait, 2005–2006)

Source	Distance (in km)	Capacity (in MLD)	Supply (in MLD)
Bilawali	5	9	4
Yashwant Sagar	21	38	27
Tubewells, etc.	–	27	13
Narmada River	70	210	160

wells, bawdis, and tanks. This can be done by cleaning them up, removing silt and debris, and repairing damaged structures. Promote rainwater harvesting: Rainwater harvesting is a simple and effective way to collect rainwater and store it for later use. The government can provide incentives and subsidies to encourage people and businesses to install rainwater harvesting systems. By recycling wastewater: Recycling and treating wastewater can offer an opportunity for its reuse in non-potable purposes like irrigation and industrial activities. This practice could potentially release additional drinking water from the reliance on the Narmada River (Fig. 11.4).

The groundwater situation in the northern parts of Indore, as depicted in Fig. 11.6 and based on data from the Madhya Pradesh State Wetland Authority in 2023, reveals a concerning trend of over-exploitation. This issue is particularly pronounced given the rapid urban expansion observed between 2001 and 2011, with the urban expansion index in the North East area surging to 2.74 during that decade, in stark contrast to the more modest index of 0.64 (Rajput 2021) in the preceding decade (Fig. 11.5).

What makes this situation intriguing is that the North–East region of Indore is endowed with more than 100 traditional sources of water, refer Table 11.4. These sources, which have historically been relied upon, underline the presence of ample water resources in the area. Yet, despite the presence of these traditional water reservoirs, the state of groundwater remains dismal. This puzzling contradiction can be linked to several factors, such as excessive groundwater extraction to meet urban and industrial demands, ineffective water management methodologies, and the absence of sustainable strategies to replenish these resources. Addressing this disparity between water availability and groundwater depletion is critical to ensure the sustainable growth and development of this rapidly expanding urban region.

Table 11.3 Indore's traditional water

S. No.	Count	Total	Particulars	No. of sources
1	Total number of sources	629	Total no. of wells greater than 3 m dia./greater than 7 m^2 plan area	389
			Total no. of wells less than 3 m dia./less than 7 m^2 plan area	204
			Closed sources	36
2	Total number of open wells	557	Total no. of open wells greater than 3 m dia./greater than 7 m^2 plan area	328
			Total no. of open wells less than 3 m dia./less than 7 m^2 plan area	204
			Closed open-wells	25
3	Total number of step wells	72	Total no. of step-wells greater than 3 m dia./greater than 7 m^2 plan area	61
			Total no. of step-wells less than 3 m dia./less than 7 m^2 plan area	0
			Closed step-wells	11

Source Environmental Planning and Coordination Organisation (EPCO) (2012)

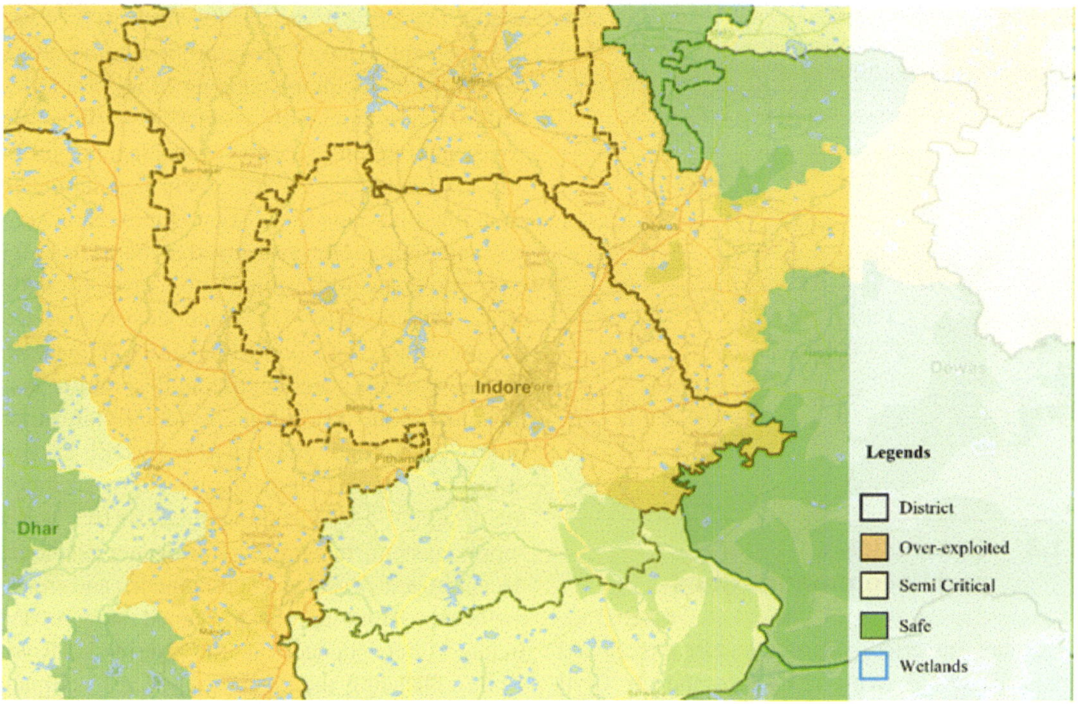

Fig. 11.4 Ground water status of Indore city. *Image Source* Madhya Pradesh State Wetland Authority (2023)

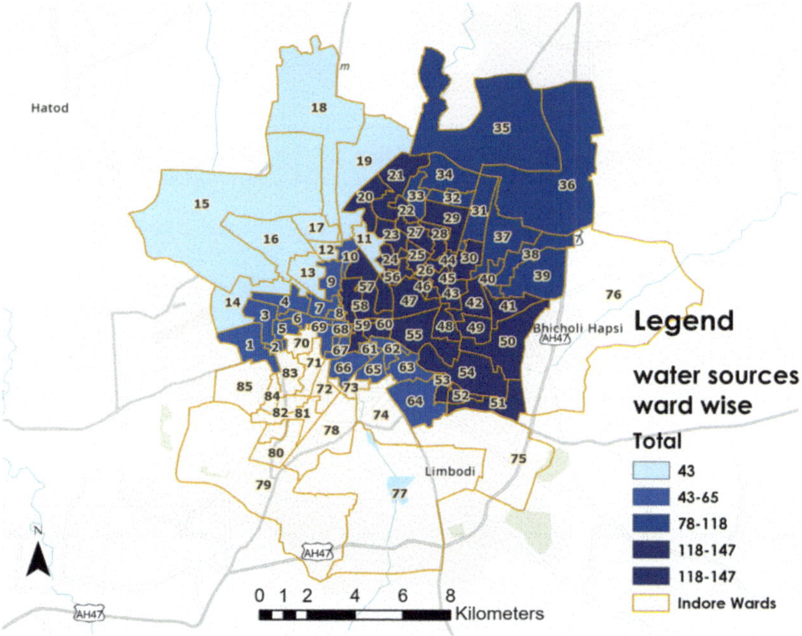

Fig. 11.5 Indore's traditional water sources, ward wise. *Image Source* Author, Data Source-Environmental Planning and Coordination Organisation (EPCO) (2012)

Table 11.4 Indore's traditional water sources, ward wise (Environmental Planning and Coordination Organisation (EPCO) 2012)

Ward No.	Open well	Step well	Defunct source	Total
1–10	52	8	4	64
11–20	30	13	0	43
21–30	97	4	0	101
31–40	69	9	0	78
41–50	114	4	0	118
51–60	132	14	1	147
61–70	59	6	0	65
	553	58	5	616

11.3 Conclusion and Recommendations

Dr. Rahul Banerjee, an urban water expert and graduate of IIT with a background in Civil Engineering, conducted a thorough study in 2012 focusing on the water supply and sanitation scenario in Indore. His study yielded a set of comprehensive recommendations aimed at addressing critical aspects of the city's water management. Among these proposals, he advocated for a meticulous mapping of all properties within the municipality, categorized based on zones and building quality, to establish fair property tax rates. This strategic approach aimed to equitably distribute the cost burden of water supply and sanitation, ensuring that heavier water consumers would contribute proportionally more. Additionally, Dr. Banerjee emphasized the importance of conducting a comprehensive inventory of all water supply and sanitation systems across the city, aiming to identify areas for improvement and enhance the overall system efficiency. His recommendations also included the design of a hybrid ground and surface water supply system, integrating stormwater recharge, decentralized wastewater treatment, and reuse methods. This approach aimed to reduce Indore's dependence on the Narmada River while enhancing the city's resilience to climate change impacts. Furthermore, he highlighted the necessity of formulating sustainable financial models for cost recovery and regularly evaluating progress to ascertain the effectiveness of the

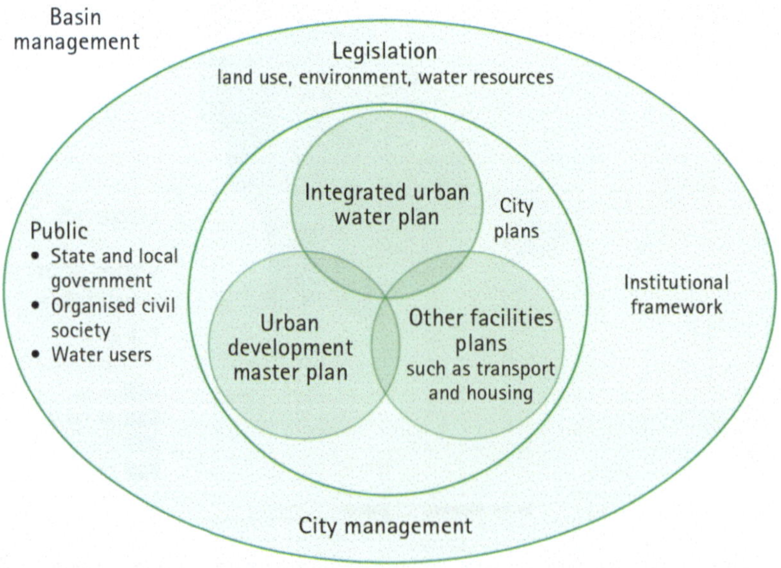

Fig. 11.6 A diagram of city management. *Image Source* Global Water Partnership, 20139140

alternative plan. Dr. Banerjee also proposed the implementation of the Central Ground Water Board's artificial recharge plan in the Gambhir and Kshipra River Basins as an integral part of Indore's water management strategy (Fig. 11.6).

Implementing these measures would enhance overall water availability within Indore's catchment area. Leveraging solar power for water supply from the Narmada aims to diminish the city's dependence on fossil fuels, fostering a more sustainable water supply system. Dr. Banerjee's proposals serve as a blueprint for a fairer and environmentally conscious water supply and sanitation framework in Indore. Their execution holds the potential to curtail the city's water footprint, bolster resilience against climate change, and ensure universal access to clean water and sanitation. As highlighted on the India Water Portal website (India Water Portal 2023), Dr. Banerjee envisions a city where clean water and sustainable, climate-resilient water supply are accessible to all. By adopting these recommendations, Indore can transform into a model city, setting an example for global urban centers. Notably, the Narmada serves as a crucial water source for millions in India, warranting its protection. Indore can alleviate its demand on the Narmada by adopting rainwater harvesting and treated wastewater reuse, ensuring sustainable resource management. Furthermore, these initiatives would strengthen Indore's resilience against climate change impacts on water resources, mitigating the anticipated effects of droughts and floods. Finally, by equitably distributing the expenses of water supply and sanitation among all residents, Indore can establish a fairer system, ensuring universal access to these fundamental services.

References

A qualitative framework to evaluate the extent of integrated urban water management in Indian cities & applying the framework to Delhi. NIUA. https://en.unesco.org/sites/default/files/unesco_ndl_and_niua_white_paper_iuwm.pdf

Amarasinghe UA et al (2007) India's water future to 2025–2050: business-as-usual scenario and deviations. Research report. International Water Management Institute, Colombo, Sri Lanka

Annual Report, CPCB (1999) Chrome-extension: //efaidnbmnnnibpcajpcglclefindmkaj/https://cpcbenvis.nic.in/annual_report/AnnualReport_4_annualreport1999-00.pdf

Bahri A, Asociación Mundial para el Agua (2012) Integrated urban water management

Capodaglio AG, Ghilardi P, Boguniewicz-Zablocka J (2016) New paradigms in urban water management for conservation and sustainability. Water Pract Technol 11(1):176–186

Environmental Planning & Coordination Organisation (EPCO) (2012) Detailed project report for climate change action programme support, enhancing adaptive capacity to climate change through conservation of traditional water supply sources (wells and bawdies) of Indore city. Environmental Planning & Coordination Organisation (EPCO), Bhopal

https://timesofindia.indiatimes.com/blogs/voices/integrated-urban-water-management/

India Water Portal (2023) Solutions to Indore's water problems don't have to be so expensive, 03 Oct 2023. Retrieved from indiawaterportal: https://www.indiawaterportal.org/articles/solutions-indores-water-problems-dont-have-be-so-expensive (2005–2006). The Water-Waste Portrait, Indore, Centre for Science and Environment, New Delhi

Leigh NG, Lee H (2019) Sustainable and resilient urban water systems: the role of decentralization and planning. Sustainability 11(3):918. https://doi.org/10.3390/su11030918

Madhya Pradesh State Wetland Authority (2023) Environmental Planning & Coordination Organisation (EPCO), Environment Department, Government of Madhya Pradesh. Retrieved from geoportal.mp.gov.in: https://geoportal.mp.gov.in/epco/login.aspx

Mitchell VG (2006) Applying integrated urban water management concepts: a review of Australian experience. Environ Manage 37:589–605

Policy Brief, Integrated Urban Water Management IUWM (2013) Toward diversification and sustainability. Global Water Partnership, Global Secretariat, Stockholm, Sweden

Rajput A (2021) Analysing spatial growth in peri-urban areas using spatial metrics: case study of Indore. Nagarlok J LIII:1–15

Rao (2022) https://timesofindia.indiatimes.com/blogs/voices/integrated-urban-water-management/

Rashid H, Manzoor MM, Mukhtar S (2018) Urbanization and its effects on water resources: an exploratory analysis. Asian J Water Environ Pollut 15(1):67–74

Salehi M (2022) Global water shortage and potable water safety; today's concern and tomorrow's crisis. Environ Int

Saleth RM (2011) Water scarcity and climatic change in India: the need for water demand and supply management. Hydrol Sci J 56(4):671–686

Scott Wilson India Pvt. Ltd. (2009) Final plan report—perspective plan for DMIC region. Scott Wilson India Pvt. Ltd., New Delhi. Retrieved from http://www.scottwilsonindia.com/

Tool kit for integrated urban water management (2010) Institute of Resource Analysis and Policy, Hyderabad, India

UNDESA World Social Report (2020) https://www.un.org/development/desa/dspd/world-social-report/2020-2.html

UNDP Annual Report (2006) https://www.undp.org/publications/undp-annual-report-2006

Wu JS (2002) Modern water resources management methodology. China Water Power Press, Beijing (in Chinese)

Part III
Surface Water Resource Modeling and Geospatial Technologies

12 Impact of Hydrological Modifications on Floodplains Wetland: Lower Atreyee River Basin of India and Bangladesh

Rajesh Sarda, Tamal Kanti Saha, and Swades Pal

Abstract

This study explores how damming affects floodplain wetlands in the lower Atreyee River basin (ARB) of India and Bangladesh. We analyzed hydrological shifts in both the river and its adjacent riparian wetland by examining three-hour interval discharge and water level data from 1993 to 2021. Our study specifically concentrated on the periods before and after the construction of a dam in 2012, encompassing pre-monsoon and post-monsoon seasons. To achieve this, we used a variety of methods to understand how damming affects water flow. These methods include box plots, heat maps, and wavelet transformation. We also used the Normalized Difference Water Index (NDWI) was carried out to delineate the water bodies from satellite images of same phase. We checked the accuracy of our wetland maps using the Kappa coefficient. Findings highlight significant differences from pre- to post-dam periods. After damming, the average water discharge decreased (70–168 m^3/sec to 36–68 m^3/sec). The variation in monsoon flow also reduced, indicating a changed natural flow pattern. Discharge rates dropped significantly, and the flow pattern was altered after dam construction. Wavelet power spectrum analysis detected a prominent shift in flow variance after damming. Flow failure rates increased post-dam, especially during lean months. This reduction was due to water diversion through the dam and other extractions, leading to an ecological deficit. After damming, the wetland area decreased from 412.63 km^2 in 2013 to 300.74 km^2 in 2021, indicating a broader loss. Seasonal wetland areas decreased from 361.15 km^2 (1991) to 62.58 km^2 (2021). This study highlights the substantial ecological impact of damming, shown through changed flow dynamics and substantial wetland loss.

Keywords

Hydrological change · Dam construction · Discharge · NDWI · Wetland · Atreyee river

R. Sarda
Department of Geography, Delhi School of Economics, New Delhi 110025, India

T. K. Saha (✉)
Department of Geography, Faculty of Sciences, Jamia Millia Islamia, New Delhi 110025, India
e-mail: tamalkantisaha999@gmail.com

S. Pal
Department of Geography, University of Gour Banga, Malda, West Bengal 732103, India

S. Talukdar et al. (eds.), *Water Resource Management in Climate Change Scenario*, GIScience and Geo-environmental Modelling, https://doi.org/10.1007/978-3-031-61121-6_12

12.1 Introduction

Water resources are essential components that support both natural ecosystems and human activities (Song et al. 2021; Yang et al. 2021). Throughout history, water has been a critical factor in determining where people settle and how societies progress. In today's world, where we use a lot of water, it is extremely important to manage it wisely and protect our water sources for the sake of creating a sustainable and ecologically balanced civilization (Imasiku and Ntagwirumugara 2020). Efficient management of water resources becomes particularly important in regions where water is scarce. This is because the unequal distribution of water sources and the various needs from nature and human activities create challenges. As a result, effective water resource management is necessary to ensure that everyone has enough water to meet their needs. In managing water resources, the needs of agriculture, industry, and cities are often given more attention than the requirements for preserving natural systems such as river habitats, wetlands, and groundwater (Stefanakis 2019; Kookana et al. 2020; Tariq et al. 2021). At times, besides the uneven distribution of rainfall, floods, and droughts caused by nature, certain strategies for managing water resources also lead to unfair treatment of different areas in terms of their environment (Srivastav et al. 2021; Li et al. 2022). Building dams is a major method used globally to manage water resources. Around 50,000 big dams were built on many large rivers such as the Nile, Ganga, Yangtze, Mississippi, according to the World Commission on Dams in 2000. India has witnessed the construction of 5264 major dams, in addition to numerous medium to small-scale dams, as documented by the National Register of Large Dams in 2021. The construction of dams often leads to an increase in water availability in the upstream area while reducing water supply downstream from the dam (Lee et al. 2019; Wang et al. 2020). This can result in the upper region benefiting economically and ecologically at the expense of the lower region (Xu et al. 2020). Therefore, various human-made systems like aquatic environments and agriculture are significantly impacted (Eekhout et al. 2020; Yoshida et al. 2020; Eslami et al. 2021). Wetlands located in the floodplain regions of rivers are severely impacted by dam construction because they rely heavily on rainfall and river water. Therefore, it is crucial to assess the changes in river flow caused by dams and understand how they affect the wetlands in these floodplain areas. This understanding is essential to ensure the balanced relationship between the river, wetland, and human activities.

The construction of dams has not only altered a river's hydrological system but has also changed the hydro-ecological patterns and physical characteristics of floodplain wetlands (Pal and Sarda 2021). This includes decreasing the frequency and size of floods, reducing the active floodplain area, and breaking up the wetland landscape (Boulange et al. 2021; Chen et al. 2021; Kundu et al. 2021; Saha et al. 2022). Wetlands stand as remarkable ecosystems, encompassing roughly 6–8% of the Earth's land surface and accounting for approximately 45% of the total economic value across all global ecosystems (Finlayson 2013). Wetlands are facing pressures stemming from alterations in land usage, population growth, shifting lifestyles, and rising demands. Unfortunately, their value has often been overlooked compared to other economic activities (Díaz-Pinzón et al. 2022; Ostad-Ali-Askari 2022). Numerous studies have shown that dam construction negatively impacts riparian zones. While dams are not the sole cause of all these changes, they significantly contribute to altering the hydrological and physical characteristics of riparian wetlands (Pal and Sarda 2022; Zhou et al. 2022). Lately, both the research community and the affected individuals have shown interest in studying this problem and seeking reasonable solutions, given its close connection to ecological survival and sustaining people's livelihoods (Meng et al. 2022; Singha and Pal 2023). Zheng et al. (2019) found that changes in water flow after dam construction along the Nenjiang River in China led to a significant 44% decrease in riparian wetlands. The Atreyee River basin also saw a loss of 45% of its active floodplain due to damming, as reported by Saha and Pal (2019). Similarly, Talukdar and Pal (2020) observed a reduction of the active floodplain by 39.72% following damming in the Punarbhaba river area in India and Bangladesh, pushing the wetland's stress beyond its natural limit. When a wetland's hydrology is negatively affected, its ability to function ecologically, its strength as a natural resource, and its physical traits all undergo changes (Yang et al. 2023). Kundu et al. (2021) pointed out that alterations in water flow have a noticeable impact on the level of wetland fragmentation and the services it provides. Furthermore, shallower water depth encourages the expansion of agriculture within wetlands, contributing to their fragmentation (Islam et al. 2021).

From reviewing existing literature, it is evident that many studies have focused on changes in river flow and their effects on downstream rivers' hydro-ecological and geomorphological aspects (Kuriqi et al. 2021; Vassoney et al. 2021). Some research has also looked into the hydrological changes occurring in riparian wetlands as a result of altering river dynamics (Chen et al. 2021; Pal and Sarda 2022; Zhou et al. 2022). However, as far as current knowledge goes, there's a gap in research connecting changes in river dynamics with alterations in wetland hydrology, particularly using relevant hydrological indicators such as detailed water depth, consistency of water presence, hydroperiod, floodwater connectivity, and more. Understanding the effects of these changes triggered by damming on both river and riparian wetland at a spatial level (pixel scale) is crucial for effective wetland management planning. While some case studies have used on-site data for this purpose, there's a lack of research covering larger geographical areas. Nonetheless, such comprehensive research is vital for effective planning. Hence, this study aimed to investigate the extent of river and adjacent wetland hydrological changes due to damming and their interconnection. The study also questioned whether these hydrological changes could significantly impact wetland morphology and contribute to various transformations.

12.2 Site Description

The Atreyee River (Fig. 12.1), commencing its journey in West Bengal's Jalpaiguri district, stretches across a distance of 390 km before eventually joining the Brahmaputra River, known as the Jamuna in Bangladesh (Adel 2013). The Atreyee River basin encompasses various regions, including parts of the Teesta mega fan, Old Himalayan Piedmont Plain, the Barind and the Ganga floodplain. Covering a significant portion of the basin is the elevated Barind tract (11–48 m above mean sea level), which is a relic of the Pleistocene era. The annual average precipitation within this basin totals around 1500 mm, with an uneven distribution across various seasons. Notably, about 80% of this precipitation takes place during the monsoon season, primarily influenced by the south-westerly monsoon winds. This region, characterized by its reliance on agriculture and the essential role of water, is presently grappling with water shortage (Rashid et al. 2019). With the aim of increasing irrigation resources and improving agricultural productivity in Bangladesh, a Rubber Dam was built in 2012 at Mohanpur, located at latitude 25.8320°N and longitude 88.8450°E. This Rubber Dam spans 135 m in length, stands at a height of 4.5 m, and includes a 12-km river reservoir with a capacity of 729,000 cubic meters (Fig. 12.1). The establishment of this structure has induced substantial shifts in the hydrological patterns, affecting both the regions upstream and downstream of the reservoir. Essentially, the reservoir operates akin to a dam, redirecting water to support irrigation needs.

12.3 Materials and Methods

12.3.1 Materials Used

Data spanning a period from 1993 to 2021, consisting of discharge and water level measurements at three-hour intervals, have been gathered from the Joda bridge gauge station situated

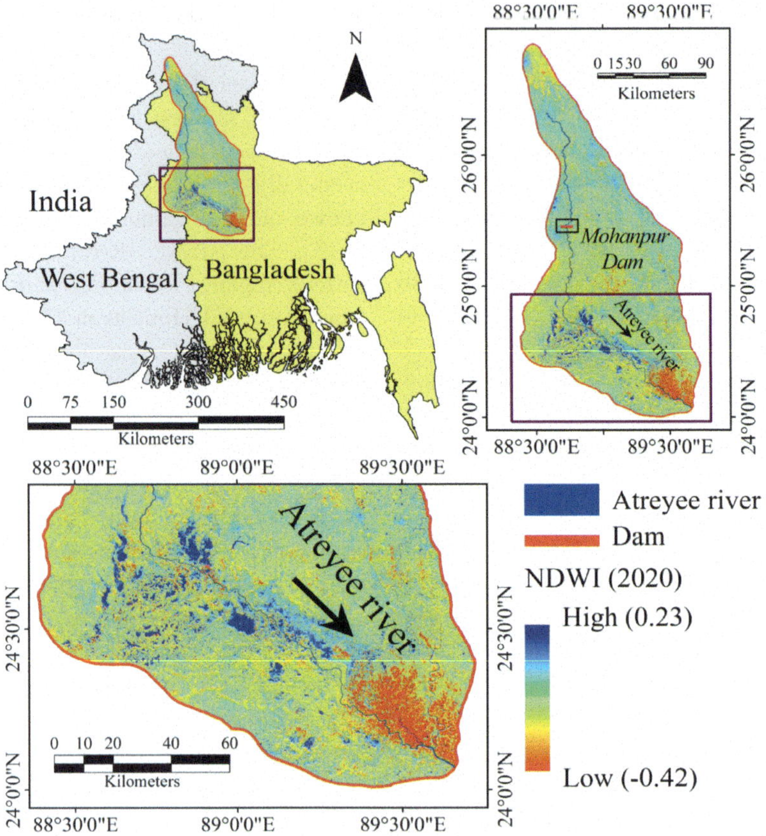

Fig. 12.1 Location map of the study region

in Balurghat, under the Irrigation & Waterways Department of the West Bengal Government. This data are being used to analyze the hydrological and ecological patterns across various temporal scales, ranging from very minute to larger timeframes. Considering the seasonal fluctuations characterizing the wetlands in the downstream riparian zone of the Atreyee River basin, a collection of multi-temporal Landsat images has been chosen to cover both pre-monsoon and post-monsoon periods. These images aim to capture the changing characteristics of the wetlands in the periods before dam construction (prior to 2012) and after the dam was built (after 2012). This involves a total of 51 Landsat images for the both seasons, acquired from the United States Geological Survey's Global Viewer (GloVis). To mitigate the influence of seasonal variations, this study employed cloud-free satellite images captured by Landsat TM and OLI sensors, ensuring a more accurate representation of the data (Table 12.1). A few satellite images that had substantial cloud cover were excluded from the study due to their limited usability.

12.3.2 Hydrological State of the River

Through the utilization of a box plot and whisker range, the river's hydrological fluctuations are visually depicted. To assess the extent of flow changes before and after dam construction, the coefficient of variation (CV) for the average discharge is computed for each month across the considered years. A heat map is employed to illustrate the shifts in river flow caused by the dam. This map displays the alteration in discharge data over time as a percentage

Table 12.1 The dataset description utilized to extract the wetland in the lower Atreyee river basin

Season	Period	Path/Row	Satellite	Sensor	Acquisition date
Pre-monsoon (March to May)	Pre-dam	138/43	Landsat-5	TM/ETM	16–03–1989, 22–03–1991, 08–03–1992, 02–04–1995, 03–03–1996, 06–03–1997, 09–03–1998, 28–03–1999, 09–03–2004, 13–04–2005, 18–03–2007, 04–03–2008, 23–03–2009, 02–03–2010
	Post-dam				05–03–2014, 24–03–2015, 11–04–2016, 14–04–2017, 01–04–2018, 03–03–2019, 06–04–2020, 25–04–2021
Post-monsoon (October to December)	Pre-dam		Landsat-8	OLI	08–12–1987, 24–11–1988, 27–11–1989, 29–10–1990, 08–12–1993, 28–11–1995, 16–10–1997, 04–11–1998, 25–12–1999, 17–11–2000, 20–11–2001, 23–11–2002, 18–11–2003, 04–11–2004, 07–11–2005, 25–10–2006, 01–12–2008, 04–12–2009, 23–12–2010, 08–11–2011
	Post-dam				29–11–2013, 31–10–2014, 21–12–2015, 20–10–2016, 08–11–2017, 11–11–2018, 30–11–2019, 16–11–2020, 03–11–2021

(%). Using a color scale, the year-wise percentage change during the post-dam phase is calculated in relation to the average discharge during the pre-dam phase. Additionally, by comparing the river's yearly flow with the mean flow for the same phase in the pre-dam phase, anomalies in flow can be identified.

12.3.3 Flow Periodicity Analysis

The Wavelet transformation technique proves to be immensely valuable for analyzing long-term changes, trends, and patterns in time series data (Pal and Talukdar 2020). This mathematical signal processing method is effective in offering insights into both frequency and time domain aspects of data sets, whether they are stationary or non-stationary. This is a capability that standard methods struggle to provide (Santos et al. 2018). The wavelet transformation method was developed as an enhancement over the Fourier transformation, offering the advantage of simultaneously providing details in both the time and frequency domains (Sifuzzaman et al. 2009). Moreover, this method enables multi-resolution analysis. For instance, at a lower scale of wavelet transformation, it offers precise time resolution, while at a higher scale, the emphasis shifts toward the opposite, namely enhanced frequency resolution. This versatility is essential for time series analyses like changes in time series discharge (Yaseen et al. 2015). The Wavelet transformation method proves especially valuable when handling non-stationary time series data, characterized by shifts in mean, variance, covariance, and autocorrelation over time without returning to their initial states. This approach enjoys broad acceptance and application, particularly in the analysis of hydro-meteorological data series.

The initial wavelets were defined by Goupillaud et al. (1984) as a set of functions created through the translation and expansion of a single function, referred to as the "mother wavelet." The wavelet transform is characterized by Eq. (12.1).

$$\psi_{x,y}(t) = \frac{1}{\sqrt{|x|}}\psi\left(\frac{t-y}{x}\right); x, y \in R; x \neq 0, \quad (12.1)$$

where the scale parameter is indicated by 'x' that represents the degree of compression, whereas, 'y' represent translation parameter. The 'x' parameter in the mother wavelet will be conditioned by the higher frequencies when $|x|$ will be less than 1. When $|x|$ more than 1 is, then $\psi_{x,y}(t)$ has a larger time width than $\psi(t)$ that will correspond to lower frequencies. Hence, wavelets possess time spans that are adjusted according to their frequencies.

12.3.4 Hydrological Threshold Estimation and Calculation Failure Rate

To determine the crucial water flow threshold, Richter et al. (1996) proposed various dispersion measures like ± 1 or ± 2 standard deviations (SD), 20th and 80th percentiles, 25th and 75th percentiles, and others. We analyzed a comprehensive set of 33 hydrological alteration indicators (IHA) that fall into five distinct categories: (1) duration and magnitude of daily extreme flow conditions, (2) magnitude of monthly flow conditions, (3) frequency and duration of high and low flow pulses, (4) timing of extreme flow conditions, and (5) rate and frequency of changes in flow conditions. These indicators were calculated using IHA software. In this study, the critical flow (threshold flow) was determined using the 25th and 75th percentile methods. For both the periods before and after dam construction, daily data were examined to determine the percentage that falls within the range of threshold flow, above and below the high and low thresholds. This provides insights into the rate of flow failures.

12.3.5 Degree of the Hydrological Alteration in the River Flow

The degree of hydrological alteration indicates the percentage of years during the post-dam phase that align with the river's discharge threshold range. This was determined by assessing the disparity between the observed and anticipated frequency of years falling within the threshold boundaries. Utilizing this data, we computed the extent of hydrological modification using Eq. (12.2).

$$DFA(\%) = \left(\frac{O - E}{E}\right) \times 100, \qquad (12.2)$$

where "O" stands for observed and "E" stand for expected. The degree of modification may assume one of three states: zero, positive, or negative. A positive change implies that the observed value frequently surpasses the expected hydro-ecological threshold limits. A value of zero indicates that the observed value matches the anticipated hydro-ecological threshold limits in the post-dam condition. Conversely, a negative change signifies that observed values less frequently fall within the hydro-ecological threshold limits than expected.

12.3.6 Measuring the Eco-Hydrological States of the River

Eco-deficit and eco-surplus are two holistic measurements used to evaluate the overall effects of modifying water flow in a river and wetland ecosystems. Vogel et al. (2007) introduced these non-dimensional metrics, which are based on the flow duration curve (FDC). To assess the conditions of eco-hydro-deficit and surplus, we constructed Flow Duration Curves (FDCs) for both unaltered natural flow and dam-regulated flow. The space between the natural FDC and the controlled FDC quantifies the reduction in water caused by dam operations. When the natural FDC surpasses the controlled FDC, it is denoted as an eco-surplus. Conversely, if the controlled FDC is above the natural FDC, it's termed an eco-deficit.

12.3.7 Methods of Water Body Delineation from Satellite Images

Numerous spectral indices based on satellite imagery are available for depicting water bodies. Several remote sensing indices are employed for extracting water bodies, such as NDWI by

McFeeters (1996), MNDWI by Xu (2006), Water Ratio Index (Shen and Li 2010), AWEI (Feyisa et al. 2014), among others. These indices are also used for mapping wetlands from satellite images. Considering the suitability of NDWI in the Barind plain of India and Bangladesh, as indicated by studies conducted by Das and Pal (2016), Pal and Sarda (2020) and Singha and Pal (2022) this research employed NDWI (Eq. 3) as the water index for generating wetland inventory maps.

$$NDWI = \frac{Green - NIR}{Green + NIR} \quad (12.3)$$

Conceptually, NDWI values lower than 0 are not indicative of a wetland, values near zero signify a saturated or shallow water body, and values close to one suggest a water body with abundant water content.

12.3.8 Methods of Wetland Map Validation

To validate the accuracy of the wetland maps, we calculated the Kappa coefficient and overall accuracy (Eq. 12.4) using Google Earth images (262 sites) and ground control points (138 sites).

$$k = \frac{P\sum_{i=1}^{N} xii - (J_i + *J_{i+i})}{P^2 \sum_{i=1}^{N} - (J_i + *J_{i+i})}, \quad (12.4)$$

where N = number of rows in the matrix; P = total number of pixels Xii = number of observations in row i and column i; J^{i+} and J_{+i} are the marginal totals for row i and column i.

12.4 Results

12.4.1 Hydrological Modification

12.4.1.1 Hydrological Variability of the River

Figure 12.2 presents a comparative analysis through box plots depicting the time series of mean, minimum and maximum discharge for both periods, categorized into monsoon and summer months. The insights drawn from these graphs are quite clear: Firstly, aligning with the skewed rainfall pattern (0.38) in the region, the discharge also exhibited a high level of skewness (ranging from 0.79 to 0.92). Notably, during the post-dam phase, monsoon months exhibited significantly higher minimum (60.28 m^3/sec), maximum (531 m^3/sec), and mean (143.54 m^3/sec) discharges compared to summer months (43.72 m^3/sec, 66.76 m^3/sec, and 53.86 m^3/sec). Secondly, the degree of fluctuation and the interquartile range (IQR) for monsoon rainfall were notably greater than those observed in dry months. Furthermore, the identification of certain extreme discharge conditions was observed in both monsoon and non-monsoon months, although such instances were less frequent during the post-dam phase. Another noteworthy finding was that the mean discharge volume during the post-dam phase (37 to 69 m^3/sec) was significantly lower than that observed in the pre-dam phase (71 to 169 m^3/sec). Additionally, the variation in monsoon flow, as calculated using the coefficient of variation (CV), showed a distinct decrease in the post-dam phase (36 to 91%) compared to the pre-dam phase (54 to 133%). This reduction in variation pointed toward a flattening of the natural flow rhythm. The loss of such natural rhythm, coupled with discharge reduction, raises ecological concerns. Moreover, this phenomenon of flow flattening extended even to non-monsoon seasons, where the variation in the pre-dam was recorded at 24 to 53%, while in the post-dam phase, it was as low as 0.49 to 13.5%.

12.4.1.2 Changes in the River's Hydrology

The heat map illustrates the patterns of average, minimum, and maximum, discharge along with the rate of discharge change. Notably, in the phase before damming, variations in discharge (both positive and negative) were observed, with a more pronounced impact on minimum discharge, particularly during monsoon months. However, after the dam's construction, a distinct and consistent reduction in discharge occurred, without any anomaly incidents (Fig. 12.3).

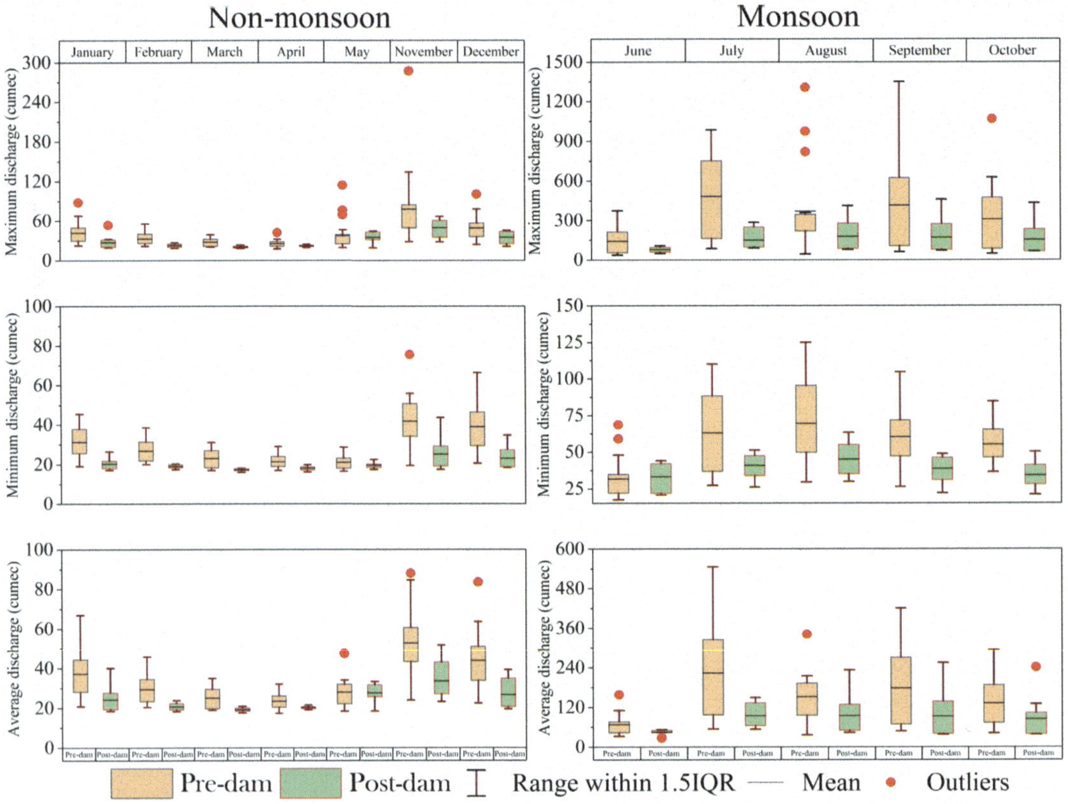

Fig. 12.2 Comparative dispersion of average, minimum, and maximum discharge in both phases in seasonal scale

Moreover, the post-dam conditions displayed a significant decline in discharge rates. Specifically, these rates reached values of up to 91.9% (max), 66.49% (min), and 74.9% (average) respectively (Fig. 12.3). This trend was particularly noticeable during the monsoon months. Interestingly, the heat maps also revealed instances of discharge increment rates; however, these were exclusively observed in the pre-dam phase. This implies that there were several years when the average, minimum, and maximum discharge surpassed their respective averages.

12.4.1.3 Periodicity of Flow

Figure 12.4 displays the continuous wavelet power spectrum (wps) of average discharge across different seasons: pre-monsoon, post-monsoon, monsoon and winter. After the installation of the dam in 2012, significant periodic patterns were identified in the 3–5 years' frequency range, particularly during various seasons. The wavelet power spectrum indicated the highest power, representing the variability in flow, within the 3–4 years' frequency range from 2013 to 2015. This suggests a change in the behavior of river flow, albeit with varying degrees. During the summer months, we observed notable power patterns in the 2–3 years frequency range between 2013 and 2015 (see Fig. 12.4). Moving on to the monsoon season, distinct wavelet power patterns emerged within the 2–8 years frequency range from 2000 to 2009 and the 16–30 years frequency range from 2013 to 2017 (refer to Fig. 12.4). Shifting our focus to the post-monsoon season, we identified three noteworthy patterns, with particular emphasis on a relatively stronger and significant pattern in the 2–2.8 years frequency range between 1995 and 2000, as well as the 4–7 years frequency range from 2013 to 2015

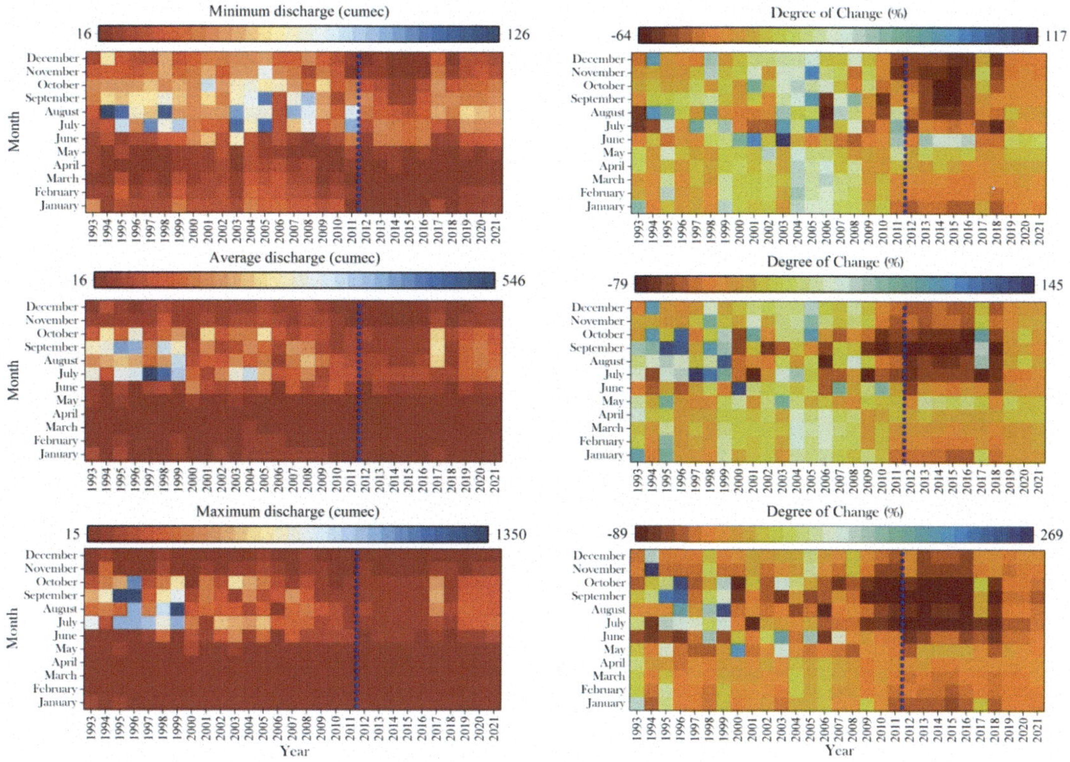

Fig. 12.3 Flow alteration showing by heat map about absolute minimum, average, and maximum discharge, and their change rate

(see Fig. 12.4). Lastly, during the winter season, several robust patterns began to manifest themselves after the year 2012.

12.4.1.4 The River's Hydrological Threshold and Its Rate of Failure

The river flow's eco-hydrological threshold for each month was categorized into three groups: (1) Optimal natural or ecological flow conditions, situated between the 25th and 75th percentiles, (2) occurrence of failure rates surpassing the upper threshold (beyond the 75th percentile), and (3) instances of failure rates falling below the lower threshold (below the 25th percentile). Figure 12.3 provides a visual representation of the monthly river flow patterns spanning from 1993 to 2021. This figure emphasizes the delineation of upper and lower threshold boundaries to distinguish flows within the optimal range from occurrences of flow failure. Notably, the post-dam phase (after 2012) indicates a notable rise in the flow failure rate. Examining seasonal shifts in flow failure rates, pre-and post-monsoon months in the lean phase consistently exhibit a high failure rate (> 70%) with minimal variance (Table 12.2). This heightened failure rate can be attributed to factors such as inadequate rainfall and substantial water diversion from dam storage units, primarily for irrigation purposes.

In Table 12.1, the computed thresholds (both upper and lower limits) for discharge based on RVA are displayed for monthly average discharge. In the phase before the dam was constructed, the monthly average discharges consistently fell within the defined threshold limits. However, in the phase after the dam's construction, a noticeable trend emerged where many months exhibited maximum discharge values either below the lower threshold limit or very near to it.

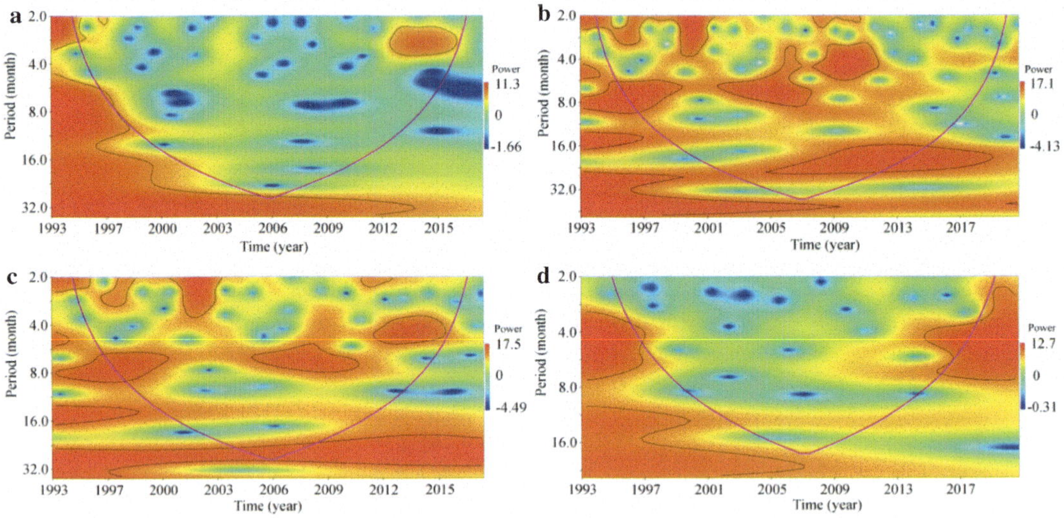

Fig. 12.4 Flow periodicity analysis of the Atreyee river using wavelet transformation

Table 12.2 Monthly average discharge with threshold, failure rate, and degree of alteration

Month	Mean discharge (m^3/sec)		RVA thresholds (m^3/sec)		Flow failure rate (%)	Hydrological alteration (HA)
	Pre-dam	Post-dam	Low	High		
January	37.20	24.19	28.37	43.08	90	– 78.89
February	29.42	20.66	24.00	33.44	100	– 100
March	25.11	19.18	21.30	28.77	100	– 100
April	23.40	20.34	20.53	25.85	60	– 15.56
May	28.01	27.56	22.19	31.90	40	– 26.67
June	67.48	44.55	49.11	75.16	60	– 15.56
July	223.42	94.22	99.62	316.44	70	– 36.67
August	151.80	94.35	100.22	190.90	70	– 36.67
September	177.53	92.35	68.39	270.08	60	– 15.56
October	131.87	84.05	77.18	170.83	70	– 36.67
November	52.64	33.54	43.49	60.05	80	– 57.78
December	43.90	26.64	33.87	50.50	70	– 36.67

Figure 12.5 depicts that higher failure rates (60–70%) were observed during monsoon months. This is unusual without some level of climatic change and variability. However, the unpredictable nature of monsoons can often lead to misleading dam regulation during this phase, potentially contributing to this situation. Similarly, the early cultivation of paddy crops may lead to dependency on dam water. In the phase after the dam's construction, flow failures below the lower RVA threshold became quite common throughout all months. Particularly in February and March, a 100% failure rate was recorded. This highlights the adverse impact of damming on hydro-ecological conditions. Consequently, these findings serve as a significant warning for riparian stakeholders, emphasizing the need to consider both the biotic and

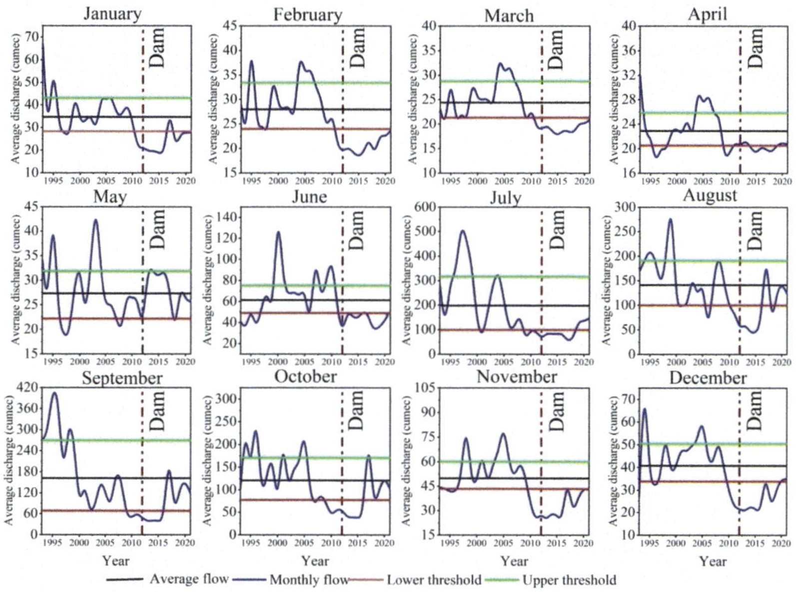

Fig. 12.5 Monthly discharge conditions in relation to the upper and lower RVA thresholds

abiotic aspects of hydro-ecological subsystems, as well as socio-economic factors, as highlighted by various well-known scholars.

12.4.1.5 Magnitude of Hydro-Ecological Alteration Effects

The hydrological indicators displayed specific negative hydrological alterations, with impact degrees ranging from − 0.14 to − 1. A negative HA degree in the range of 0 to − 1 indicates a more significant level of negative alteration. In the context of rivers, the degree of HA for different IHA groups ranged from − 0.66 to − 0.82 (Table 12.3). All of these findings suggest potential adverse eco-hydrological impacts, indicating the potential for increasing ecological distress.

Table 12.2 presents a spectrum of negative degrees of hydrological alteration (ranging from 15.57 to 100%) in the monthly average discharge of the Atreyee River. These alterations contribute to the escalating eco-deficit within the river ecosystem. Pal et al. (2019) underscored the significance of stream flow modification, which not only affects the river's ecosystem but also leads to hydro-ecological deficit in the riparian wetlands and floodplain. Consequently, it is imperative to take precautionary measures to prevent further degradation of hydro-ecological conditions.

Table 12.3 Average level of hydrological alteration (HA) within various IHA (Indicators of Hydrologic Alteration) categories

IHA group/parameters	Hydrological alteration	Remarks
Group 1	− 0.69	High HA
Group 2	− 0.70	
Group 3	− 0.74	
Group 4	− 0.82	
Group 5	− 0.66	

Note Based on Richter et al. (1997) suggested classes

12.4.2 The Ecological Condition of the River's Hydrology

Figure 12.6 illustrates the monthly flow duration curves (FDCs) under two distinct conditions: natural/unregulated (pre-dam phase) and controlled/regulated (post-dam phase). These

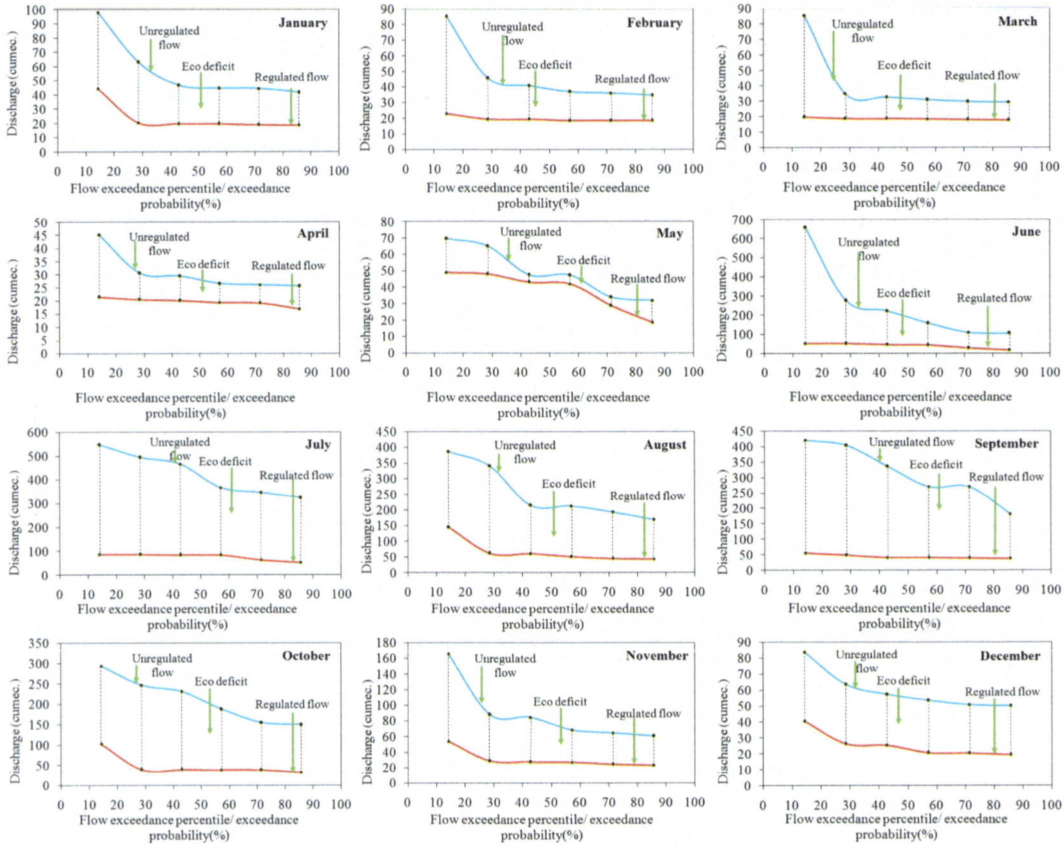

Fig. 12.6 Ecological condition of the Atreyee river using FDC curve

curves highlight the ecological surplus and deficit in river flow. Across all months, it is consistently evident that the FDC during the pre-dam phase consistently exceeded that of the post-dam phase (indicating an eco-hydro-deficit). This means that at each level of flow probability, the discharge during the pre-dam phase was higher compared to the post-dam flow levels. This difference was most pronounced during the monsoon and post-monsoon seasons (as depicted in Fig. 12.6). For instance, in July, the discharge exceeded 300 m^3/sec in the unregulated condition, but dropped to less than 100 m^3/sec under regulated conditions. Such alterations often lead to species loss, reduced habitat suitability, and compromised breeding and spawning areas, as previously discussed by Gain and Giupponi (2014).

The monthly flow duration curves (FDCs) serve as a visual representation of ecological deficit and surplus in river flow. In these curves, the blue line represents unregulated flow, while the red line depicts regulated flow. The vertical black line demarcates the disparity between pre-dam and post-dam flow conditions. This observable difference between unregulated and regulated flow, as illustrated, symbolizes the ecological deficit.

Alterations in flow that surpass the ecological threshold can result in vulnerabilities within habitats and ecosystems, causing heightened stress and hydrological impoverishment (Belmer et al. 2013; Sabater and Tockner 2010). The observed eco-deficit across all months, depicted in Fig. 12.6 with reference to the pre-dam flow conditions, underscores the emerging state of hydro-ecological impoverishment. A notable decline in discharge, attributed to water diversion through the dam and other extraction practices, emerges as a primary driver behind the increasing eco-deficit condition. The pattern of ecological deficit resulting from flow modifications is consistent with research outcomes from various studies, including those by Lin et al. (2014) studying the Dongjiang River, Li et al. (2017) regarding the Mekong River, and Wang et al. (2017) concerning the Yangtze River. Sima et al. (2021) have also reported that reduced flow and the consequent growth of eco-deficit have contributed not only to ecological degradation within the river but also within the riparian floodplain wetlands. The consequences of this increasing ecological deficit are evident in the form of reduced wetland habitats, shallower water depths, shortened availability of water, and heightened uncertainty in wetland hydrological dynamics, as noted by Pal and Sarda (2022), Pal and Singha (2023), and Wang et al. (2023). While this study doesn't directly investigate the impact on ecological species, previous research suggests a simplification of species diversity and species loss. For instance, Gain and Giupponi (2014) highlighted how the diversion of water through the Farakka barrage in 1975 on the Ganga River resulted in the destruction of breeding and hunting grounds for 109 species of Gangetic fishes, amphibians, and other aquatic species. Additionally, Hossain and Haque (2005) discovered that over 50 species became rare in the Bangladesh region in the aftermath of the post-Farakkaphase.

12.4.3 Seasonal Change of Wetland Area

In Fig. 12.7, you can observe the wetland areas for specific years captured through NDWI during both pre and post-monsoon seasons. This provides insight into how wetland areas change over time and across seasons. The analysis of wetland area changes before the dam was constructed showed a decrease in wetland coverage. Interestingly, after the dam was built, the rate of wetland reduction significantly intensified.

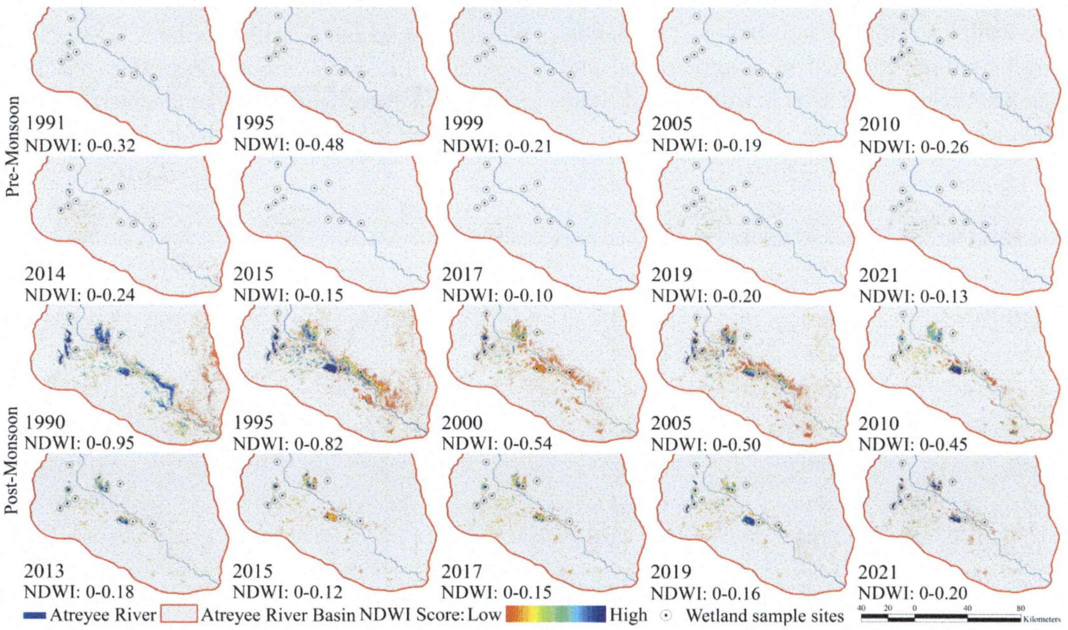

Fig. 12.7 Seasonal change of Wetland area of the lower Atreyee river basin

For example, during the year 1990, the post-monsoon wetland area measured 924.79 km^2, but by 2010, it had decreased to 541.66 km^2. This indicated a loss of 383.12 km^2 of wetland within this pre-dam phase (Table 12.4). Following the construction of the dam, the wetland area was 412.62 km^2 in 2013 and further reduced to 300.73 km^2 by 2021, reflecting a broader decline in wetland coverage.

It's worth noting that while the monsoon season typically witnesses the most significant expansion of floodplain seasonal wetland area, this study primarily focused on the post-monsoon season due to the limited availability of cloud-free images during the monsoon. The trend of wetland loss during the pre-monsoon season remained consistent in both periods, as illustrated in Fig. 12.7. Seasonal wetland areas measured 361.14 km^2 in 1991, 197.9 km^2 in 2010, 154.22 km^2 in 2014, and 62.57 km^2 in 2021, indicating considerable seasonal fluctuations. This also highlighted that the pre-dam phase experienced more significant fluctuations in wetland area compared to the post-dam phase. Significantly, the wetland areas with greater depth exhibited lower seasonal fluctuations. Among the regions affected by wetland reduction, the ones most severely impacted included fringe wetlands, smaller and shallower wetlands, as well as those situated near agricultural and urbanized areas, all of which witnessed a more pronounced rate of decline.

12.4.4 Validation of the Wetland Map

Table 12.4 presents the validation results for the NDWI images depicting water bodies, including overall accuracy and the Kappa coefficient. The findings indicate that across all years, the overall accuracy exceeded 88%, with Kappa coefficients exceeding 0.81 (Table 12.5). These values suggest a remarkably strong agreement between the generated maps and the actual ground conditions (Monserud and Leemans, 1992). Consequently, the wetland maps generated using NDWI can be considered highly valid and reliable.

Accurate delineation of wetlands plays a critical role in implementing effective management strategies to mitigate the rapid changes observed in wetland areas, especially within floodplain regions (Garg 2015). The size of floodplain or seasonally flooded wetlands tends to vary annually during both the pre- and post-monsoon periods. This underscores the importance of ongoing monitoring to enhance management efforts. Previous studies (Li et al. 2021; Mao et al. 2021) have primarily focused on quantifying wetland fluctuations in numerical terms, which might not fully capture the spatial complexities associated with wetland loss. Such numerical analyses might not offer the best foundation for effective management strategies. Thus, assessing wetland areas at the pixel level

Table 12.4 Season-specific wetland area in pre and post-damming year

Season	Year	Area	Season	Year	Area
Pre-monsoon	1991	361.14	Post-monsoon	1990	924.79
	1995	260.25		1995	780.8
	1999	401.45		2000	1070.85
	2005	245.85		2005	813.25
	2010	197.9		2010	541.66
	2014	154.22		2013	412.62
	2015	112.98		2015	358.96
	2017	89.98		2017	384.12
	2019	77.85		2019	321.76
	2021	62.57		2021	300.73

Table 12.5 Overall accuracy and Kappa coefficient values of NDWI water indices in pre and post-monsoon seasons

Phase	Year	Pre-monsoon			Year	Post-monsoon		
		Overall accuracy	Kappa coefficient	Degree of agreement		Overall accuracy	Kappa coefficient	Degree of agreement
Pre-dam	1991	88	0.81	Very good	1990	94.25	0.91	Excellent
	1995	94	0.88	Excellent	1995	94	0.88	Excellent
	1999	91.33	0.82	Very good	2000	92	0.82	Very good
	2005	93.77	0.87	Excellent	2005	94.67	0.89	Excellent
	2010	95.67	0.91	Excellent	2010	96.33	0.92	Excellent
Post-dam	2014	92.6	0.85	Excellent	2013	93.67	0.87	Excellent
	2015	91	0.81	Very good	2015	95	0.9	Excellent
	2017	92.33	0.85	Excellent	2017	93	0.87	Excellent
	2019	96.46	0.95	Excellent	2019	98.4	0.96	Excellent
	2021	94.44	0.92	Excellent	2021	95.6	0.91	Excellent

Note If kappa coefficient value is < 0.4 signifies poor agreement; 0.40–0.55 indicates fair agreement; 0.55–0.70 signifies good agreement; 0.7–0.85 indicates very good agreement; > 0.85 depicts Excellent agreement (Monserud and Leemans 1992)

can provide more accurate and reliable results, benefiting both numerical precision and spatial understanding, which is of great value in terms of management considerations.

12.5 Discussion

The findings of this study highlight the significant discharge attenuation (ranging from 16 to 75%) following damming, which has had a notable impact on river systems, extending its influence to nearby wetlands. This alteration has resulted in flow failure rates exceeding 85%, a consistent trend observed both before and after the monsoon seasons. Furthermore, the variability in flow has diminished in the post-dam phase, reflecting a more regulated flow regime. The degree of alteration within different Hydrological Indicator-Specific Hydrological Alteration (IHA) groups ranged from − 0.67 to − 0.83, indicating a consistent negative alteration trend. Moreover, in the wetland areas, water depth failure rates experienced a significant increase of 28% after damming, contributing to the growing Hydrological Alteration (HA) in those regions. Both the river and wetland ecosystems exhibited eco-hydro-deficit conditions, unmistakably pointing to an increasing ecological distress. Yet, upon conducting a comparative assessment of the influence of hydrological alteration (HA) and its ensuing outcomes on rivers and wetlands, it becomes apparent that the ramifications of dam construction were more prominently observed within the river ecosystem, surpassing their impact on wetlands.

The phenomenon of flow attenuation observed in the Atreyee River in this research is consistent with findings from analogous studies, including those conducted by Pal (2016) on the Mayurakshi River in India, as well as the Atreyee River in both India and Bangladesh, as well as by Nowreen et al. (2020) on a river in a yet unspecified location. The impact of damming has brought about significant alterations in both the river and the surrounding riparian environment's hydrological patterns (Chen et al. 2021; Pal and Sarda 2022). While the general trend of findings corresponds with these studies, there are variations in the extent of alteration observed. The pace of change is chiefly determined by factors like inflow, outflow, reservoir storage levels, and the specific strategies employed in dam control (Yu et al. 2023). Within floodplain regions, characterized by fertile soil and ample moisture for agriculture, particularly in the context of paddy cultivation, a significant amount of water is frequently redirected for irrigation needs. This diversion plays a role in the reduction of water flow in downstream sections of the river (Pal and Sarda 2021; Ji et al. 2022). The timing and extent of this impact are heavily contingent on the ratio between inflow and outflow, as well as the tactics employed for managing stored water. Furthermore, the increasing demand for water usage in urban and industrial settings also plays a role in diminishing water flow within rivers (Hoque et al. 2022; Zehtabian et al. 2023), although this study did not specifically analyze the proportion of water used for industrial, municipal, and agricultural purposes apart from dam diversion. In this study, it was observed that the degree of the coefficient of variation (CV) during the pre-dam phase was significantly greater than that during the post-dam phase, suggesting a flattening of the flow dynamics. A parallel discovery was reported by Pal et al. (2019) in their research on the Tangon River in the Barind tract, which encompasses both India and Bangladesh. Conversely, studies by Gierszewski et al. (2020) suggested that the range of flow variability increased in the post-dam phase. This discrepancy primarily hinges on the flow regulation policies implemented by dam control agencies and the natural flow-contributing factors.

Recurring occurrences of flow reduction below the lower threshold flow limit were conspicuous in the Atreyee River, leading to the expansion of an ecological deficit and subsequent adverse repercussions on the survival of ecological species. In a similar vein, Latrubesse et al. (2021) documented a decline in biota vulnerability resulting from dam construction in Amazonian rivers. Intriguingly, there is currently no existing research indicating an increase in ecological surplus following dam construction. In proximity to the Atreyee River, Talukdar and Pal (2019) reported a significant rise in ecological deficit in the Punarbhaba River as a consequence of damming. Likewise, the Yangtze River in China, the Mekong River in Myanmar, and the Tangon River in India and Bangladesh also showed a consistent trend of ecological deficit, as observed by Wang et al. (2017), Li et al. (2017), and Pal et al. (2019).

Numerous research endeavors, including Gain and Giupponi's study (2014) on the Ganga River downstream of the Farakka barrage, Ren et al. investigation (2018) on the Wei River in China's Guanzhong Plain, and Dong et al.'s research (2019) on the upper Yangtze River, have brought attention to cases of eco-hydro-deficit within river systems. In these cases, water scarcity disrupts crucial ecological processes like breeding, spawning, and maintaining suitable habitats. In a study conducted by Kumar and Jayakumar (2020), they noted ecological degradation in the Krishna River attributed to reduced water availability. Similarly, the decrease in water availability resulted in the total loss of critical ecological species within Hakaluki Haor in Northeastern Bangladesh, as documented by Aziz et al. (2021). Although this study didn't quantitatively assess species loss or frequency, the insights from local inhabitants pointed toward the reduction in fish and biodiversity due to decreased flow. While numerous studies have reported eco-deficit resulting from damming, only a limited number have delved into the impact of damming on wetland eco-hydrological conditions. Importantly, the findings of the present study align with those of Biswas and Pani (2021), Khatun et al. (2021), and Wu et al. (2021), further emphasizing the significance of this issue.

Riparian wetlands share a strong interdependency with the hydrological regime of their associated rivers. Consequently, any reduction in water flow caused by dams can significantly impact these wetlands. In the specific study area under consideration, the wetlands primarily receive water from rainfall and the adjacent river. As a result, the decrease in water supply from the river after damming directly leads to a decline in water levels within the dependent wetlands. However, there exists a notable distinction in the impact experienced by the river and the wetlands. The river bears a higher level of impact, which is to be expected as it directly encounters the consequences of dam construction. The degree of alteration and impact was notably more pronounced in the river, exceeding

0.7. The wetlands, too, experienced significant effects, although in certain areas the rate of depth modification wasn't as drastic. In terms of quantifying the alteration, the study assessed all five IHA groups for the river, whereas the wetlands were evaluated solely in terms of magnitude (group 1). Consequently, the comparison was limited to group 1, showcasing how the impact was distributed across both the river and the wetland ecosystems.

In the typical scenario, the primary conduits through which water is conveyed from the river to the wetlands encompass tie channels, subterranean seepage pathways, and surface water spillage (Acreman et al. 2022; Pal et al. 2023). A reduction in river flow levels and discharge plays a significant role in limiting the exchange of water between the river and its neighboring regions, which becomes evident through phenomena such as seepage and flooding, as elucidated by Pradhan et al. (2021). The diminished frequency and magnitude of floods result in the contraction of the active floodplain area, thereby curtailing the provision of water services to the riparian wetlands. Sustaining a steady water supply to wetlands can positively impact water availability, increase water depth, and extend the wetland's hydrological duration, as discussed in studies by Pal and Sarda (2021) and de Morais et al. (2022). This, in turn, fosters a more favorable ecological environment and amplifies ecological productivity. In the aftermath of dam construction, a situation arose where river flow failure rates were consistently observed, often surpassing 80%. The consequences of this diminished water supply extended to the wetlands nourished by the river. The decline in water availability within the wetland gave rise to a water crisis, resulting in an increasing failure rate across substantial portions of the wetland area to achieve ecologically significant threshold water depths. As a result, this escalated the influence of hydrological alteration on the eco-hydrological system, intensifying the eco-hydro-deficit across significant sections of the present wetlands. In a related context, Kundu et al. (2021) discerned a substantial connection between the hydrological condition and the ecological functionality of wetlands. The level of impact is significant, to the extent that no single pixel exhibited an eco-hydro-surplus condition, emphasizing that none of the pixels achieved the ecologically relevant threshold water depth. Tie channels as highlighted by Pal et al. (2023) plays a crucial role in supplying water to the surrounding wetlands. In this particular instance, the decrease in water levels in the main river not only diminished the flow of water to the adjacent channel but also completely halted it. Consequently, this led to the unavailability of water in the wetland and its subsequent ecological deterioration. A similar situation of hydrological and ecological deprivation was documented by Khatun et al. (2021).

Additionally, it is vital to recognize that water from the river, in its multiple facets, serves not only as a direct water source but also fulfills a pivotal role in delivering a wide array of water-related services. These services encompass the provision of fish seeds, nutrient inputs, facilitation of species exchange, and the purification of pollutants from wetlands, among other functions (Funk et al. 2019; Grizzetti et al. 2019; Sarkar et al. 2020). These processes establish a mutually beneficial relationship between river and wetland ecosystems, supporting the preservation of natural ecological equilibrium and functioning (Grizzetti et al. 2019; Meng et al. 2019). The decrease in discharge, reduced occurrence of floods, shrinking flood-spilling regions, and the weakening of hydrological connectivity between the river and the wetland in significant parts of the study area have all played a part in the eco-hydrological changes observed within the wetlands. Particularly, wetlands located farther away from the main rivers, which are hydrologically isolated, have experienced a more pronounced transformation in their eco-hydrological dynamics.

12.6 Conclusion

In conclusion, this study delved into the intricate relationship between hydrological modifications resulting from damming and their impact on the Lower Atreyee River Basin's floodplain wetlands in India and Bangladesh. The post-dam phase demonstrated a significant reduction in mean discharge volume compared to the pre-dam phase, signifying the substantial impact of damming on river flow. Notably, the variation in monsoon flow, as quantified by the coefficient of variation (CV), indicated a flattening of the natural flow rhythm in the post-dam era. The rate of discharge attenuation in the post-dam phase was prominent. Wavelet power spectrum analysis revealed changes in flow nature. Even the flow failure rate was consistently high during the lean months of the post-monsoon and pre-monsoon periods, indicating a pronounced effect of damming. The wetland area exhibited a noticeable decline in both pre-dam and post-dam periods, with a higher rate of reduction observed after damming. These findings underscore the vulnerability of wetlands to hydrological modifications. In summary, this study highlights how changes in water flow caused by building dams can have a big impact on both rivers and the nearby wetlands. These changes affect the environment a lot and need to be taken seriously when making plans to manage these important ecosystems. The information we learned from this research helps us better understand how water and nature work together in the Lower Atreyee River Basin. It also gives us important ideas about how changes in water flow can affect other places like this all over the world.

References

Acreman M, Casier R, Salathe T (2022) Evidence-based risk assessment of ecological damage due to groundwater abstraction; the case of Doñana natural space, Spain. Wetlands 42(7):63

Adel MM (2013) Upstream water piracy, the strongest weapon of cornering a downstream nation. Environ Ecol Res 1(3):85–128

Aziz MSB, Hasan NA, Mondol MMR, Alam MM, Haque MM (2021) Decline in fish species diversity due to climatic and anthropogenic factors in Hakaluki Haor, an ecologically critical wetland in northeast Bangladesh. Heliyon 7(1)

Belmar O, Bruno D, Martínez-Capel F, Barquín J, Velasco J (2013) Effects of flow regime alteration on fluvial habitats and riparian quality in a semiarid Mediterranean basin. Ecol Ind 30:52–64

Biswas SS, Pani P (2021) Changes in the hydrological regime and channel morphology as the effects of dams and bridges in the Barakar river, India. Environ Earth Sci 80:1–20

Boulange J, Hanasaki N, Yamazaki D, Pokhrel Y (2021) Role of dams in reducing global flood exposure under climate change. Nat Commun 12(1):417

Chen L, Wu Y, Xu YJ, Zhang G (2021) Alteration of flood pulses by damming the Nenjiang river, China-implication for the need to identify a hydrograph-based inundation threshold for protecting floodplain wetlands. Ecol Ind 124:107406

Das RT, Pal S (2016) Identification of water bodies from multispectral landsat imageries of Barind tract of West Bengal. Int J Innovative Res Rev 4(1):26–37

de Morais M, Abdo MSA, dos Santos C, Sander NL, da Silva Nunes JR, Lázaro WL, da Silva CJ (2022) Long-term analysis of aquatic macrophyte diversity and structure in the Paraguay river ecological corridor, Brazilian Pantanal wetland. Aquat Bot 178:103500

Díaz-Pinzón L, Sierra L, Trillas F (2022) The Economic value of wetlands in urban areas: the benefits in a developing country. Sustainability 14(14):8302

Dong Q, Fang D, Zuo J, Wang Y (2019) Hydrological alteration of the upper Yangtze river and its possible links with large-scale climate indices. Hydrol Res 50(4):1120–1137

Eekhout JP, Boix-Fayos C, Perez-Cutillas P, de Vente J (2020) The impact of reservoir construction and changes in land use and climate on ecosystem services in a large Mediterranean catchment. J Hydrol 590:125208

Eslami V, Ashofteh PS, Golfam P, Loáiciga HA (2021) Multi-criteria decision-making approach for environmental impact assessment to reduce the adverse effects of dams. Water Resour Manage 35(12):4085–4110

Feyisa GL, Meilby H, Fensholt R, Proud SR (2014) Automated water extraction index: a new technique for surface water mapping using Landsat imagery. Remote Sens Environ 140:23–35

Finlayson CM (2013) Climate change and the wise use of wetlands: information from Australian wetlands. Hydrobiologia 708(1):145–152

Funk A, Martínez-López J, Borgwardt F, Trauner D, Bagstad KJ, Balbi S, Magrach A, Villa F, Hein T (2019) Identification of conservation and restoration priority areas in the Danube river based on the multi-functionality of river-floodplain systems. Sci Total Environ 654:763–777

Gain AK, Giupponi C (2014) Impact of the Farakka dam on thresholds of the hydrologic flow regime in the lower Ganges river basin (Bangladesh). Water 6(8):2501–2518

Garg JK (2015) Wetland assessment, monitoring and management in India using geospatial techniques. J Environ Manage 148:112–123

Gierszewski PJ, Habel M, Szmańda J, Luc M (2020) Evaluating effects of dam operation on flow regimes and riverbed adaptation to those changes. Sci Total Environ 710:136202

Goupillaud P, Grossmann A, Morlet J (1984) Cycle-octave and related transforms in seismic signal analysis. Geoexploration 23(1):85–102

Grizzetti B, Liquete C, Pistocchi A, Vigiak O, Zulian G, Bouraoui F, De Roo A, Cardoso AC (2019) Relationship between ecological condition and ecosystem services in European rivers, lakes and coastal waters. Sci Total Environ 671:452–465

Hoque MM, Islam A, Ghosh S (2022) Environmental flow in the context of dams and development with special reference to the Damodar valley project, India: a review. Sustain Water Resour Manage 8(3):62

Hossain MA, Haque MA (2005) Fish species composition in the river Padma near Rajshahi. J Life Earth Sci 1(1):35–42

Imasiku K, Ntagwirumugara E (2020) An impact analysis of population growth on energy-water-food-land nexus for ecological sustainable development in Rwanda. Food Energy Secur 9(1):e185

Islam ARMT, Talukdar S, Mahato S, Ziaul S, Eibek KU, Akhter S, Pham QB, Mohammadi B, Karimi F, Linh NTT (2021) Machine learning algorithm-based risk assessment of riparian wetlands in Padma river basin of Northwest Bangladesh. Environ Sci Pollut Res 28:34450–34471

Ji H, Chen S, Jiang C, Fan Y, Fu Y, Li P, Liu F (2022) Damming-Induced hydrogeomorphic transition in downstream channel and delta: a case study of the yellow river, China. Water 14(13):2079

Khatun R, Talukdar S, Pal S, Kundu S (2021) Measuring dam induced alteration in water richness and eco-hydrological deficit in flood plain wetland. J Environ Manage 285:112157

Kookana RS, Drechsel P, Jamwal P, Vanderzalm J (2020) Urbanisation and emerging economies: Issues and potential solutions for water and food security. Sci Total Environ 732:139057

Kumar AU, Jayakumar KV (2020) Hydrological alterations due to anthropogenic activities in Krishna river basin, India. Ecol Indic 108:105663

Kundu S, Pal S, Talukdar S, Mandal I (2021) Impact of wetland fragmentation due to damming on the linkages between water richness and ecosystem services. Environ Sci Pollut Res 28(36):50266–50285

Kuriqi A, Pinheiro AN, Sordo-Ward A, Bejarano MD, Garrote L (2021) Ecological impacts of run-of-river hydropower plants—current status and future prospects on the brink of energy transition. Renew Sustain Energy Rev 142:110833

Latrubesse EM, d'Horta FM, Ribas CC, Wittmann F, Zuanon J, Park E, Dunne T, Arima EY, Baker PA (2021) Vulnerability of the biota in riverine and seasonally flooded habitats to damming of Amazonian rivers. Aquat Conserv: Mar Freshw Ecosyst 31(5):1136–1149

Lee G, Lee HW, Lee YS, Choi JH, Yang JE, Lim KJ, Kim J (2019) The effect of reduced flow on downstream water systems due to the kumgangsan dam under dry conditions. Water 11(4):739

Li D, Long D, Zhao J, Lu H, Hong Y (2017) Observed changes in flow regimes in the Mekong river basin. J Hydrol 551:217–232

Li M, Cao X, Liu D, Fu Q, Li T, Shang R (2022) Sustainable management of agricultural water and land resources under changing climate and socio-economic conditions: a multi-dimensional optimization approach. Agric Water Manag 259:107235

Li Z, Sun W, Chen H, Xue B, Yu J, Tian Z (2021) Interannual and seasonal variations of hydrological connectivity in a large shallow wetland of north China estimated from Landsat 8 images. Remote Sens 13(6):1214

Lin K, Lian Y, Chen X, Lu F (2014) Changes in runoff and eco-flow in the Dongjiang river of the Pearl river basin, China. Front Earth Sci 8:547–557

Mao D, Tian Y, Wang Z, Jia M, Du J, Song C (2021) Wetland changes in the Amur river basin: differing trends and proximate causes on the Chinese and Russian sides. J Environ Manage 280:111670

McFeeters SK (1996) The use of the normalized difference water index (NDWI) in the delineation of open water features. Int J Remote Sens 17(7):1425–1432

Meng B, Liu JL, Bao K, Sun B (2019) Water fluxes of Nenjiang river basin with ecological network analysis: conflict and coordination between agricultural development and wetland restoration. J Clean Prod 213:933–943

Meng Y, Wu H, Guan Q, Lu K, Kang Y, Dai G (2022) Responses of riparian insect communities to flow regulation in riverine wetlands of Northeastern China. Hydrobiologia 1–13

Monserud RA, Leemans R (1992) Comparing global vegetation maps with the Kappa statistic. Ecol Model 62(4):275–293

Nowreen S, Haque P, Mondal MS, Zzaman RU (2020) Hydrological assessment for the availability of water for off-stream uses of Karatoa-Atrai river in Bangladesh. Water Policy 22(1):70–84

Ostad-Ali-Askari K (2022) Review of the effects of the anthropogenic on the wetland environment. Appl Water Sci 12(12):260

Pal S (2016) Impact of Massanjore dam on hydro-geomorphological modification of Mayurakshi river, Eastern India. Environ Dev Sustain 18:921–944

Pal S, Sarda R (2020) Damming effects on the degree of hydrological alteration and stability of wetland in lower Atreyee River basin. Ecol Ind 116:106542

Pal S, Sarda R (2021) Measuring the degree of hydrological variability of riparian wetland using hydrological attributes integration (HAI) histogram comparison approach (HCA) and range of variability approach (RVA). Ecol Ind 120:106966

Pal S, Sarda R (2022) Modelling water richness in riparian flood plain wetland using bivariate statistics and machine learning algorithms and figuring out the role of damming. Geocarto Int 37(19):5585–5608
Pal S, Singha P (2023) Linking river flow modification with wetland hydrological instability, habitat condition, and ecological responses. Environ Sci Pollut Res 30(5):11634–11660
Pal S, Chowdhury P, Singha P, Let M (2023) Role of tie channel on wetland hydrological security and sustenance. J Clean Prod 418:138162
Pal S, Saha A, Das T (2019) Analysis of flow modifications and stress in the Tangon river basin of the Barind tract. Inter J River Basin Manage 17(3):301–321
Pal S, Talukdar S (2020) Modelling seasonal flow regime and environmental flow in Punarbhaba river of India and Bangladesh. J Clean Prod 252:119724
Pradhan NS, Das PJ, Gupta N, Shrestha AB (2021) Sustainable management options for healthy rivers in South Asia: the case of Brahmaputra. Sustainability 13(3):1087
Rashid MB, Islam MB, Sultan-Ul-Islam M (2019) Causes of acute water scarcity in the Barind Tract, Bangladesh. Int J Econ Environ Geol 5–14
Ren W, Yao T, Xie S (2018) Stable isotopic composition reveals the spatial and temporal dynamics of discharge in the large river of Yarlungzangbo in the Tibetan Plateau. Sci Total Environ 625:373–381
Richter BD, Baumgartner JV, Powell J, Braun DP (1996) A method for assessing hydrologic alteration within ecosystems. Conserv Biol 10(4):1163–1174
Richter B, Baumgartner J, Wigington R, Braun D (1997) How much water does a river need? Freshw Biol 37(1):231–249
Sabater S, Tockner K (2010) Effects of hydrologic alterations on the ecological quality of river ecosystems. Water Scarcity Mediterr: Perspect Under Global Change 15–39
Saha TK, Pal S (2019) Emerging conflict between agriculture extension and physical existence of wetland in post-dam period in Atreyee river basin of Indo-Bangladesh. Environ Dev Sustain 21:1485–1505
Saha TK, Pal S, Sarda R (2022) Impact of river flow modification on wetland hydrological and morphological characters. Environ Sci Pollut Res 29(50):75769–75789
Santos CAG, Kisi O, da Silva RM, Zounemat-Kermani M (2018) Wavelet-based variability on streamflow at 40-year timescale in the Black sea region of Turkey. Arab J Geosci 11:1–11
Sarkar UK, Mishal P, Borah S, Karnatak G, Chandra G, Kumari S, Meena DK, Debnath D, Yengkokpam S, Das BK (2020) Status, potential, prospects, and issues of floodplain wetland fisheries in India: synthesis and review for sustainable management. Rev Fish Sci Aquac 29(1):1–32
Shen L, Li C (2010) Water body extraction from Landsat ETM+ imagery using adaboost algorithm. In: 2010 18th international conference on geoinformatics. IEEE, pp 1–4
Sifuzzaman M, Islam MR, Ali MZ (2009) Application of wavelet transform and its advantages compared to Fourier transform
Sima S, Kavousi K, Saed B (2021) Application of satellite data for determining environmental water requirement of wetlands in data-sparse regions (the case of Kanibrazan wetland, Iran). Iran J Ecohydrol 8(1):109–125
Singha P, Pal S (2022) Predicting wetland area and water depth in Barind plain of India. Environ Sci Pollut Res 29(47):70933–70949
Singha P, Pal S (2023) Wetland transformation and its impact on the livelihood of the fishing community in a flood plain river basin of India. Sci Total Environ 858:159547
Song F, Su F, Mi C, Sun D (2021) Analysis of driving forces on wetland ecosystem services value change: a case in Northeast China. Sci Total Environ 751:141778
Srivastav AL, Dhyani R, Ranjan M, Madhav S, Sillanpää M (2021) Climate-resilient strategies for sustainable management of water resources and agriculture. Environ Sci Pollut Res 28(31):41576–41595
Stefanakis AI (2019) The role of constructed wetlands as green infrastructure for sustainable urban water management. Sustainability 11(24):6981
Talukdar S, Pal S (2019) Effects of damming on the hydrological regime of Punarbhaba river basin wetlands. Ecol Eng 135:61–74
Talukdar S, Pal S (2020) Modeling flood plain wetland transformation in consequences of flow alteration in Punarbhaba river in India and Bangladesh. J Clean Prod 261:120767
Tariq MAUR, Wangchuk K, Muttil N (2021) A critical review of water resources and their management in Bhutan. Hydrology 8(1):31
Vassoney E, Mochet AM, Bozzo M, Maddalena R, Martinet D, Paternoster C, Quiriconi C, Rocco R, Comoglio C (2021) Definition of an indicator assessing the impact of a dam on the downstream river landscape. Ecol Indic 129:107941
Vogel RM, Sieber J, Archfield SA, Smith MP, Apse CD, Huber-Lee A (2007) Relations among storage, yield, and instream flow. Water Resour Res 43(5)
Wang H, Wang B, Yang H, Zhou H, Chen H, Guo W (2023) Analysis of hydrological regime evolution and ecological response in the Min River, China. J Water Clim Change
Wang Y, Feng L, Liu J, Hou X, Chen D (2020) Changes of inundation area and water turbidity of Tonle Sap Lake: responses to climate changes or upstream dam construction?. Environ Res Lett 15(9):0940a1
Wang Y, Wang D, Lewis QW, Wu J, Huang F (2017) A framework to assess the cumulative impacts of dams on hydrological regime: a case study of the Yangtze river. Hydrol Process 31(17):3045–3055

Wu Y, Zhang G, Xu YJ, Rousseau AN (2021) River damming reduces wetland function in regulating flow. J Water Resour Plan Manag 147(10):05021014

Xu C, Xu Z, Yang Z (2020) Reservoir operation optimization for balancing hydropower generation and biodiversity conservation in a downstream wetland. J Clean Prod 245:118885

Xu H (2006) Modification of normalised difference water index (NDWI) to enhance open water features in remotely sensed imagery. Int J Remote Sens 27(14):3025–3033

Yang D, Yang Y, Xia J (2021) Hydrological cycle and water resources in a changing world: a review. Geogr Sustain 2(2):115–122

Yang R, Chen Y, Qiu Y, Lu K, Wang X, Sun G, Liang Q, Song H, Liu S (2023) Assessing the landscape ecological health (LEH) of wetlands: research content and evaluation methods (2000–2022). Water 15(13):2410

Yaseen ZM, El-Shafie A, Jaafar O, Afan HA, Sayl KN (2015) Artificial intelligence based models for stream-flow forecasting: 2000–2015. J Hydrol 530:829–844

Yoshida Y, Lee HS, Trung BH, Tran HD, Lall MK, Kakar K, Xuan TD (2020) Impacts of mainstream hydropower dams on fisheries and agriculture in lower Mekong basin. Sustainability 12(6):2408

Yu C, Xu Z, Li Y, Yang Y, Cai Y, Yang Z (2023) Rethinking environmental flow management strategies in reservoir operations from an integrated water quality perspective. J Hydrol 130196

Zehtabian E, Masoudi R, Yazdandoost F, Sedghi-Asl M, Loáiciga HA (2023) Investigation of water allocation using integrated water resource management approaches in the Zayandehroud river basin, Iran. J Cleaner Prod 395:136339

Zheng Y, Zhang G, Wu Y, Xu YJ, Dai C (2019) Dam effects on downstream riparian wetlands: the Nenjiang river, Northeast China. Water 11(10):2038

Zhou Z, Tuo Y, Li J, Chen M, Wang H, Zhu L, Deng Y (2022) Effects of hydrology and river characteristics on riverine wetland morphology variation in the middle reaches of the Yarlung Zangbo–Brahmaputra river based on remote sensing. J Hydrol 607:127497

Effect of Hydrological Modification on Wetland Morphology in Reference to Parts of the Bagri Region of Eastern India

13

Swades Pal and Manabendra Let

Abstract

The previous existing literature concentrated majorly on how mounting anthropogenic pressure invited hydrological evolution and morphological insecurity in the wetlands. In the floodplain region where the wetlands confronted severe threats in the form of agricultural expansion, built-up encroachment, etc. and resulting hydrological evolution is immensely obvious, how far the hydrological parameters can control landscape insecurity was not well addressed. Therefore, the present work endeavored to identify the association between hydrological parameters and wetland landscape fragmentation. Results exhibited on average 55% of the wetland area was vulnerable to fragmentation, whereas a huge areal coverage, i.e., nearly 69% for water presence frequency (WPF) and > 90% in the case of Wetland depth comes under an insecure hydrological area. A large area converted from wetland area to land area. A negative association was found between hydrological stability and morphological instability. Poor hydrological condition reinvigorates the morphological insecurity of the wetland as the anthropogenic pressure made the outer area more vulnerable and prevented the wetland to maintain desired hydrological stability.

Keywords

Floodplain wetlands · Hydrological dynamics · Morphological evolution · Association between hydrological and morphological condition

13.1 Introduction

Wetland, also known as the "kidney" of the earth, is one of the world's most productive ecosystems and offers multidimensional functions incorporating biodiversity conservation (Bobbink et al. 2006; Menbere et al. 2018; Mitsch et al. 2015; Sundar et al. 2013), materials provision, climate regulation (Mander et al. 2014; Wong et al. 2017; Xu et al. 2019), water purification (Li et al. 2021; Trepel 2010), erosion control (Grenfell et al. 2022; Yoon 2009), groundwater recharge (Bullock and Acreman 2003) as well as educational services (Kingsford et al. 2016; Wachenheim et al. 2018; Xu et al. 2019). Despite their importance, due to continuous human disturbance and economic development, nearly 50% of the world's wetlands have

S. Pal · M. Let (✉)
Department of Geography, University of Gour Banga, Malda, West Bengal, India
e-mail: manabendralet20@gmail.com

S. Talukdar et al. (eds.), *Water Resource Management in Climate Change Scenario*, GIScience and Geo-environmental Modelling, https://doi.org/10.1007/978-3-031-61121-6_13

already disappeared, and the rest are threatened by degradation (Davidson 2014).

Floodplains are comparatively low and periodically inundated areas adjacent to rivers, oceans, and lakes that are rich in wetlands due to rich hydrological conditions. However, floodplain wetlands are the world's most dynamic region (Singh and Sinha 2020). The floodplain wetlands are directly related to surface and groundwater, terrestrial ecosystems, riparian zones and soils, simultaneously their interactions serve to maintain the health and resilience of the floodplain (Krause et al. 2007). Wetlands in the floodplain region are very conducive to wide communities like planktons, fishes, insects and macrophytes including other fauna and flora bestowed with economic value (Lair 2006; Sarkar et al. 2021). Besides, it fosters and secures security of water, food security, and the condition of human health. Seasonal floodplain wetlands are very important like the perennial ones cause these mark the zone of interaction between surface water and groundwater (Paul and Pal 2020; White 1993). These seasonal wetlands, because of their ephemeral nature often overlooked and avoided. Interestingly, their hydro-ecological significance is as important as the perennial wetlands (Calheiros and Oliveira 2016; Wu et al. 2017) and socio-economic value (Let and Pal 2023; Terer et al. 2004) as the perennial wetlands do. Despite being one of the conducive and profitable land covers for biodiversity, ecosystem services and ecological profits, these landscapes are continuously transforming into other landscapes. Even though these are critically vulnerable to the path of their endangerment. More specifically, the changing water surface resulting from climate change has converted the floodplain region into agricultural land (Pal et al. 2022).

Many research work has already investigated the drivers of floodplain wetland loss. Sica et al. (2016) carried out a work concerning the area loss of wetland in deltaic part of Parana River in Argentina and established that settlement construction was the major driver of wetland loss in this region. Yu et al. (2018) conducted a study in floodplain wetlands in north–east China and found that agricultural activity was the prominent cause of wetland loss in this region. Pal et al. (2018) investigated the major causes of wetland loss in the lower reach of Atreyee River basin and found agricultural activity as the principal driver of wetland loss. Barbier et al. (2008) investigated and found agricultural encroachment as the driving force behind wetland area loss. Twilley et al. (2016) studied that urbanization is the main cause of coastal floodplain wetland loss in the Mississippi river delta region. Burgin et al. (2016) evaluated how the growing urbanization triggered the concomitant wetland loss in western Sydney of Australia. According to Bosma et al. (2017), the urbanization impacted by the growing population has remained the major causes of wetland loss in many parts of the world. Das et al. (2022) studied the floodplain wetlands in the lower Gangetic plain, where they found continuous urbanization was the leading factor in floodplain wetland loss in English Bazar, Manikchak, Ratua-II, and Kaliachak-II blocks within the study region.

Some researchers, along with identifying the wetland area loss worldwide, investigated on the hydrological changes in wetlands keeping their interest on ecological fate regarding those wetlands. Pal and Singha (2022) indicated lowering of depth in wetland and increasing inconsistency concerning water presence within the floodplain wetlands experienced and this happened because of dam construction on the river in the Barind region of Bangladesh and India. Similar kind of findings also recorded by Sarda and Pal (2022) in the basin area of the river Atreyee. Singha and Pal (2023) documented the change experienced regarding hydro-period in the floodplain region of the Tangon river after the construction of dam on this river. Khatun et al. (2021) revealed how water richness undergo changes as per the hydrological alteration in neighboring master stream. The flood water is an important source of water of the floodplain wetlands. How important the flood frequency, flood magnitude, etc. are for hydrological security and stability been rightly deciphered through the work of Talukdar and Pal (2019), Saha et al. (2022). The Flow reduction in rivers induced by dam construction is a major reason for the reduction of active floodplain domain.

Fragmentation is one of the major reasons behind wetland transformation. It corroborates the shattering of natural wetland landscape unit into many fragmented patches that are comparatively smaller than the original landscape (Fahirg 2003). Fragmentation of any kind of ecological landscape increases patch frequency, edge area ratio, edge length, reduces patch size etc. Expanding agricultural land (Saha and Pal 2019), settlement area (Brinkmann et al. 2020) encroachment, and construction of roads (Bennett 2017) and railways (Barrientos and Borda-de-Água 2017) bifurcating the wetland are mainly responsible for the fragmentation of the wetlands.

Although a number of research works have already been conducted regarding the drivers of wetland transformation as well as the hydrological security and morphological insecurity of the wetlands in different parts of the world at different levels, very few works have been attempted on how the hydrological security and morphological insecurity are interrelated with each other and how far the changes in one aspect changes the other. Keeping in mind the research gap mentioned above the present work tried to show the spatio-temporal changes in the hydrological and morphological characteristics of the wetlands. Further, we attempted to assess the interaction between hydrological and morphological transformation.

13.2 Study Area

The study area of this work comes under Moribund deltaic part of river Ganga. More precisely the region stands in the northeastern part of the moribund Gangetic delta. This region lying east of the Bhagirathi River is known as the Bagri region (Sen and Sen 1990). The Bagri is characterized by alluvial fertile soil having a huge thickness of recent to sub-recent sediments of the Ganga River system (Fig. 13.1). The Moribund deltaic part is the one along with the active delta and mature delta, together which form the Ganges delta (Bagchi 1944; Umitsu 1993). The moribund deltaic part is the region where the delta formation is completed. Here, an interconnected network formed by the rivers like Bhagirathi, Jalangi, Mathabhanga, Ichamati, Churni, Bhairabi are found along with numerous oxbow lakes, meandering channels, left channels, marshy swamps. This is among the three types of deltaic parts where the delta formation is completed (Rudra 2014). The wetlands within the region are mainly rain-fed or receives its water from the rivers, hence, mostly seasonal in nature (Bala and Mukherjee 2010).

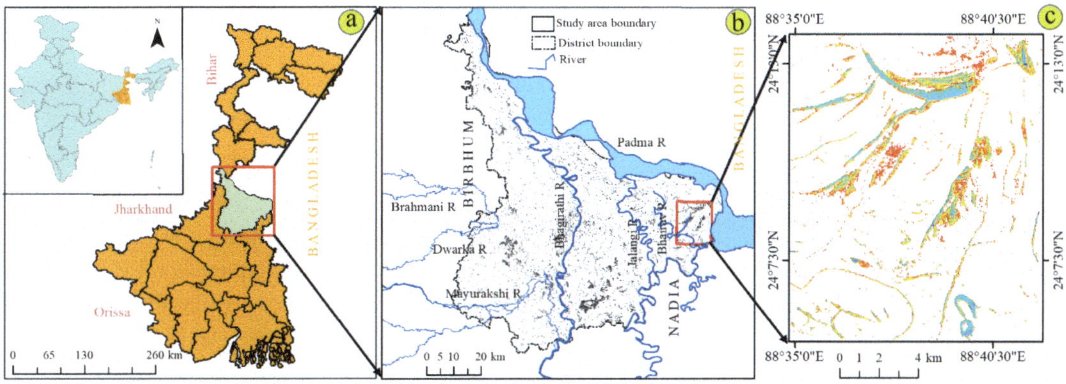

Fig. 13.1 Geographical location of the study area along with floodplain wetlands

13.3 Materials

The materials used for this work include Landsat-TM (1991 to 2012) and Landsat-OLI (2013–2021) satellite images which availed from USGS platform. One mentionable thing is while downloading the images we did not get the same date image for all seasons. Therefore, we choose images from the month March to May on consideration of pre-monsoon and month October to December in case of post-monsoon time.

13.4 Methods

13.4.1 Spectral Water Indices-Based Wetland Mapping

Different spectral water index like Normalized Differences Water index (NDWI) developed by Mcfeeters (1996), Modified Normalized Differences Water index (MNDWI) developed by Xu (2006), Remodified Normalized Differences Water Index (RmNDWI) developed by Debanshi and Pal (2020) are used for wetland area identification (Eq. 13.1, 13.2, and 13.3). As these indices are very effective in surface water detection, wetland mapping, and vegetation water content. Considering the usability of these indices, in this work NDWI, MNDWI, RmNDWI have been used for wetland area mapping. The methods for calculating NDWI, MNDWI, and RmNDWI is of the following.

$$NDWI = (b_G - b_{MIR}) / (b_G + b_{MIR}) \quad (13.1)$$

Here, NDWI = Normalized Differences Water Index; b_G = TM band 2 and OLI band 3; b_{MIR} = Tm band 4 and OLI band 5.

$$MNDWI = (b_G - b_{MIR}) / (b_G + b_{MIR}) \quad (13.2)$$

Here, MNDWI = Modified Normalized Differences Water Index; b_G = TM band 2 and OLI band 3; b_{MIR} = Tm band 4 and OLI band 5.

$$RmNDWI = (b_R - b_{MIR}) / (b_R + b_{MIR}) \quad (13.3)$$

Here, *RmNDWI* = Remodified Normalized Differences Water Index; Br = TM band 3 and OLI band 4; BMIR = TM band 5 and OLI band 6.

13.4.2 Validation of Wetland Maps for Best Method Selection

The indices used for wetland identification do not exhibit the same results. Therefore, validation of these indices is mandatory. On account of validation, 108 field based and 205 Google image-based reference sites were taken into account. A total of 111 Google Earth-based reference points were considered for the validation of the wetland maps in 1991. Kappa coefficient, True positive rate (TPR), and True negative rate (TNR) were applied to compute the accuracy of wetland maps.

$$Kappa\ Statistics\ (K) = \frac{(TS \times TSC) - \sum (CT \times RT)}{TS^2 - \sum (CT \times RT)} \times 100 \quad (13.4)$$

Here, TS is the total reference pixel, TSC is the total corrected classified pixel, CT is the column total and RT is the row total.

$$TPR = \frac{TP}{TP + FN} \quad (13.5)$$

$$TNR = \frac{TN}{TN + FP} \quad (13.6)$$

Here, *TP* means true positive, *TN* means true negative, *FP* means false positive, *FN* means false negative.

13.4.3 Hydrological Evolution

13.4.3.1 Hydrological Consistency Gradient Mapping

The Water Presence Frequency (WPF) denotes the frequency of water availability in a pixel to the total counted year (Das and Pal 2017).

To show the hydrological consistency gradient, WPF maps were produced for the Phase-I (1991–2000) and the Phase-II (2012–2021) addressing the post-monsoon season. The MNDWI maps were changed into the binary maps providing '1' to waterbody and '0' to non-waterbody. Finally, phase-wise MNDWI maps were integrated by applying the following equation (Eq. 13.7).

$$WPF_P = \frac{\sum_{I=1}^{NS} BTK_L}{N_O} \tag{13.7}$$

Here, WPF_P is WPF for pixel P for a specific period, BTK_L denotes Lth pixels with water in the chosen MNDWI maps; N_O is the number of the images. The value lies in between 0 and 1. Value closer to '1' signify high water consistency and nearer to '0' indicates high water inconsistency.

13.4.3.2 Water Depth Mapping

The intensity of spectral water indices demonstrates the thickness of the water (Gao 1996). Li et al. (2015) also found such a relationship between spectral water indices and water depth. For water depth mapping, MNDWI maps of 1991 and 2021 were considered. A total of 31 field sites-based depth data were collected and linear regression was computed between MNDWI intensity value and field-based depth data. The resultant linear regression equation (Eq. 13.8) was further calibrated with the MNDWI map of 1991 and 2021, respectively. For our convenience, we categorized the wetland area under three depth classes viz. low class (< 1.35 m), moderate class (1.35–2 m), and high class (> 2 m).

$$\mathrm{Y} = 0.50 + 6.38\mathrm{x} \tag{13.8}$$

13.4.4 Wetland Morphological Mapping

In landscape ecology, landscape fragmentation types are very important while focusing on it. Likewise, the forest fragmentation, the wetland landscape also fragmented into four major types such as the patch area, edge area, perforated area, and core area (Lopez and Frohn 2017). The core area was further divided into three sub categories namely small core, medium core, and large core. The 'core area' shows that part of the wetland which is separated from the non-wetlands (Zhang et al. 2013). This is generally the area expressed as wetlands with large area and high concentrations. The 'Patch' means wetland area much smaller than the core and encompassed by non-wetlands. 'Edge' and 'Perforated' area are actually the buffer zone between the core and non-wetlands. In the majority of the cases, an increment in the 'patch' area and 'perforated' area directly exhibits an increase in fragmentation.

The landscape fragmentation carried out for the year 1991 and 2021 to understand the wetland fragmentation over the course of time. To perform this, at first binary raster data for both water and non-water areas was created and henceforth using an edge width of 50 m, wetland fragmentation performed with the aid of landscape fragmentation tool over the GIS platform (Festus et al., 2020; Rahaman et al., 2016; Li et al., 2013).

13.4.5 Association Between Hydrological and Morphological Modifications

To show the association between hydrological and morphological parameters we correlated the parameters with each other. The raster layers for both morphological modifications and hydrological parameters (WPF and depth) were overlaid over the GIS platform. Thereafter, with respect to individual morphological units (like patch, small core, edge, etc.) how many pixels of a particular class under hydrological parameters (WPF, depth) came into existence were counted. Later, on the basis of these counted pixels, total area were calculated for that respective class.

13.5 Result and Analysis

13.5.1 Wetland Mapping

Wetland maps of different years were produced with the help of frequently used indices like NDWI, MNDWI, and RmNDWI. Areal extension of wetlands were depicted by these indices differs from each other. A noticeable decline was observed in terms of area coverage from the year 1991 to 2021 regarding post-monsoon season (Fig. 13.2). In the year 1991, the wetland area shown through NDWI, MNDWI and RmNDWI were 9.48 km^2, 19.04 km^2 and 15.93 km^2, respectively, which changed into 4.22 km^2, 9.11 km^2 and 14.68 km^2 in 2021 (Table 13.1). Hence, choosing the best spectral water index for delineating the wetland area is of due importance. The very same procedure being applied by Saha et al. (2021); Pal and Singha (2022) in their respective work on floodplain wetlands.

Kappa coefficient, TPR, TNR induced accuracy results exhibited among all the indices MNDWI showed good consistency between the map and the ground (Eqs. 13.4, 13.5 and 13.6). The Kappa coefficient value, TPR, and TNR for MNDWI were 0.72, 0.70, and 0.96, respectively, for the post-monsoon season in 2021. Henceforth, MNDWI was used for further works like mapping different hydrological parameters (Table 13.2).

13.5.2 Hydrological Evolution

13.5.2.1 Evolution of Hydrological Consistency

The hydrological consistency maps in relation to the post-monsoon season for 1991 to 2000 displayed the area under low category (< 0.33), moderate category (33.33–66.66), and high category (> 66.66) categories were 22.37 km^2, 10.26 km^2 and 9.28 km^2, respectively, which

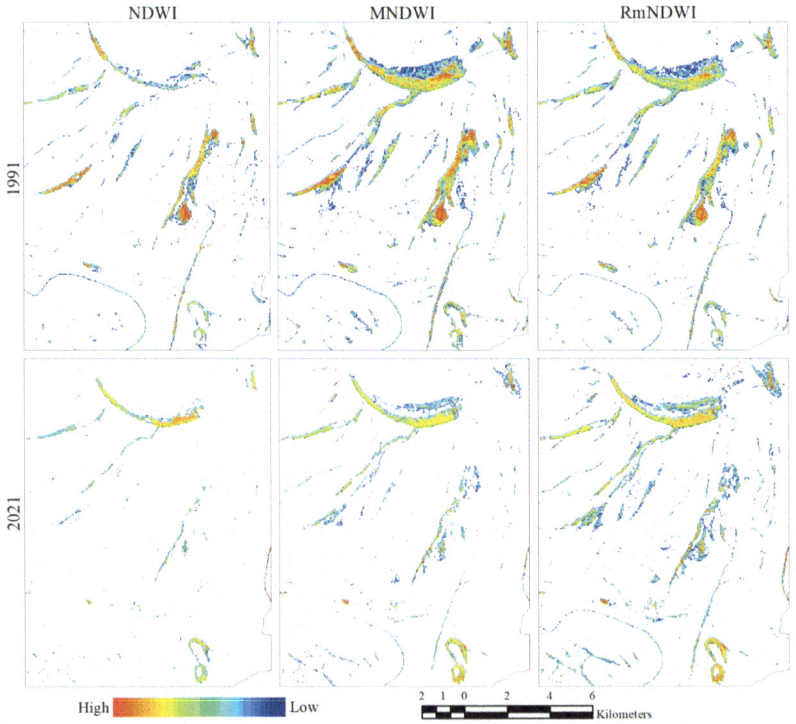

Fig. 13.2 Wetland maps produced using three spectral indices regarding post-monsoon season (1991 to 2021)

Table 13.1 Wetland area under different water index for the year 1991 and 2021

Indices	1991		2021	
	Area (km^2)	Area (%)	Area (km^2)	Area (%)
NDWI	9.48	21.34	4.22	15.07
MNDWI	19.04	42.84	9.11	32.53
RmNDWI	15.93	35.83	14.68	52.40

Table 13.2 TPR, TNR, and Kappa coefficient values for different spectral water index

Validation methods	NDWI	MNDWI	RmNDWI
TPR	0.62	0.70	0.81
TNR	0.98	0.96	0.89
Kappa coefficient	0.57	0.72	0.69

converted into 12.36 km^2, 5.21 km^2 and 7.68 km^2 in the time period of 2012 to 2021 (Table 13.3). Increasing areal coverage under the lower WPF class category clearly indicates mounting inconsistency. In the later phase, WPF area simultaneously declined with the overall wetland area. The area under low (< 0.33) and moderate (0.33–0.66) WPF classes not only declined for the later phase the areal shrinkage was striking in nature (Fig. 13.4).

13.5.2.2 Water Depth Dynamics

The depth of the wetlands experienced huge changes from 1991 to 2021. The area under higher wetland depth was 11.23 km^2 in 1991which changed to 0.03 km^2 in 2021. Similarly, the area under the moderate class and low class showed raising profoundly from 1991 to 2021. In 1991 area under the moderate depth class was 3.35 km^2 which increased to 6.16 km^2 (Table 13.4). Additionally, the area under low class was 23.39% in 1991 and this changed to 32.09 5 in 2021. The area under the higher depth category changed drastically from 1991 to 2021 (Fig. 13.5).

13.5.3 Morphological Evolution of Wetland Area

Morphologically, the wetland area witnessed a paramount transformation since the year 1991 (Fig. 13.6). Here, morphological transformation encompassed changes in shape, size, expansion or shrinkage in the core, patch and perforated area. In 1991 the high core area was 6.56 km^2 which vanished in 2021. Simultaneously, no area was under the medium core category in 1991 but 1.43 km^2 area became prominent in 2021. The patch and edge area also declined drastically in between 1991 and 2001, the area declined to 2.16 km^2 and 3.78 km^2 in 2021(Table 13.5). The area under perforated class also noticeably squeezed into 0.19 km^2 from 0.76 km^2 in1991 (Table 13.5).

Table 13.3 Different WPF class wise area for the year 1991 and 2021

WPF class	1991		2021	
	Area (km^2)	Area (%)	Area (km^2)	Area (%)
<0.33	22.37	53.36	12.36	48.93
0.33–0.66	10.26	24.48	5.21	20.63
>0.66	9.29	22.16	7.68	30.43

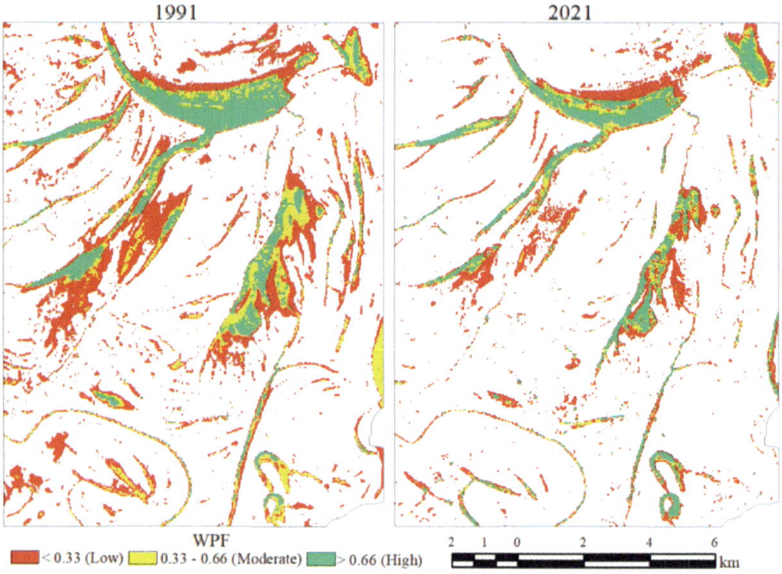

Fig. 13.4 The WPF in two different phases (1991 to 2000, 2012 to 2021) for the post-monsoon season

Table 13.4 Area under different wetland depth class

Depth class	1991		2021	
	Area (Km^2)	Area (%)	Area (Km^2)	Area (%)
Low (< 1.35 m)	4.45	23.39	2.92	32.09
Moderate (1.35–2 m)	3.35	17.61	6.16	67.58
High (> 2 m)	11.23	59.00	0.03	0.33

13.5.4 Association Between Hydrological and Morphological Parameters

A strong association could be found between hydrological and morphological parameters within the study region. In the year 1991, the area under the Core 3 was 17.70 km^2 whereas in that phase area under the high WPF class was as high as 9.29 km^2 which changed simultaneously in 2021 to 4.29 km^2 and 7.68 km^2, respectively (Table 13.6). In 1991 existence of larger portion of areas in the high WPF class and Core 3 category revealed the interrelation between wetland fragmentation and WPF (Fig. 13.6). The decreasing core area with the decreasing area under the high WPF showed how the WPF was influenced by wetland fragmentation. The area under patch, edge and core area was 2.30 km^2, 8.45 km^2, and 0.76 km^2 in 1991 and at that time the area under the low WPF class category was 22.37 km^2. This area squeezed to 12.36 km^2 in 2021. The association not only existed between wetland fragmentation class and WPF rather an intimate association could also find between wetland depth and wetland fragmentation. Wetland areas under high-depth, changes over time with diminishing core area (Table 13.7).

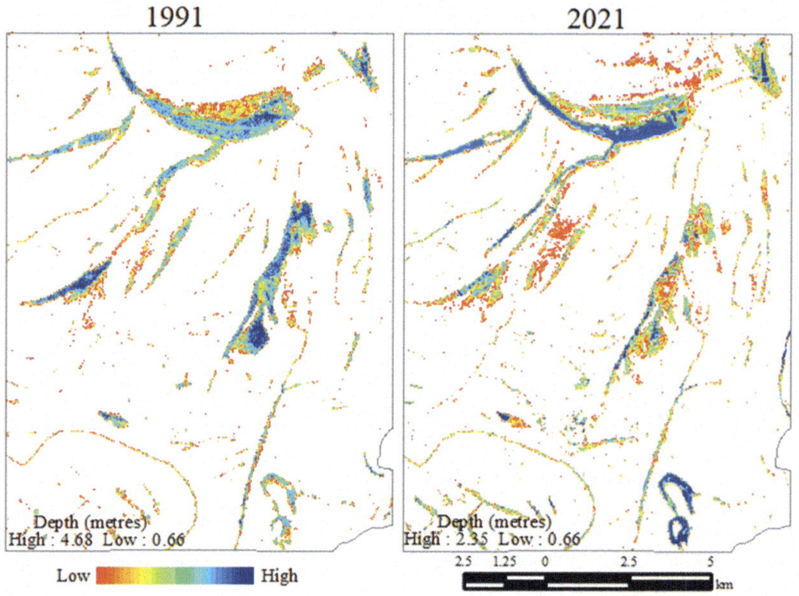

Fig. 13.5 Wetland water depth (1991–2021)

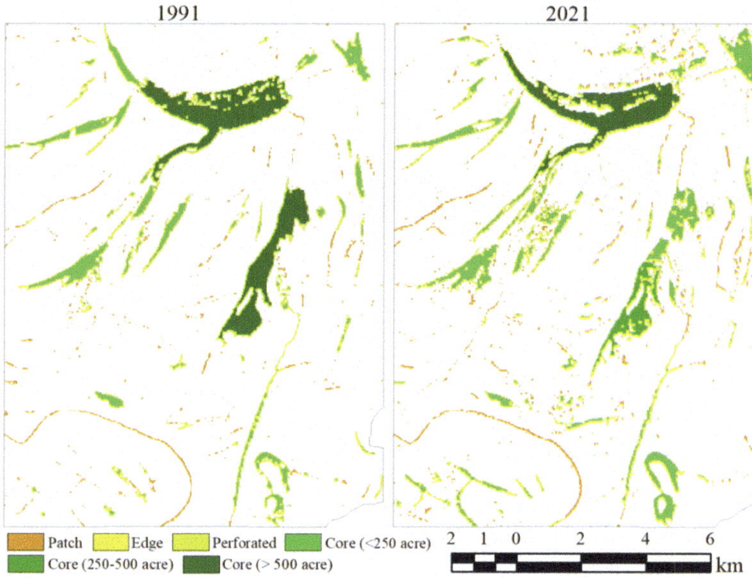

Fig. 13.6 Wetland fragmentation map showing patch, edge, core and perforated area for the year 1991 and 2021

13.6 Discussion

The areal shrinkage of wetlands along with hydrological and morphological interrelationships could also be found in the works of (Kundu et al. 2022). This has been the result of widespread hydrological change, land use change and other forms of human activities (Sica et al. 2016). It is worth mentioning that the major drivers of floodplain wetland loss

Table 13.5 Areal coverage of different fragmentation classes of wetlands

Wetland fragmentation class	1991		2021	
	Area (km^2)	Area (%)	Area (km^2)	Area (%)
Patch	2.30	6.81	2.88	22.29
Edge	8.45	25.02	4.13	31.89
Perforated	0.76	2.25	0.16	1.25
Core 1 (< 250 acre)	3.46	10.24	1.48	11.40
Core 2 (250–500 acre)	1.10	3.26	0.00	0.00
Core 3 (> 500 acre)	17.70	52.41	4.29	33.17

Table 13.6 Interrelation between WPF and morphology of the wetland

Wetland landscape class	The area under WPF classes (km^2)							
	1991				2021			
	<0.33	0.33–0.66	>0.66	Total	<0.33	0.33–0.66	>0.66	Total
Patch	0.90	0.94	0.26	2.1	0.42	0.84	0.87	2.13
Edge	1.15	3.40	1.52	6.07	0.50	1.21	2.06	3.77
Perforated	0.02	0.32	0.42	0.76	0.03	0.06	0.10	0.19
Core 1 (< 250 acre)	0.08	1.28	2.20	3.56	0.18	0.39	0.99	1.56
Core 2 (250–500 acre)	0.00	0.00	0.00	0	0.00	0.17	1.26	1.43
Core 3 (> 500 acre)	0.01	1.85	4.70	6.56	0.00	0.00	0.00	0

Table 13.7 Interrelation between wetland depth and wetland morphology

Wetland landscape class	The area under depth class (km^2)							
	1991				2021			
	<1.35 m	1.35–2 m	>2 m	Total	<1.35 m	1.35–2 m	>2 m	Total
Patch	1.25	0.33	0.51	2.09	0.25	0.16	0.01	0.42
Edge	2.4	1.01	2.66	6.07	0.95	0.9	0.07	1.92
Perforated	0.38	0.15	0.24	0.77	0.06	0.03	0	0.09
Core 1 (< 250 acre)	0.12	0.21	3.24	3.57	0.92	1.91	0.33	3.16
Core 2 (250–500 acre)	0	0	0	0	0.2	0.33	0.08	0.61
Core 3 (> 500 acre)	0.31	0.41	5.83	6.55	0.57	2.31	0.02	2.9

across the world identified agricultural activities (Banda et al. 2023; Schuyt 2005; Yu et al. 2018). The floodplain wetlands are treated as an underrated resource in many less-developed and developing countries and henceforth their value is overlooked (Barbier and Thompson 1998). Additionally, in the name of mere reclamation, vast numbers of land development programs are being incorporated which itself hampering the wetlands. Hydrological transformation of the wetlands, in terms of WPF and wetland depth, was also clearly witnessed in the wetlands. Previous works like Golden et al. (2014); Pal and Singha (2023), also supported this phenomenon regarding the floodplain wetlands across different parts of the world. According to Schneider et al. (2017), the dam construction across the river is the major reason for the hydrological transformation in floodplain wetlands. This has also been reflected in the works of Dudgeon (2000); Khatun et al. (2021); Pal and Sarda (2020), etc. in tropical Asian rivers and their associated wetlands, Punarbhaba River Basin of Bangladesh and India in the lower reach of the Atreyee River Basin. Li et al., (2009) condemned how changing the river course brings changes in the hydrological characteristics of the wetland. A number of scholars like Erwin (2009); Junk et al. (2014) opined widespread climate change is one of the major reasons for the hydrological degradation of the floodplain wetlands. Declining groundwater level leads to the hydrological transformation, as envisaged by Dwire et al. (2018).

The morphological changes in the floodplain region have also been discussed in the works of Pal et al. (2022) regarding the floodplain wetlands in different scales of various nature. Our present study results exhibited the wetland area experienced fragmentation with the passing of time (Fig. 13.6). Around the year 1991, the existence of a wider core area exclusively showed the then wetlands were more regular and compact. Later, the fragmentation of these initial wetlands over time showed the widespread core area squeezed into either small patches or perforated areas. This kind of result could also be found in the works of (Pal et al., 2022). These alterations or diversification of the core area denote nothing but the fragmentation of the previous pristine wetland. This type of wetland fragmentation is mainly the immediate result of hydrological transformation in the same (Pal and Debanshi 2021). Whether decreasing the water flow from the master river or decreasing in rainwater, the hydrological setup of the wetland is hampered followed by morphological configuration (Pal et al. 2023; Wei et al., 2023; Zhang et al. 2022). In the floodplain region, the main source of water is rainwater and floodwater (Weilhoefer et al. 2008). Fluctuation of water from these sources brings about changes in the hydrological characteristics of wetland. Increasing water scarcity within wetland means decreasing the wetted area. In such cases, the fringe part of the wetland which is shallower in nature, becomes more vulnerable to getting affected (Pal and Talukdar 2018; Khatun et al. 2021). Majority of the cases this shallower fringe part is converted into agricultural land. In connection with the fragmentation of wetlands, it was found instead of the interior part, the fringe and edge areas do get affected primarily. This mechanism of morphological transformation was also reported in the work of Debanshi and Pal (2020); Pal et al., (2022); Ralph and Hesse (2010). They clearly revealed hydrological transformation stands as the prime cause behind the morphological changes in the wetland. A number of studies strongly supported this finding. So, the morphological transformation witnessed in the present study region further cemented how changes in the hydrological scenario of the wetland, invited the morphological transformation concerning the floodplain wetland.

13.7 Conclusion

This work focused on the spatio-temporal transformation of the wetland area as well as tried to find out how far the hydrological and morphological transformations interacted with each other. The result clearly showed wetland area declined substantially between 1991 and 2021. The result is actually an overall representation of the overall change that happened in the wetland landscape. Morphological transformation triggered by hydrological changes explicitly unraveled in this study. On one hand, encroachment of agricultural land, infrastructural development, clogging of tie channels, damming across the master river, lowering of groundwater, etc. have directly influenced the hydrological characters, which in turn craved the path for morphological transformation on the other hand. Results also corroborated how the two aspects of the wetland acted dynamically and impacted each other.

The hydrological security along with morphological stability of the wetland is inevitably necessary for the ecological viability of the same. The increasing fragmentation of these floodplain wetlands further leads to degradation and as a consequence of this, only about 15% of wetlands are under in the stable condition. So, required proper attention must be given to prevent the loss and degradation of the wetland. Simultaneously, due attention should be given in maintaining optimum hydrological condition of the remnant wetlands. All the stakeholders, therefore, need to take proper initiative in this concern.

References

Bagchi K (1944) The Ganges delta. (No Title)

Bala G, Mukherjee A (2010) Inventory of wetlands of Nadia district, West Bengal, India and their characterization AS. J Environ Sociobiol 7(2):93–106

Banda AM, Banda K, Sakala E, Chomba M, Nyambe IA (2023) Assessment of land use change in the wetland of Barotse floodplain, Zambezi river sub-basin, Zambia. Nat Hazards 115(2):1193–1211

Barbier EB, Koch EW, Silliman BR, Hacker SD, Wolanski E, Primavera J, ... Reed DJ (2008) Coastal ecosystem-based management with nonlinear ecological functions and values. Sci 319(5861):321–323

Barbier EB, Thompson JR (1998) The value of water: floodplain versus large-scale irrigation benefits in northern Nigeria. Ambio 434–440

Barrientos R, Borda-de-Água L (2017) Railways as barriers for wildlife: current knowledge. Railway Ecol 43–64

Bennett JW (2017) The ecological transition: cultural anthropology and human adaptation. Routledge

Bobbink R, Whigham DF, Beltman B, Verhoeven JT (2006) Wetland functioning in relation to biodiversity conservation and restoration. Springer Berlin Heidelberg, pp. 1–12

Borda-de-Água L, Barrientos R, Beja P, Pereira HM (2017) Railway ecology. Springer Nature, p 320

Bosma C, Glenk K, Novo P (2017) How do individuals and groups perceive wetland functioning? Fuzzy cognitive mapping of wetland perceptions in Uganda. Land Use Policy 60:181–196

Brinkmann K, Hoffmann E, Buerkert A (2020) Spatial and temporal dynamics of urban wetlands in an Indian megacity over the past 50 years. Remote Sens 12(4):662

Bullock A, Acreman M (2003) The role of wetlands in the hydrological cycle. Hydrol Earth Syst Sci 7(3):358–389

Burgin S, Franklin MJ, Hull L (2016) Wetland loss in the transition to urbanisation: a case study from Western Sydney, Australia. Wetlands 36:985–994

Das RT, Pal S (2017) Exploring geospatial changes of wetland in different hydrological paradigms using water presence frequency approach in Barind tract of West Bengal. Spat Inf Res 25:467–479

Das S, Bhunia GS, Bera B, Shit PK (2022) Evaluation of wetland ecosystem health using geospatial technology: evidence from the lower Gangetic flood plain in India. Environ Sci Pollut Res 29(2):1858–1874

Davidson NC (2014) How much wetland has the world lost? Long-term and recent trends in global wetland area. Mar Freshw Res 65(10):934–941

Debanshi S, Pal S (2020) Wetland delineation simulation and prediction in deltaic landscape. Ecol Ind 108:105757

Dudgeon D (2000) The ecology of tropical Asian rivers and streams in relation to biodiversity conservation. Annu Rev Ecol Syst 31(1):239–263

Dwire KA, Mellmann-Brown S, Gurrieri JT (2018) Potential effects of climate change on riparian areas, wetlands, and groundwater-dependent ecosystems in the Blue mountains, Oregon, USA. Clim Serv 10:44–52

Erwin KL (2009) Wetlands and global climate change: the role of wetland restoration in a changing world. Wetlands Ecol Manage 17(1):71–84

Fahrig L (2003) Effects of habitat fragmentation on biodiversity. Annu Rev Ecol Evol Syst 34(1):487–515

Festus O, Ji W, Zubair OA (2020) Characterizing the landscape structure of urban wetlands using terrain and landscape indices. Land 9(1):29

Gao BC (1996) NDWI—a normalized difference water index for remote sensing of vegetation liquid water from space. Remote Sens Environ 58(3):257–266

Golden HE, Lane CR, Amatya DM, Bandilla KW, Kiperwas HR, Knightes CD, Ssegane H (2014) Hydrologic connectivity between geographically isolated wetlands and surface water systems: a review of select modeling methods. Environ Model Softw 53:190–206

Grenfell S, Grenfell M, Tooth S, Mehl A, O'Gorman E, Ralph T, Ellery W (2022) Wetlands in drylands: diverse perspectives for dynamic landscapes. Wetlands Ecol Manage 30(4):607–622

Junk WJ, Piedade MTF, Lourival R, Wittmann F, Kandus P, Lacerda LD, Bozelli RL, Esteves FDA, Nunes da Cunha C, Maltchik L, Schöngart J, Agostinho AA (2014) Brazilian wetlands: their definition, delineation, and classification for research, sustainable management, and protection. Aquat Conserv: Mar Freshw Ecosyst 24(1):5–22

Khatun R, Talukdar S, Pal S, Kundu S (2021) Measuring dam induced alteration in water richness and eco-hydrological deficit in flood plain wetland. J Environ Manage 285:112157

Kingsford RT, Basset A, Jackson L (2016) Wetlands: conservation's poor cousins. Aquat Conserv Mar Freshwat Ecosyst 26(5):892–916

Krause S, Bronstert A, Zehe E (2007) Groundwater–surface water interactions in a North German lowland floodplain–implications for the river discharge dynamics and riparian water balance. J Hydrol 347(3–4):404–417

Kundu S, Pal S, Mandal I, Talukdar S (2022) How far damming induced wetland fragmentation and water richness change affect wetland ecosystem services? Remote Sens Appl: Soc Environ 27:100777

Lair N (2006) A review of regulation mechanisms of metazoan plankton in riverine ecosystems: aquatic habitat versus biota. River Res Appl 22(5):567–593

Let M, Pal S (2023) Socio-ecological well-being perspectives of wetland loss scenario: a review. J Environ Manage 326:116692

Li X, Xing Y, Jiang Y, Ding Y, Li W (2009) Antimicrobial activities of ZnO powder-coated PVC film to inactivate food pathogens. Int J Food Sci Technol 44(11):2161–2168

Li X, Yang L, Xu K, Bei K, Zheng X, Lu S, ... Jin Z (2021) Application of constructed wetlands in treating rural sewage from source separation with high-influent nitrogen load: A review. World J Microbiol Biotechnol 37:1–8

Li Z, Ren W, Liu X, Xie L (2013) Distributed consensus of linear multi-agent systems with adaptive dynamic protocols. Automatica 49(7):1986–1995

Li W, Cao T, Ni L, Zhu G, Zhang X, Fu H, Song X, Xie P (2015) Size-dependent C, N and P stoichiometry of three submersed macrophytes along water depth gradients. Environ Earth Sci 74:3733–4373

Lopez R, Frohn R (2017) Remote sensing for landscape ecology: new metric indicators. CRC Press

Mander Ü, Tournebize J, Kasak K, Mitsch WJ (2014) Climate regulation by free water surface constructed wetlands for wastewater treatment and created riverine wetlands. Ecol Eng 72:103–115

McFeeters SK (1996) The use of the normalized difference water index (NDWI) in the delineation of open water features. Int J Remote Sens 17(7):1425–1432

Menbere IP, Menbere TP (2018) Wetland ecosystems in Ethiopia and their implications in ecotourism and biodiversity conservation. J Ecol Nat Environ10(6):80–96

Mitsch WJ, Bernal B, Hernandez ME (2015) Ecosystem services of wetlands. Int J Biodivers Sci, Ecosyst Serv Manage 11(1):1–4

Pal S, Chakraborty S, Datta S, Mukhopadhyay SK (2018) Spatio-temporal variations in total carbon content in contaminated surface waters at East Kolkata Wetland Ecosystem, a Ramsar Site. Ecol Eng 110:146–157

Pal S, Chowdhury P, Singha P, Let M (2023) Role of tie channel on wetland hydrological security and sustenance. J Cleaner Prod 418:138162

Pal S, Sarda R (2020) Damming effects on the degree of hydrological alteration and stability of wetland in lower Atreyee river basin. Ecol Ind 116:106542

Pal S, Sarkar R, Saha TK (2022) Exploring the forms of wetland modifications and investigating the causes in lower Atreyee river floodplain area. Ecol Informatics 67:101494

Pal S, Singha P (2022) Image-driven hydrological parameter coupled identification of flood plain wetland conservation and restoration sites. J Environ Manage 318:115602

Pal S, Singha P (2023) Linking river flow modification with wetland hydrological instability, habitat condition, and ecological responses. Environ Sci Pollut Res 30(5):11634–11660

Pal S, Talukdar S (2018) Application of frequency ratio and logistic regression models for assessing physical wetland vulnerability in Punarbhaba river basin of Indo-Bangladesh. Hum Ecol Risk Assess Int J 24(5):1291–1311

Paul S, Pal S (2020) Exploring wetland transformations in moribund deltaic parts of India. Geocarto International 35(16):1873–1894

Rahaman T, Vasiljevic T, Ramchandran L (2016) Effect of processing on conformational changes of food proteins related to allergenicity. Trends Food Sci Technol 49:24–34

Ralph TJ, Hesse PP (2010) Downstream hydrogeomorphic changes along the Macquarie River, southeastern Australia, leading to channel breakdown and floodplain wetlands. Geomorphology 118(1–2):48–64

Rudra K (2014) Changing river courses in the western part of the Ganga–Brahmaputra delta. Geomorphology 227:87–100

Saha TK, Pal S (2019) Emerging conflict between agriculture extension and physical existence of wetland in post-dam period in Atreyee river basin of Indo-Bangladesh. Environ Dev Sustain 21:1485–1505

Saha TK, Pal S, Sarkar R (2021) Prediction of wetland area and depth using linear regression model and artificial neural network based cellular automata. Eco Inform 62:101272

Saha TK, Pal S, Sarda R (2022) Impact of river flow modification on wetland hydrological and morphological characters. Environ Sci Pollut Res 29(50):75769–75789

Sarda R, Pal S (2022) Evaluating damming effect on eco-hydrological alteration in river and wetland using indicators of hydrological alteration. Geocarto Int 37(27):16132–16156

Sarkar UK, Mishal P, Karnatak G, Lianthuamluaia L, Saha S, Bandopadhyay A, Das Ghosh B (2021) Regional climatic variability and fisher's adaptation to climate-induced risks in an impacted tropical floodplain wetland: a case study. Sustain Water Resour Manage 7:1–12

Schneider T, Lan S, Stuart A, Teixeira J (2017) Earth system modeling 2.0: a blueprint for models that learn from observations and targeted high-resolution simulations. Geophys Res Lett 44(24):12–396

Schuyt KD (2005) Economic consequences of wetland degradation for local populations in Africa. Ecol Econ 53(2):177–190

Sen S, Sen J (1990) Regional geography of the moribund ganga delta, West Bengal. Patterns Reg Geogr: Int Perspect 2:87

Sica YV, Quintana RD, Radeloff VC, Gavier-Pizarro GI (2016) Wetland loss due to land use change in the lower Paraná river delta, Argentina. Sci Total Environ 568:967–978

Singh M, Sinha R (2020) Distribution, diversity, and geomorphic evolution of floodplain wetlands and wetland complexes in the Ganga plains of north Bihar, India. Geomorphology 351:106960
Singha P, Pal S (2022) Predicting wetland area and water depth in Barind plain of India. Environ Sci Pollut Res 29(47):70933–70949
Singha P, Pal S (2023) Wetland transformation and its impact on the livelihood of the fishing community in a flood plain river basin of India. Sci Total Environ 858:159547
Sundar KG, Kittur S (2013) Can wetlands maintained for human use also help conserve biodiversity? Landscape-scale patterns of bird use of wetlands in an agricultural landscape in north India. Biol Conserv 168:49–56
Talukdar S, Pal S (2019) Effects of damming on the hydrological regime of Punarbhaba river basin wetlands. Ecol Eng 135:61–74
Terer T, Ndiritu GG, Gichuki NN (2004) Socio-economic values and traditional strategies of managing wetland resources in Lower Tana River, Kenya. Hydrobiologia 527:3–15
Trepel M (2010) Assessing the cost-effectiveness of the water purification function of wetlands for environmental planning. Ecol Complex 7(3):320–326
Twilley RR, Bentley SJ, Chen Q, Edmonds DA, Hagen SC, Lam NSN, Willson CS, Xu K, Braud D, Hampton Peele R, McCall A (2016) Co-evolution of wetland landscapes, flooding, and human settlement in the Mississippi river delta plain. Sustain Sci 11:711–731
Umitsu M (1993) Late quaternary sedimentary environments and landforms in the Ganges delta. Sed Geol 83(3–4):177–186
Wachenheim CJ, Roberts DC, Addo NS, Devney J (2018) Farmer preferences for a working wetlands program. Wetlands 38:1005–1015
Wei C, Wang X, Cai J, Liao Z, Li C, Liu Q (2023) A new framework combining hydrological connectivity metrics and morphological spatial pattern analysis for the hydrological connectivity evaluation of wetlands. Integr Environ Assess Manag 19(4):1064–1078
Weilhoefer CL, Pan Y, Eppard S (2008) The effects of river floodwaters on floodplain wetland water quality and diatom assemblages. Wetlands 28:473–486
White WA, Wermund EG, Handley LR (1993) Trends and status of wetland and aquatic habitats in the Galveston Bay system. Texas (Vol. 31). Galveston Bay National Estuary Program
Wong CP, Jiang B, Bohn TJ, Lee KN, Lettenmaier DP, Ma D, Ouyang Z (2017) Lake and wetland ecosystem services measuring water storage and local climate regulation. Water Resour Res 53(4):3197–3223
Wu G, Liu Y (2017) Assessment of the hydro-ecological impacts of the Three Gorges dam on China's largest freshwater lake. Remote Sens 9(10):1069
Xu H (2006) Modification of normalised difference water index (NDWI) to enhance open water features in remotely sensed imagery. Int J Remote Sens 27(14):3025–3033
Xu W, Fan X, Ma J, Pimm SL, Kong L, Zeng Y, ... Ouyang Z (2019) Hidden loss of wetlands in China. Curr Biol 29(18):3065–3071
Yoon CG (2009) Wise use of paddy rice fields to partially compensate for the loss of natural wetlands. Paddy Water Environ 7(4):357–366
Yu T, Zhang R, Yang K, Yasunaga M, Wang D, Li Z, Ma J, Li I, Yao Q, Roman S, Zhang Z, Radev D (2018) Spider: a large-scale human-labeled dataset for complex and cross-domain semantic parsing and text-to-sql task. arXiv preprint arXiv:1809.08887
Zhang Y, Cao B, Zhang Q, Cui S, Cui B, Du J (2022) Responses of urban wetland to climate change and human activities in Beijing: a case study of hanshiqiao wetland. Sustainability 14(8):4530
Zhang Z, Solaiman ZM, Meney K, Murphy DV, Rengel Z (2013) Biochars immobilize soil cadmium, but do not improve growth of emergent wetland species Juncus subsecundus in cadmium-contaminated soil. J Soils and Sediments 13:140–151

Retreating Gya Glacier and Expansion of Gya Lake in the Ladakh Region of North-Western Himalaya

14

Rayees Ahmed, Abid Farooq Rather, Gowhar Farooq Wani, Syed Towseef Ahmad, Taha Shamim, Pervez Ahmed, and Riyaz Ahmad Mir

Abstract

Glacial lake outburst floods (GLOF's) and its associated hazards in the mountainous areas are a matter of great concern worldwide due to ongoing warming climate and glacier recession. The region has witnessed heavy damages to property, agricultural and other assets during recent years e.g. Gya GLOF event in 2014 is the recent manifestation of the GLOF hazard in this region. The present study has analysed the recent evolution of Gya glacial lake and its mother glacier using high resolution satellite data sources from 1990 to 2022 and geospatial techniques. The results show that the glacier (Gya) has lost an area of about 0.4 km^2 from 1990 (0.86 km^2) to 2022 (0.46 km^2), which is about 46% of the surface area. While the Gya glacial lake on the other hand revealed a significant growth in the area from 0.048 km^2 in 1990 to 0.11 km^2 in 2022, therefore indicating a growth of 126% since the past 32 years. The snout of the Gya glacier exhibited a considerable retreat during a 32-year period, similar to other glaciers in the Himalayan area. During the years 1990 and 2022, snout of the glacier receded by 381 m at an average pace of 11.9 m/yr. The study also identified several potential factors that may have contributed to the rapid expansion of the Gya lake, including glacier retreat, climate change, geology, and geomorphology. The key findings of the present case study have significant implications for the management of glacial lakes in the Himalayan region. The findings of this study will be also useful for decision makers and to formulate the strategies in this direction to mitigate and reduce the risks from hazards associated with glacial lakes. However, an in-depth field-based study for mass balance of the glacier and to investigate the other geological and geotechnical properties of the moraine-dam sediments impounding the rapidly increasing glacial melt water is highly recommended.

Keywords

Climate change · Glacier recession · Snout retreat · GLOF · Gya lake expansion · Trans-Himalayan region · Ladakh

R. Ahmed (✉) · A. F. Rather · G. F. Wani · S. T. Ahmad · T. Shamim · P. Ahmed
Department of Geography and Disaster Management, School of Earth and Environmental Sciences, University of Kashmir, Srinagar 190006, India
e-mail: rayeesrashid84@gmail.com

R. A. Mir
Geological Survey of India, UT: Jammu and Kashmir, Srinagar 190008, India

S. Talukdar et al. (eds.), *Water Resource Management in Climate Change Scenario*, GIScience and Geo-environmental Modelling, https://doi.org/10.1007/978-3-031-61121-6_14

14.1 Introduction

High mountains are dynamic systems with multiple regimes governed by interactions between various processes and elements. Because of this complexity, high-mountain systems are vulnerable to external factors/impacts. In the present scenario, climate change/variability is major concern before the scientific community (Gadek 2014). The increase in temperatures is a conclusively observed global phenomenon that has impacted all major systems including oceans and cryosphere, and is predicted to rise continuously in the future (IPCC 2013). The Himalayan region is home to thousands of glaciers, which are vital sources of freshwater for millions of people living in the region (Dasgupta and Shyamoli 2021; Mountford 2022). However, due to changing climate, these glaciers are melting and receding at an alarming rate, which has resulted in the development expansion and formation of several glacial lakes across the high mountain regions (Ahmed et al. 2021a, b; Taylor et al. 2023). These lakes are growing rapidly, which might have disastrous effects on the downstream infrastructure, wildlife, and local populations (Sherpa et al. 2019; Riaz et al. 2014; Khan 2022). According to ICIMOD report 2007, there are around 15,000 glaciers and 9000 glacial lakes in the Himalayan region (Bajracharya et al. 2007; Jain et al. 2012). The glaciers in the Himalayan region are the source of major river systems of Asia like Indus, Ganga, Brahmaputra, Mekong, Yangtze, and Yellow (Yang et al. 2019; Yao et al. 2022). These rivers are a lifeline for the people living in the region, providing water for agriculture, industry, and domestic use (Yang et al. 2019). Geologically and ecologically the Himalayas are one of the most important regions on the planet, providing millions of people in South Asia with a crucial supply of freshwater (Bandyopadhyay and Gyawali 1994).

The Himalayan region is facing a significant threat due to the climate change, which is causing the glaciers to melt at an alarming rate (Verma 2021; Young 2021). As a consequence of this, glacial lakes are developing and growing quickly as a result of the rising temperatures (Sattar et al. 2021; Ahmed et al. 2021a, b). There is an immediate need to monitor the growth of these glacial lakes and put effective mitigating measures in place since they pose a serious threat to the communities, wildlife, and infrastructure downstream (Sharma et al. 2022; Hussain and Khan 2023).

In the last few decades, volume area and number of these lakes is increasing significantly in different catchments of the Himalayan region as reported by the previous studies (Taylor et al. 2023; Veh et al. 2019; Zheng et al. 2021). However, the rate of change differs from one catchment to other owing to the different climatic and environmental conditions prevailing in these regions. The third pole region is home to hundreds of PDGLs. Few of these lakes are highly susceptible to breach and cause flash floods in the downstream regions (Ashraf et al. 2021; Saifullah et al. 2020; Singh et al. 2023; Rawat et al., 2023b; Mir et al., 2023). For instance, the 2013 Kedarnath flood in the Uttarakhand (state of India), which resulted in substantial destruction and fatalities, was attributed to the sudden breaching of a glacial lake (Gupta et al. 2022; Naithani and Saha 2021; Rafiq et al. 2019). Similarly, in 2014 a GLOF event was reported in the Trans-Himalayan region of Ladakh, which resulted in the destruction of Agriculture, bridges, roads and houses of Gya village (Majeed et al. 2021; Schmidt et al. 2020).

In order to address the possible threats, it is necessary to monitor and assess the growth of these lakes in the Himalayan region regularly and to put in place proper mitigation strategies. A useful technique for analysing the behaviour of glacier and lakes through satellite data, and several studies have utilised such data to monitor the rapid expansion of these lakes (Rawat et al., 2023a; Wang et al., 2022; Mir et al. 2022; Wangchuk et al. 2022; Zhang et al. 2021; Ahmed et al. 2024; Rather et al. 2024). Keeping in view the above discussion, this research study utilized multi-temporal satellite imagery to examine the growth of the Gya lake and its mother glacier. In addition to highlighting the need for improved monitoring and risk mitigation measures to address the possible threats posed by growing glacier lakes in the Himalayan region the study provides valuable understandings into the expansion of the Gya lake in the Trans-Himalayan region of Ladakh.

14.2 Study Area Description

The Gya lake is located in the "Trans-Himalayan region of Ladakh". The Gya lake is a moraine dammed type of lake located at the snout of the glacier. It is positioned between 33° 39′ 12″ N Longitude and 77° 44′ 07″ E Latitude (Fig. 14.1). The Gya Glacier has an area of approximately 0.46 km^2 (2022), and its terminus is located at an elevation of around 5462 m above sea level. The nearest settlement i.e. Gya village is situated at a distance of 12.83 km from the outlet of the lake. The lake has been formed due to the melting of the glacier, and it has been expanding rapidly since the recent past as a result of climate warming in the region. Gya glacial lake is approximately 0.65 km long and 0.25 km wide, with a depth of around 17.19 m. The lake is directly fed by the melt water supply of the Gya glacier. The melt water from this glacier and lake is utilized by the people living in the downstream village (Gya) for agriculture and domestic purposes. The village comprises of 140 households with a population of approximately 658 persons, as reported in the Census of 2011 (Majeed et al. 2021).

It is very difficult to monitor both Gya glacier and its lake through the field-based techniques due to their remote location, challenging weather conditions, financial constraints and lack of instrumentation framework. However, satellite remote sensing provides a valuable tool for monitoring the changes in the glacier and the

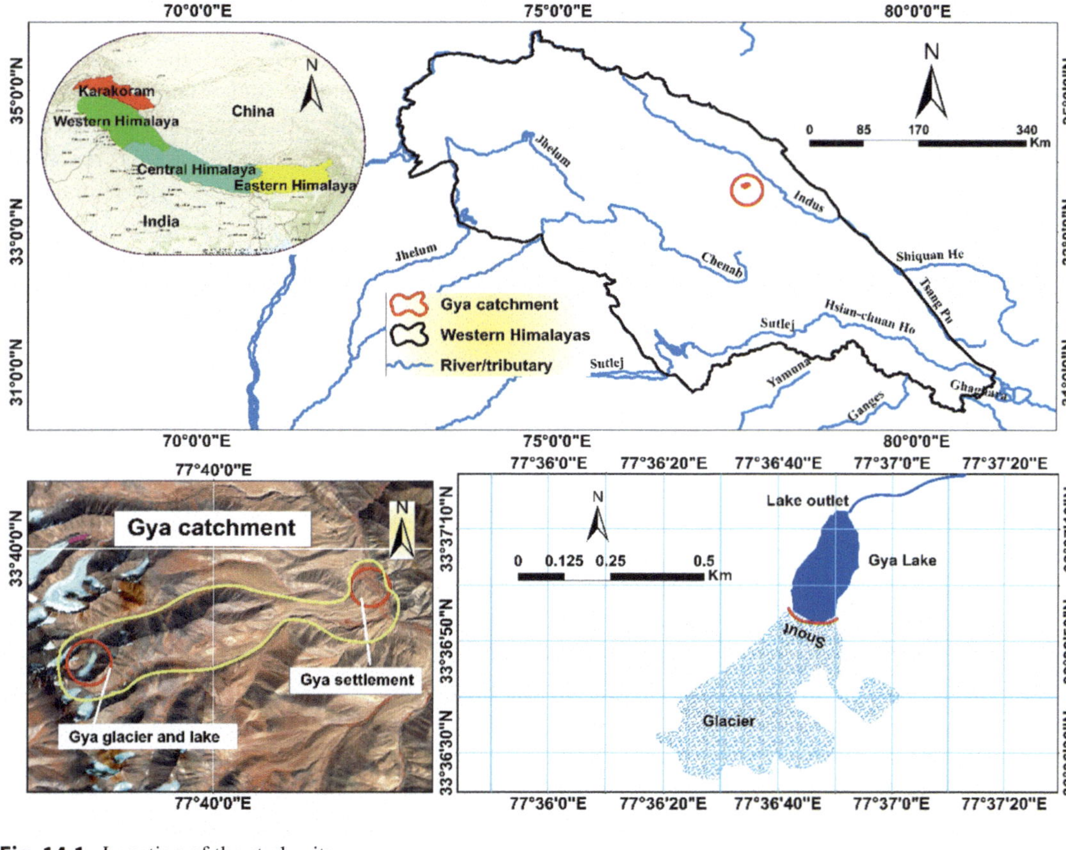

Fig. 14.1 Location of the study site

lake over time. The use of satellite data, geo spatial tools, and techniques allows for a comprehensive analysis of the changes in the lake area, volume, and other parameters, which can provide valuable insights into the potential risks posed by the expanding lake.

14.3 Datasets and Methodology

We have used satellite imageries (multi-temporal) from various source to map and analyse the changes in the Gya glacial lake and its feeding glacier. The spatio-temporal dynamics of the lake was estimated using multi-temporal Landsat data. The satellite imageries were extracted from four different time points i.e. 1990, 2000, 2010 and 2022 in order to track the growth of the Gya glacial lake and its feeding glacier over the period of 32 years. Sentinel-2 images for the Year 2021 and 2022 were also used to supplement the analysis of the lake area changes. GE images and Planet-CubeSat satellite images were also utilized for the cross check and validation purpose.

The Landsat data utilized in the current study were obtained from the US Geological Survey's (USGS) Earth Explorer platform. The platform provides access to a vast archive of Landsat data. The Sentinel-2 data were obtained from the "European Space Agency's (ESA)" Sentinel Hub platform. These types of satellite data have been widely used in the past studies for the change detection analysis.

Software tools such QGIS, ERDAS Imagine, and Google Earth Engine were utilised to process and analyse the Landsat and Sentinel-2 images that were used in the study. The lake area was extracted from the satellite images using software tools, which also produced time series graphs showing how the lake area has changed over the period of time. The methodology utilized to carry out the present study is illustrated in the Fig. 14.2.

We have used various satellite datasets and geospatial techniques to investigate the Gya lake and its feeding glacier. The methodology comprises of several steps as given below:

1. Data Collection: Satellite imagery (Landsat and Sentinel-2) were obtained from the USGS site and ESA Sentinel Hub platforms (https://www.sentinel-hub.com/), respectively, for the study area covering both Gya Glacier and associated glacial lake.
2. Pre-processing: The Landsat and Sentinel-2 images were pre-processed to remove any atmospheric effects and to enhance the contrast of the images. The pre-processing was done using the Atmospheric and Topographic Correction (ATCOR) algorithm and the Sentinel-2 Sen2 Cor algorithm.
3. Image Analysis: The pre-processed data were then analysed using a combination of GIS software's like Arc-GIS, ERDAS Imagine, and Google Earth Engine (GEE). The lake boundaries were extracted through both automated and semi-automated technique such as NDWI and manual correction. The lake extents were validated and checked using through high resolution Google Earth (GE) and Planet-CubeSat data. In this study, we have manually delineated the glacier area from the Landsat and Sentinel-2 images through visual interpretation and mapping technique as utilized by various researchers in different catchments of the Himalayan basin (Ahmed et al. 2022; Begam et al. 2018; Mir et al. 2018). The glacier area was defined as the area covered by perennial snow and ice, which was distinguished from other land cover types based on its spectral characteristics and morphology.
4. Time-Series Analysis: The lake area changes over time were analysed using Landsat images from four different time periods (1990, 2000, 2010 and 2022) and Sentinel-2 images from 2021 and 2022. The time series analysis was done using the GEE platform, and the changes in the lake area were calculated for each time period.
5. Statistical Analysis: The lake area changes were analysed statistically to determine the rate of expansion of the lake. The statistical analysis was carried out in Microsoft Excel to determine the trend in the glacier and lake area.

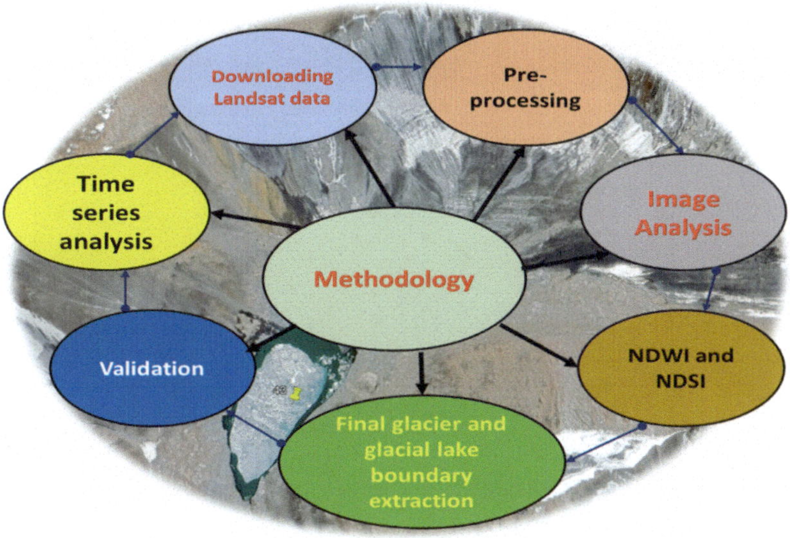

Fig. 14.2 Methodology flowchart

14.4 Results and Discussion

14.4.1 Gya Glacier Recession

The analysis of the multi-temporal Landsat data reveal that the Gya glacier has shown a remarkable decrease in area over the 32 years from 1990 to 2022. In 1990, the area of the glacier was estimated as 0.86 km^2 which has reduced to 0.46 km^2 in 2022, thus showing an overall decline of 46.51%. Between 1990 and 2000, the Gya glacier experienced a noticeable reduction in area, declining from 0.86 km^2 to 0.73 km^2, reflecting a loss of approximately 15.12%. The trend of retreat continued into the next decade, with the glacier's area further diminishing to 0.68 km^2 by 2010. This represented an additional decrease of around 6.85% compared to 2000. The most dramatic reduction occurred between 2010 and 2022, during which the Gya glacier's area shrunk from 0.68 km^2 to just 0.46 km^2, resulting in a significant loss of approximately 32.35% as depicted in the Figs. 14.3 and 14.6. This rapid retreat emphasises the accelerated pace at which the glacier has been receding in recent years. The data underline the alarming trend of the Gya glacier's retreat, highlighting the vulnerability of glaciers to climate change. The decrease in area over the years points towards the ongoing impact of global warming, with rising temperatures leading to the melting and shrinking of glaciers.

14.4.2 Snout Retreat

Like other glaciers in the Himalayan region, the snout of the Gya glacier revealed a significant retreat over the period of 32 years. The snout of the glacier has retreated by 381 m from 1990 to 2022 at an average retreat rate of 11.9 m/yr. The highest retreat i.e. 175 m at the rate of 14.5 m/yr was estimated during the years 2010 and 2022 (Fig. 14.4). Analysis of GE images reveal the presence of calving activity at the snout of the glacier. This retreat is primarily attributed to the climate change and global warming, as rising temperatures have resulted in increased melting of the glacial ice.

14.4.3 Gya Lake Expansion

The results of the study showed that the Gya Lake associated with the Gya Glacier in the

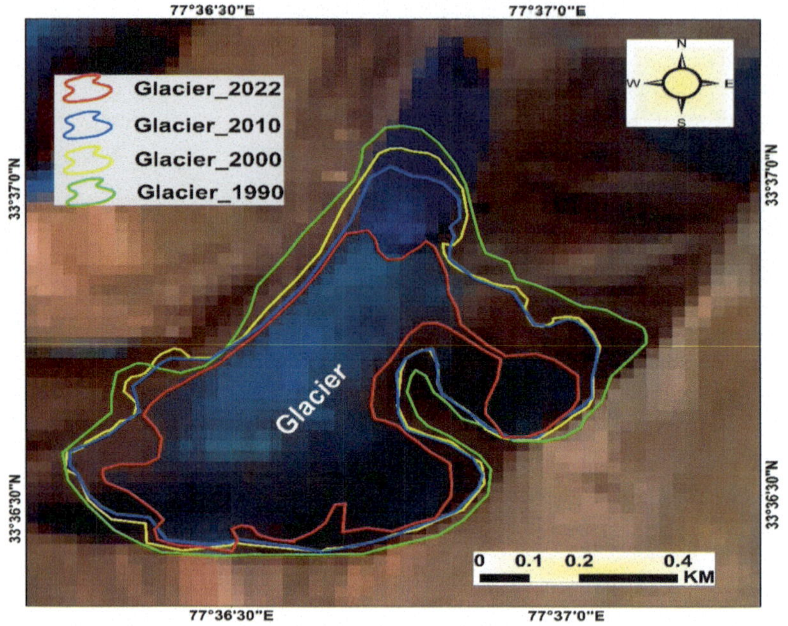

Fig. 14.3 Recession of Gya glacier from 1990 to 2022

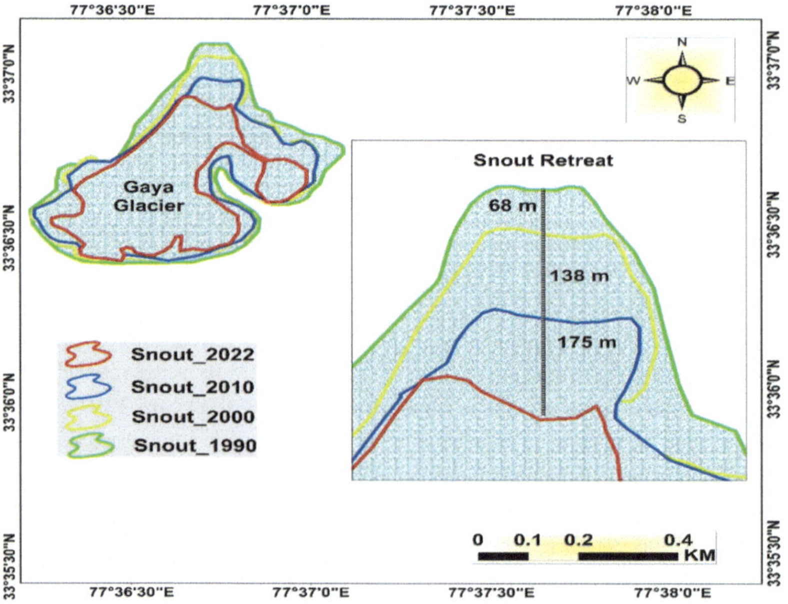

Fig. 14.4 Terminus retreat of Gya glacier from 1990 to 2022

Ladakh Himalayas has been rapidly expanding in recent decades. The analysis of multi-temporal satellite data revealed a significant increase in the surface area of lake from 1990 (0.048 km^2) to 2022 (0.11 km^2), which represents a nearly three-fold increase in the lake area

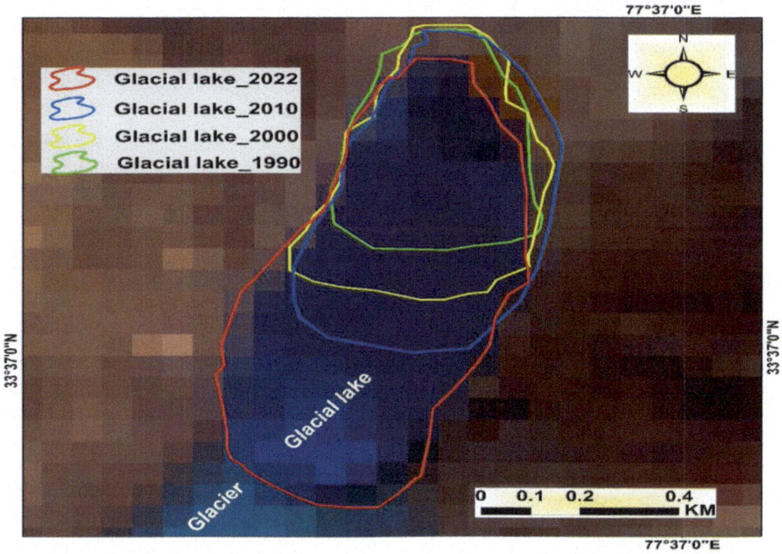

Fig. 14.5 Glacial lake area changes from 1990 to 2022

over the study period. The time series analysis of Landsat and Sentinel-2 images also showed that the lake area has been expanding at an average rate of 0.003 km^2/year between 1990 and 2022. In 1990, the Gya Lake area was recorded as 0.048 km^2. The lake area during 1990 and 2020 has increased by approximately 33.33% (Fig. 14.5). This indicates a notable expansion of the lake over this period of time. Between 2000 and 2010, there was a further increase in the Gya Lake area by approximately 28.15%. From 2010 to 2022, the lake area experienced another significant increase of around 34.14%. This demonstrates a continued trend of expansion and growth over the years. Overall, from 1990 to 2022, the Gya Lake area witnessed a substantial increase of approximately 129.16% (Fig. 14.6). This signifies a substantial overall growth and expansion of the lake during the entire period.

The study also revealed that the expansion of the Gya Lake has been accompanied by changes in the surrounding landscape, including the retreat of the Gya Glacier. These changes have important implications for the hydrology and ecology of the region, as well as for the safety of local communities and infrastructure.

14.4.4 Gya GLOF 2014

The Gya GLOF occurred in August 2014, the resultant flood hit the Gya village located in the Ladakh region (Majeed et al. 2021). The sudden flash flood washed away the agriculture fields, bridges and roads besides causing destruction to the infrastructure in the downstream village (Fig. 14.7). The flood was triggered due to the failure of the terminal moraine of the Gya glacier, which suddenly releases an enormous volume of water, debris and ice in the downstream region. The tributary was heavily impacted due to the flash flood, which also led to the deposition of sediments and debris in the downstream region. The incident brought attention to the growing concern of GLOFs in the Himalayan region which are rising in number due the ongoing climate change, rising temperatures and rapid melting of glaciers.

The Gya glacier and lake is located in the Upper Indus Basin, where the impacts of climate change on the cryosphere and associated hazards are significant. The UIB region has witnessed a remarkable increase in the temperatures with a corresponding decrease in the snowfall, which has resulted to widespread glacier melting,

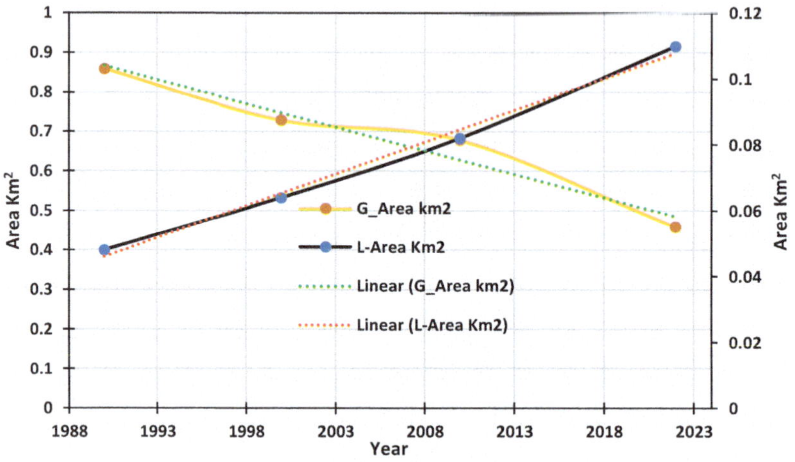

Fig. 14.6 Gya glacier retreat and expansion of associated lake from 1990 to 2022

Fig. 14.7 Damage caused due to 2014 Gya GLOF event in the downstream village of the lake (after Majeed et al. 2021)

development, formation, and expansion of several glacial lakes. The rapid expansion of these lakes is a matter of concern as they are often dammed by loose and unconsolidated ice and debris material, which can be unstable and prone to breach/failure. In the case of Gya Glacier, the collapse of the terminal moraine and the subsequent release of water and debris highlights the potential for catastrophic flood events in the region. A couple of research studies have been conducted to investigate the Gya glacier and lake as well as Gya GLOF 2014. Schmidt et al. (2020) studied the evolution of Gya lake (1969 to 2018), Gya GLOF 2014 and its implication for the risk assessment. This case study of Gya lake further illustrates that, the issue of potentially dangerous glacial lakes being missed by the inventories and therefore proposes an integrated method for finding undocumented GLOFs based on a comprehensive investigation. Another study by Majeed et al. (2021) studied the spatio-temporal evolution of the Gya lake and its feeding glacier as well reconstruction of the Gya GLOF event (2014). Further, the study revealed that, during the GLOF event approximately 25% of the water was drained from the Gya lake with a peak discharge of around 470 m^3 $s-^1$, which inundated around 4 km^2 of the Gya village.

The Gya GLOF 2014 event led to increased attention on the need for effective early warning systems, as well as for better monitoring of potentially dangerous glacial lakes and the surrounding areas. The Indian government has since taken steps to improve the monitoring, assessment and early warning systems in the Ladakh region, as well as to improve infrastructure to withstand future flood events. However, there is still much work to be done to mitigate the risks associated with GLOFs in the region, particularly given the ongoing impacts of climate change on the Himalayan glaciers.

In the HR majority of the glaciers are retreating trend, while on the other hand lakes are showing an increase in the area over the time as revealed by a boom of studies conducted in the area (Ahmed et al. 2021a, b; Shukla et al. 2018; Wang et al. 2015; Zhang et al. 2019, 2021). There are several causes which have contributed to the glacier retreat and glacial lake expansion in the Ladakh region. Climate change is considered to be one of the key reasons of glacier retreat and thereby lake expansion across the globe as well as in the Himalayan region. The rising temperatures and declining trend of precipitation is the major cause of glacier melting and retreating in the Himalayas which has resulted in the rapid expansion of the lakes. The geology and topography of the region are also considered as one of the factors responsible for the formation and expansion of glacial lakes. The presence of impermeable rock formations, for example, can lead to the accumulation of water in depressions and valleys, which can form glacial lakes.

The results of the research work have important implications for the monitoring and management of glacial lakes in the Himalayan region. The rapid expansion of the Gya Lake highlights the potential risks linked with the glacier fed lakes, including the potential of GLOFs and other hazards. The formation of new proglacial lakes downstream also increases the potential for GLOFs and other hazards, which can have substantial impacts on infrastructure and local communities.

The study underscores the need of continued monitoring and assessment of glacial lakes in the Himalayan region, and the need for effective management strategies to mitigate the potential risks and impacts of these lakes. This includes the development of early warning systems and evacuation plans in the event of a GLOF or other hazard, as well as measures to reduce the potential for GLOFs, such as lake stabilisation and drainage works.

14.5 Discussion

The findings of this research emphasise how crucial it is to keep a check on and understand the temporal dynamics of glacier lakes in the Himalayan region. The rapid expansion of Gya lake has significant ramifications for the hydro-ecology of the region as well as for the humans living in the region. The rapid melting of the glacier feeding the Gya lake since the recent past has enhanced the threat of potential

Fig. 14.8 Gya village (agriculture and built-up area) and lake. The subset 'a' and 'b' shows the zoomed view of the Gya village

lake outburst flood and other related hazards which can have the significant impacts on the downstream communities and infrastructure. Figure 14.8 depicts the Gya lake and the Gya village including roads, bridges, and settlement. One of the main findings of this study is the rapid expansion of the Gya lake and melting of the Gya glacier. Between 1990 and 2022, the lake area almost tripled, expanding at an average rate of 0.003 km^2/year. The rate at which Gya lake is expanding is in line with other studies which have reported the rapid growth of the Himalayan Glacial lakes over the last few decades. This is concerning because glacial lake expansion exaggerates the susceptibility of GLOF disaster. The study also identifies the causes which are responsible for the expansion of the Gya lake. These factors include, glacier retreat, climate change as well as geology and geomorphology of the region. The rapid melting of the Gya glacier provides the continuous meltwater supply, which makes it one of the major drivers of the lake expansion. The findings of this study have important implications for the management of glacial lakes in the Himalayan region. The rapid expansion of the Gya Lake highlights the likely hazard and the risks associated with glacial lakes, including the potential for GLOFs and other hazards.

14.6 Conclusions

The rapid melting of glaciers and expansion of lakes in the Himalayan region is a growing concern owning to the potential risk associated with these glacier and lakes, including the potential of GLOFs. The present study has utilized the multi-temporal satellite imageries to understand the evolution of Gya lake and its feeding glacier located in the Ladakh region over the past three decades. The results show that, the Gya glacier has lost an area of about 0.4 km^2 from 1990 (0.86 km^2) to 2022 (0.46 km^2), which is about 46% of the surface area. While the Gya glacial lake on the other hand revealed a significant growth in the area from 0.048 km^2 in 1990 to 0.11 km^2 in 2022, therefore indicating a growth of 126% since the past 32 years. The Gya glacier snout showed a substantial retreat during a 32-year period, similar to other glaciers in the Himalayan region. During the years 1990 and 2022, glacier snout receded by 381 m at an average pace of 11.9 m/yr. The study also identified several potential factors that may have resulted to the rapid expansion of the Gya lake, including glacier retreat, climate change, and geology/topography. The major findings of the present case study have significant implications for the management of glacial lakes across the Himalayas. Effective management strategies are needed to mitigate the potential risks and impacts of glacial lakes, including the development of EWS and evacuation plans in the event of a GLOF or other hazard, as well as measures to reduce the potential for GLOFs. The continuous monitoring and evaluation of glacier lakes in the region is critical to better understand the causes of lake extension and to create efficient management measures to reduce the risk.

References

Ahmed R, Wani GF, Ahmad ST, Sahana M, Singh H, Ahmed P (2021a) A review of glacial lake expansion and associated glacial lake outburst floods in the Himalayan region. Earth Syst Environ 5(3):695–708

Ahmed R, Wani GF, Ahmad ST, Mir RA, Almazroui M, Jain SK, Ahmed P (2021b) Spatiotemporal dynamics of glacial lakes (1990–2018) in the Kashmir Himalayas, India using remote sensing and GIS. Discov Water 1:1–17

Ahmed R, Wani GF, Rather AF, Javaid S, Ahmad ST, Sharma S, Taloor AK (2024) Modelling a glacial lake outburst flood of sona-sar lake in the jhelum basin of kashmir himalayas, India, Using HEC-RAS. In polar ice and global warming in cryosphere regions. CRC Press, pp 37–57

Ahmed R, Ahmad ST, Wani GF, Mir RA, Ahmed P, Jain SK (2022) High resolution inventory and hazard assessment of potentially dangerous glacial lakes in upper Jhelum Basin, Kashmir Himalaya, India. Geocarto Inter 37(25):10681–10712

Ashraf A, Rustam M (2020) Monitoring supraglacial lakes formation and risk of outburst flooding in the Himalayan cryosphere of Pakistan. Inter J Environ 9(1):52–67

Ashraf A, Iqbal MB, Mustafa N, Naz R, Ahmad B (2021) Prevalent risk of glacial lake outburst flood hazard in the hindu kush–karakoram–himalaya region of pakistan. Environ Earth Sci 80(12):451

Begam S, Sen D, Dey S (2018) Moraine dam breach and glacial lake outburst flood generation by physical and numerical models. J Hydrol 563:694–710

Bajracharya, S. R., Mool, P. K., & Shrestha, B. R. (2007) Impact of climate change on Himalayan glaciers and glacial lakes: case studies on GLOF and associated hazards in Nepal and Bhutan. Inter Centre Integr Mt Dev (ICIMOD)

Bandyopadhyay J, Gyawali D (1994) Himalayan water resources: ecological and political aspects of management. Mt Res Dev 1–24

Dasgupta A, Shyamoli S (2021) Himalayan glacial retreat—an interpretation of the glacial mass balance model. Int J Adv Eng Manag (IJAEM) 1054–1063

Gądek B (2014) Climatic sensitivity of the non-glaciated mountains cryosphere (Tatra Mts., Poland and Slovakia). Glob Planet Change 121:1–8

Gupta N, Das J, Paul JX (2022, Dec) Climate change—induced natural disaster: a case study of 2013 Kedarnath disaster, Uttrakhand. In: 5th world congress on disaster management. Routledge, pp 336–344

Hussain W, Khan MA (2023) Climate change-induced Glacial lake outburst floods in Hunza valley of Pakistan: an assessment of indigenous farming community perceptions and adaptation. Environ Dev Sustain 1–25

IPCC (2013) Summary for policymakers. In: Stocker TF, Qin D, Plattner GK, Tignor M, Allen SK, Boschung J, Nauels A, Xia Y, Bex V, Midgley PM (eds) Climate change 2013: the physical science basis. Contribution of working group I to the fifth assessment report of the intergovernmental panel on climate change. Cambridge University Press, Cambridge, pp 3–29

Jain SK, Lohani AK, Singh RD, Chaudhary A, Thakural LN (2012) Glacial lakes and glacial lake outburst flood in a Himalayan basin using remote sensing and GIS. Nat Hazards 62:887–899

Khan AF (2022 Dec) Increasing risk of Glacial lake outburst floods (GLOFs) as a consequence of climate change in the Himalayan region. In: 5th world congress on disaster management. Routledge, pp 405–412

Majeed U, Rashid I, Sattar A, Allen S, Stoffel M, Nüsser M, Schmidt S (2021) Recession of Gya Glacier and the 2014 glacial lake outburst flood in the Trans-Himalayan region of Ladakh, India. Sci Total Environ 756:144008

Mir RA, Jain SK, Lohani AK, Saraf AK (2018) Glacier recession and glacial lake outburst flood studies in Zanskar basin, western Himalaya. J hydrol 564:376–396

Mir RA, Jain SK, Lohani A K, Ahmed R, Ahmad ST, Wani GF (2022) Application of mike 11 for one-dimensional GLOF modeling of a rapidly expanding dalung Proglacial lake, Indus river Basin, Western Himalaya. In Challenges of Disasters in Asia:

Vulnerability, Adaptation and Resilience. Springer Nature Singapore, Singapore, pp 147–161

Mir RA, Majeed Z, Ahmed R, Jain SK, Ahmed ST, Mukhtar MA, Wani GF (2023) Heterogeneity in glacier area loss in response to climate change in selected basins of western himalaya. In: Climate change and environmental impacts: past, present and future perspective. Cham: Springer International Publishing, pp 137–174

Mountford J (2022) A comparative analysis of the Andes and the Himalayas: how climate change effects the health of Glaciers and the social impacts this has on surrounding communities

Naithani S, Saha AK (2021) Social capital and livelihood strategies in response after 2013 Kedarnath disaster (India). Disaster Prev Manage: Int J 30(2):179–193

Pandey P, Ali SN, Champati Ray PK (2021) Glacier-glacial Lake interactions and glacial lake development in the central Himalaya, India (1994–2017). J Earth Sci 1–12

Rafiq M, Romshoo SA, Mishra AK, Jalal F (2019) Modelling Chorabari lake outburst flood, Kedarnath, India. J Mt Sci 16(1):64–76

Rather AF, Ahmed R, Jun C, Bateni SM, Ahmed P, Mir RA (2024) Understanding glacial lake evolution and the associated GLOF hazard in the shyok catchment of the upper indus basin using geospatial techniques. Nat Hazards Rev 25(3):04024014

Rawat M, Ahmed R, Jain SK, Lohani AK, Rongali G, Tiwari KC (2023a) Glacier–glacial lake changes and modeling glacial lake outburst flood in upper Ganga Basin, India. Model Earth Syst Environ 9(1):507–526

Rawat M, Jain SK, Ahmed R, Lohani AK (2023b) Glacial lake outburst flood risk assessment using remote sensing and hydrodynamic modeling: a case study of satluj basin, Western Himalayas, India. Environ Sci Pollut Res 30(14):41591–41608

Riaz S, Ali A, Baig MN (2014) Increasing risk of glacial lake outburst floods as a consequence of climate change in the Himalayan region. Jàmbá: J Disaster Risk Stud 6(1):7

Saifullah M, Liu S, Adnan M, Ashraf M, Zaman M, Hashim S, Muhammad S (2020) Risks of glaciers lakes outburst flood along China Pakistan economic corridor. In: Glaciers and the polar environment. IntechOpen

Sattar A, Goswami A, Kulkarni AV, Emmer A, Haritashya UK, Allen S, ... Huggel C (2021) Future glacial lake outburst flood (GLOF) hazard of the South Lhonak Lake, Sikkim Himalaya. Geomorphology 388:107783

Schmidt S, Nüsser M, Baghel R, Dame J (2020) Cryosphere hazards in Ladakh: the 2014 Gya glacial lake outburst flood and its implications for risk assessment. Nat Hazards 104:2071–2095

Sharma RK, Kumar R, Pradhan P, Sharma A (2022) Climate-induced Glacier retreats and associated Hazards: need for robust Glaciers and Glacial lake management policy in Sikkim Himalaya, India. Climate change: impacts, responses and sustainability in the Indian Himalaya, pp 161–182

Sherpa SF, Shrestha M, Eakin H, Boone CG (2019) Cryospheric hazards and risk perceptions in the Sagarmatha (Mt. Everest) national park and buffer zone, Nepal. Nat Hazards 96:607–626

Shukla A, Garg PK, Srivastava S (2018) Evolution of glacial and high-altitude lakes in the Sikkim, Eastern Himalaya over the past four decades (1975–2017). Front Environ Sci 6:81

Singh H, Varade D, de Vries MVW, Adhikari K, Rawat M, Awasthi S, Rawat D (2023) Assessment of potential present and future glacial lake outburst flood hazard in the Hunza valley: a case study of Shisper and Mochowar glacier. Sci Total Environ 161717

Taylor C, Robinson TR, Dunning S, Rachel Carr J, Westoby M (2023) Glacial lake outburst floods threaten millions globally. Nat Commun 14(1):487

Veh G, Korup O, von Specht S, Roessner S, Walz A (2019) Unchanged frequency of moraine-dammed glacial lake outburst floods in the Himalaya. Nat Clim Chang 9(5):379–383

Verma O (2021) Climate change and its impacts with special reference to India. In: Water, cryosphere, and climate change in the Himalayas: a geospatial approach. Springer International Publishing, Cham, pp 39–55

Wang W, Xiang Y, Gao Y, Lu A, Yao T (2015) Rapid expansion of glacial lakes caused by climate and glacier retreat in the Central Himalayas. Hydrol Process 29(6):859–874

Wang W, Zhang T, Yao T, An B (2022) Monitoring and early warning system of Cirenmaco glacial lake in the central Himalayas. Inter J Disaster Risk Reduction 73:102914

Wangchuk S, Bolch T, Robson BA (2022) Monitoring glacial lake outburst flood susceptibility using Sentinel-1 SAR data, google earth engine, and persistent scatterer interferometry. Remote Sens Environ 271:112910

Yang X, Lu X, Park E, Tarolli P (2019) Impacts of climate change on lake fluctuations in the Hindu Kush-Himalaya-Tibetan Plateau. Remote Sens 11(9):1082

Yao T, Bolch T, Chen D, Gao J, Immerzeel W, Piao S, Su F, Thompson L, Wada Y, Wang L, Wang T, Zhao P (2022) The imbalance of the Asian water tower. Nat Rev Earth Environ 3(10):618–632

Young JDA (2021) Impacts of climate change on water. Inter J Humanit Soc Stud 9(3)

Zhang G, Bolch T, Allen S, Linsbauer A, Chen W, Wang W (2019) Glacial lake evolution and glacier–lake interactions in the Poiqu River basin, central Himalaya, 1964–2017. J Glaciol 65(251):347–365

Zhang M, Chen F, Zhao H, Wang J, Wang N (2021) Recent changes of Glacial lakes in the high mountain Asia and its potential controlling factors analysis. Remote Sens 13(18):3757

Zheng G, Allen SK, Bao A, Ballesteros-Cánovas JA, Huss M, Zhang G, Li J, Yuan Y, Jiang L, Yu T, Chen W, Stoffel M (2021) Increasing risk of Glacial lake outburst floods from future third Pole deglaciation. Nat Clim Change 11(5):411–417

15 Enhancing Resilience: Flood Vulnerability Assessment in the Uttarakhand Himalaya

Nitin Rathi, Sanjeev Sharma, and Susanta Mahato

Abstract

This chapter delves into the critical task of categorizing Chamoli District in Uttarakhand, India, into five distinct flood-vulnerable zones by meticulously evaluating the influence of physiographic, hydrological, and geological parameters on the dynamics of flooding. The study also seeks to unearth the impact of flooding on various land use classes within these different flood hazard zones. A comprehensive set of fifteen parameters has been identified to delineate these flood-vulnerable zones. These parameters encompass drainage density, flow accumulation, river distance, normalized difference water index (NDWI), elevation, aspect, stream power index (SPI), fault distance, soil depth, soil texture, land use/land cover, slope, rainfall, topographical wetness index (TWI), and geomorphology. To achieve this, a sophisticated fuzzy logic approach was employed to create a flood vulnerability zone map for the entire study area. The process involved assigning fuzzy membership values ranging from 0 to 1 to all thematic layers, with the aid of Arc-GIS software. The resulting flood vulnerability zone map reveals a stark reality: approximately 7.13% of the region lies within the highly precarious "very high" vulnerability zone. The significance of this map extends beyond research, offering invaluable insights to decision-makers for formulating effective flood management strategies, rapid response plans, and rescue operations. In a world grappling with increasing climate variability and its devastating consequences, this study's findings hold the promise of bolstering the region's resilience against the looming threat of floods.

Keywords

Flood hazard assessment · Hydrological parameters · Fuzzy logic modeling · Disaster resilience

N. Rathi · S. Sharma
Centre for the Study of Regional Development (CSRD), Jawaharlal Nehru University, New Delhi 110067, India

S. Mahato (✉)
Special Centre for Disaster Research (SCDR), Jawaharlal Nehru University, New Delhi 110067, India
e-mail: mahatosusanta2011@gmail.com

15.1 Introduction

Floods stand as one of the deadliest and most pervasive natural disasters on a global scale, impacting countries across the world with devastating consequences. These catastrophic events result in substantial losses, including loss of human lives, damage to properties, and

S. Talukdar et al. (eds.), *Water Resource Management in Climate Change Scenario*, GIScience and Geo-environmental Modelling, https://doi.org/10.1007/978-3-031-61121-6_15

the destruction of the natural environment. According to the 14th annual report of the International Charter Space and Major Disaster, floods are the most frequent type of hazard occurring on Earth's surface. In recent decades, floods have emerged as a paramount global concern due to their increasing frequency and escalating intensity, attributed in large part to the far-reaching impacts of global climate change. Climate-induced factors, such as abrupt changes in rainfall patterns, have substantially contributed to the rise in flood occurrences and their destructive potential (Hirabayashi et al. 2013; Ibarrarán et al. 2009; Ranger et al. 2011).

In the Himalayan region, flash floods constitute one of the most extreme and devastating events. Uttarakhand, located in northern India, is particularly susceptible to a range of natural disasters, including floods, earthquakes, landslides, forest fires, snow avalanches, road accidents, and droughts. The vulnerability of Uttarakhand to such disasters has been well-documented in various studies (Kumar et al. 2022; Mittal et al. 2019; Pande 2010; Singh et al. 2016). Over the years, the state has faced several catastrophic events, with some of the most recent and deadliest ones including the Malpa Flood in 1998, the Rudraprayag Flood in 2001, the Tehri Flood in 2001, the Pithoragarh Flood in 2002, the Govind Ghat (Chamoli) flood in 2005, the Kedarnath flood in 2013, and the Chamoli Flood in 2021, among others (Pande 2010).

Flood risk assessment typically encompasses the measurement of exposure to damage and loss resulting from flooding, taking into account physio-climatic, hydrodynamic, economic, social, and ecological aspects (Dash and Sar 2020). Of particular concern are the religious sites in Uttarakhand, such as Badrinath, which has experienced recurrent floods in recent years, including events in 2013, 2022, and 2023. For instance, the town of Badrinath, one of the four dhams in Hindu pilgrimage, has witnessed devastating floods due to torrential rainfall within short timeframes. In the broader context, Chamoli District has emerged as one of the areas most susceptible to flash flood events. In February 2021, this region experienced a deadly flood, resulting in the tragic loss of more than two hundred lives. To address the recurring threat of flooding in Chamoli and similar regions, an innovative approach is required. Fuzzy logic, known for its structured and highly adaptable nature, presents a promising avenue for addressing complex decision-making problems characterized by multiple criteria and parameters. Table 15.1 provides valuable insights into flood vulnerability and risk assessment in mountainous regions, shedding light on various aspects of multi-hazard mapping, susceptibility, and potential causes of flooding.

In recent years, the state of Uttarakhand in northern India has faced a series of devastating flood events, causing significant social, economic, and environmental repercussions. Uttarakhand's distinctive geographical features, characterized by its mountainous terrain, abundant biodiversity, and a network of rivers, including the Ganges and its tributaries, make it susceptible to flash floods and landslides. These natural disasters are often exacerbated by the region's unique geography and its vulnerability to extreme weather events. Chamoli District, although not experiencing recurrent floods, faces extreme flooding events when they occur, as witnessed during the Chamoli flash flood in February 2021. Various factors, including glacial lake outbursts, cloudbursts, dam breaches, and heavy rainfall contribute to the flood risk in this district (Siddique et al. 2022; Thakur et al. 2019). The primary objective of this case study is to identify flood-vulnerable zones within Chamoli District by considering both flood conditioning or causing factors and flood inventories. The insights generated from this study will serve as a basis for the development of a comprehensive flood management plan aimed at reducing the vulnerability of the region and enhancing its resilience to future flood events. In the subsequent sections of this research, we will delve into the methodologies employed, data collection and analysis procedures, and the application of fuzzy logic to identify flood-prone areas within Chamoli District, Uttarakhand. Through this research, we aim to contribute valuable insights that can inform flood risk reduction strategies and protect the lives and livelihoods of the region's residents.

Table 15.1 Previous studies on flood in mountainous regions

Objective	Study area	Findings	References
To prepared multi-hazard map using machine learnings	Saveh city of Iran	Bayesian generalized linear model (BGLM) is the best model for landslide susceptibility mapping	Pal et al. (2022)
To assess the multi-hazard vulnerability of Persian and Oman Gulf	Southern coasts of Iran	Some districts were identified as the most vulnerable to multi-hazard based on the nine parameters	Mafi-Gholami et al. (2019)
To determine the flash flood potential index and identify vulnerable zones	Mandakini river basin	Approximately 40% area lies in the very high susceptibility zones	Singh and Pandey (2021)
To know the impact of seismic vulnerable zones on floods	Kedarnath	Uncontrolled and unplanned growth of urban places is the main reason of multi-hazards	Sati and Gahalaut (2013)
To find the gaps and implementation issues in policies and plans in context of land use in disaster-prone areas	Bhagirathi Valley, Uttarakhand	There is a need to improve multi-institutional coordination and implementation of policies and plans in flood-vulnerable zones	Dash and Punia (2019)
To assess the potential causes and process of flooding	Chamoli District, Uttarakhand	Earthquake triggered landslide was the main causes of the flood in downstream	Pandey et al. (2021)

15.2 Study Area

Chamoli District, nestled in the picturesque Garhwal region of Uttarakhand, is a land of astounding natural beauty and geographical significance (Fig. 15.1). Encompassing an expansive area of 7820 km^2, it ranks as the second-largest district in Uttarakhand (Singh et al. 2010). Situated in the heart of the central Himalayas, Chamoli boasts two prominent mountain ranges, namely the Dhauliganga and the iconic Nanda Devi range. Among its towering peaks, the district proudly cradles Nanda Devi, India's second-highest summit, soaring majestically at 7816 m above sea level. Other formidable peaks that grace the Chamoli skyline include Kamet (7756 m), the Elephant peak, fondly known as "Hathi Parbat" (6727 m), Abi Gamin (7355 m), Mukut Parbat (7242 m), Dunagiri (7066 m), and Saraswati Parbat (6940 m). This region is administratively divided into ten tehsils: Joshimath, Chamoli, Ghat, Karnaprayag, Pokhari, Jilasu, Gairsain, Adibadri, Tharali, and Narayanbagar. Chamoli's pristine landscape serves as the cradle for two significant glaciers, Gangotri and Pindari, which are the lifelines of major rivers in the area (Anthwal et al. 2006). The district is graced by the presence of the Alaknanda, Pindar, and Ramganga rivers, with the Alaknanda serving as the primary tributary of the Ganga and covering the largest expanse of the district. Additional tributaries such as Dhauli Ganga and Birhi Ganga enrich the river network of Chamoli.

The region's rich biodiversity is a testament to its ecological significance, featuring subtropical forests in the low-altitude areas and lush coniferous forests in the higher elevations. Agriculture forms the backbone of the local economy, with crops like rice, wheat, potatoes, pulses, millets, and seasonal vegetables thriving in the fertile soil (Diduck et al. 2007; Tiwari and Rawat 2017). Notably, Chamoli is home to a diverse range of wildlife, including the Himalayan black bear, snow leopard, musk deer, barking deer, sambar, tiger, and brown bear (Badola et al. 2018). The district also houses the renowned Kedarnath Wildlife

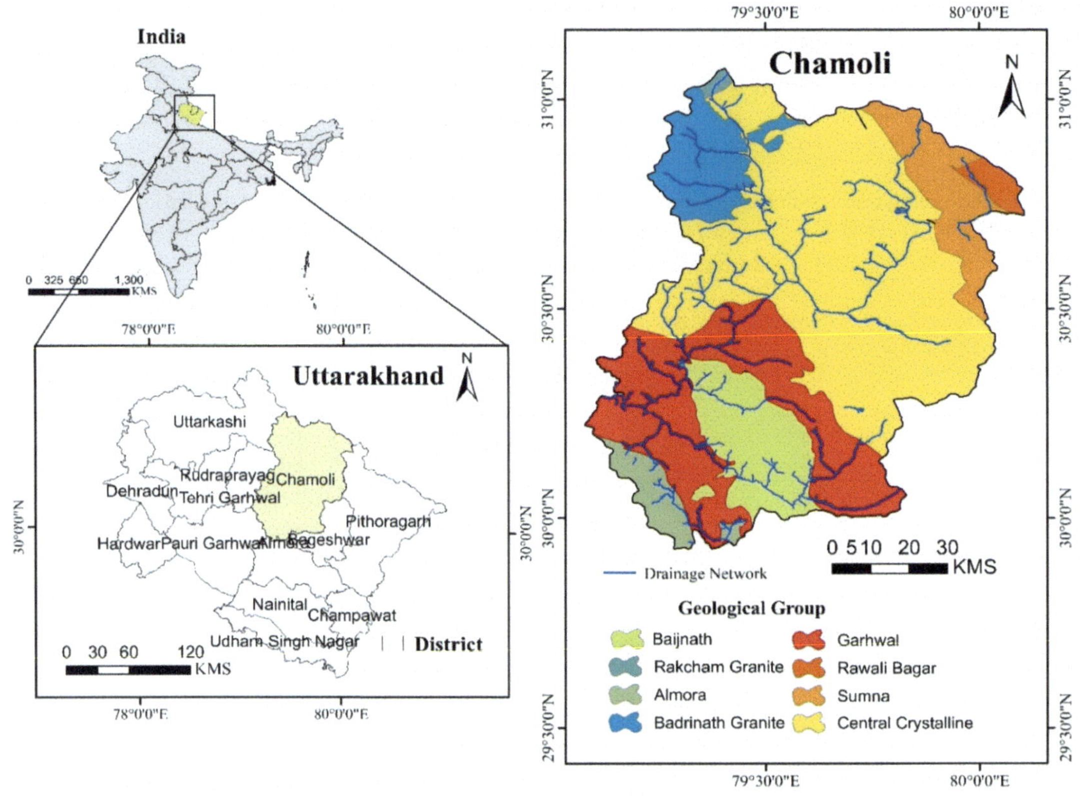

Fig. 15.1 Study area—Chamoli district, Uttarakhand

Sanctuary, spanning 975 km^2, extending into both Chamoli and Rudraprayag Districts (Malik and Bhatt 2015). Beyond its natural wonders, Chamoli beckons tourists and pilgrims from around the world to its sacred and tourist destinations. Among the most frequented are the Badrinath temple, Valley of Flowers National Park, Hemkund Sahib, Auli, Govind Ghat, Rudraprayag, and Nanda Devi National Park, each offering a unique blend of spirituality and natural splendor.

15.3 Material and Methods

This chapter relies exclusively on secondary data meticulously sourced from various reputable outlets and official datasets furnished by government entities. Employing a rigorous process of multi-collinearity analysis, a selection of 12 flood-condition factors was meticulously identified for inclusion in this study. To enhance clarity and precision, these factors were thoughtfully categorized into five distinct types based on their inherent physical attributes: topographical, hydrological, soil-related, geological, and environmental (Table 15.2).

15.3.1 Identification of Flood Conditioning Factors

The flood vulnerability zone map was developed based on the comprehensive identification of fifteen parameters derived from the physiographic, hydrologic, and geological attributes of the study area. In the context of the fuzzy logic methodology, the selection of parameters or criteria holds paramount importance, as the entire outcome of the analysis hinges on the influence

Table 15.2 Flood conditioning factors

Type	Factors	Spatial resolution	Sources
Topographical	Elevation	12.5 m	Alaska satellite
	Slope	12.5 m	DEM
	Aspect	12.5 m	DEM
	SPI	12.5 m	DEM
Hydrological	Rainfall	0.25 m	IMD
	NDWI	30 m	Landsat
	Flow accumulation	12.5 m	DEM
	TWI	12.5 m	DEM
	Drainage density	12.5 m	DEM + OSM
Soil	Soil texture	Vector layer	Soil survey of India
	Soil depth	Vector layer	Soil survey of India
Geological	Fault distance	Vector layer	Geological survey of India
	Geomorphology	Vector layer	Geological survey of India
Environmental	LU/LC	30 m	Landsat-9 (USGS)
	River distance	Vector layer	Survey of India

of these chosen factors. Drawing upon the insights gleaned from previous research endeavors employing fuzzy logic techniques for flood vulnerability zone mapping, it is evident that the selection of criteria has consistently aligned with the unique geographical characteristics influencing the physical processes underlying flood mechanisms (Kazakis et al. 2015; Papaioannou et al. 2015; Kabenge et al. 2017; Das et al. 2018; Dash and Sar 2020).

Considering the geographical context of Chamoli, a comprehensive set of fifteen criteria was thoughtfully employed, spanning hydrologic, lithological, and physiographic characteristics, among others. These fifteen criteria encompass drainage density, flow accumulation, river distance, normalized difference water index (NDWI), elevation, aspect, stream power index (SPI), fault distance, soil depth, soil texture, land use/land cover, slope, rainfall, topographical wetness index (TWI), and geomorphology.

The parameter drainage density is primarily contingent on the density of the drainage network and plays a pivotal role in influencing flood intensity. Areas characterized by higher drainage network density are more susceptible to flooding due to their proximity to water bodies, although other factors such as absorption capacity and slope also contribute significantly (Ogden et al. 2011). The raster layer depicting drainage density was meticulously derived from a vector layer of the drainage network, which was sourced from OpenStreetMap.

Elevation, another critical parameter, significantly impacts water flow, drainage capabilities, and runoff coefficients. Low-lying regions with gentle slopes are particularly vulnerable to floods and waterlogging, and many flood risk assessment and prediction models rely heavily on elevation data (Garrote and Bras 1995; Manfreda et al. 2011; McClean et al. 2020). The raster layer illustrating elevation zones was generated using the ALOS PALSAR Digital Elevation Model (DEM) dataset with a resolution of 12.5 m.

Slope is a direct and influential flood conditioning factor, crucially governing water flow, drainage capacity, and infiltration rates (Aryal et al. 2020). The degree of steepness of the land surface, represented by slope percentage, is particularly pertinent, with higher slopes correlating to an increased risk of flash floods, especially in mountainous regions. The thematic layer representing slope was meticulously created using the ALOS PALSAR DEM dataset with a resolution of 12.5 m.

Rainfall is intrinsically linked to flooding and stands out as one of the most critical parameters for flood occurrence in mountainous regions. Excessive rainfall often leads to a high volume of water, and if this water exceeds the normal flow capacity, it results in flooding (Arnaud et al. 2002; Guhathakurta et al. 2011). Understanding rainfall patterns, monitoring weather forecasts, and implementing sustainable land management practices are essential steps in managing the risks associated with rainfall and floods. The rainfall map was meticulously prepared using model data from the tropical rainfall measuring mission (TRMM) for the year 2020.

Geomorphology assumes a pivotal role in influencing flood causation in any area. Factors such as drainage pattern, water movement and storage, channel morphology, and sediment transportation collectively impact the intensity of flooding (Sofia and Nikolopoulos 2020). Geomorphology significantly influences floodplain size, elevation, and slopes, ultimately affecting the extent and severity of flooding (Beven 1987). The geomorphology map was sourced from the Bhukosh portal of the Geological Survey of India.

Soil texture emerges as a critical factor that influences the intensity of floods. It determines variables such as infiltration capacity, soil saturation, surface runoff, soil erosion, and slope stability, all of which are dependent on soil types. Soil with high infiltration capacity reduces the likelihood of flash floods, whereas soil with low infiltration capacity increases the risk of flooding. Soil erosion, on the other hand, is directly linked to soil type and indirectly contributes to flooding by potentially blocking drainage systems downstream. The soil texture map was prepared using the map provided by the Soil Survey of India.

Normalized difference water index (NDWI) relies on satellite data and is primarily employed to differentiate water pixels from all other types of pixels. While NDWI serves as an indirect factor for delineating flood zones, it specifically identifies water pixels, which, when combined with historical data, can be used to calculate inundated areas. NDWI is a valuable tool for time-series analysis of water accumulation, and many flood assessment methodologies incorporate NDWI in conjunction with other direct factors for water detection, flood mapping, and monitoring (Jain et al. 2005; Singh and Kansal 2022). NDWI calculations were carried out using Landsat satellite imagery based on the formula: NDWI = (Green − NIR)/(Green + NIR) (Gao 1996; McFeeters 1996).

Topographic wetness index (TWI) is a significant indicator influencing flood dynamics. It is based on flow accumulation and slope angle and quantifies topographic controls on hydrological processes (Besnard et al. 2013; Pourali et al. 2016). TWI also offers a highly cost-efficient approach to flood mapping and zoning (Pourali et al. 2016). The formula for calculating TWI is: TWI = Ln (Flow Accumulation Scaled/Tan(slope)).

Flow accumulation assumes a pivotal role in influencing flooding, with its value contingent on the density of the drainage network within the study area. ALOS POLSAR data with a spatial resolution of 12.5 m was used to create the spatial layer depicting flow accumulation. Flood occurrences are contingent upon the values of flow accumulation, with higher values in a given area indicating a greater likelihood of inundation and vice versa (Dash and Sar 2020).

Fault distance plays a significant role in shaping the topography and hydrology of the region (Pham et al. 2020). Fault distance can impact hydrological processes at the local level, influencing factors such as fissures and fractures that affect water flow and storage. Faults can lead to flash flooding through the failure of levees and earthen dams (Shafapour Tehrany et al. 2017; Pham et al. 2020). To create the thematic layer depicting fault distance, fault line data was sourced from the Geological Survey of India and processed using Arc-GIS software.

River distance is a critical factor in flood mapping and the sustenance of life. Areas located at a certain distance from rivers are generally less vulnerable to flooding, while those in close proximity to river channels are more prone to flooding (Yamazaki et al. 2011; O'Neill et al. 2016). Human interventions, such as construction on riverbanks and alterations to the topography, play a significant role in determining the distance between a particular location and a river. The thematic layer depicting river distance was prepared using data from the Geological Survey of India and processed using Arc-GIS tools.

Land use land cover (LU/LC) is a key determinant of infiltration capacity, water movement, runoff, and water retention. The presence of impervious surfaces and concrete increases surface runoff while decreasing infiltration, thereby elevating the risk of flooding (Posthumus et al. 2008; Saghafian et al. 2008; Swanson et al. 1998). Anthropogenic activities like urbanization, deforestation, and other development practices directly contribute to flooding by encroaching on floodplains, reducing infiltration, and altering evapotranspiration patterns. The land use/land cover map of Chamoli was prepared using Erdas Imagine software through supervised classification of Landsat-9 data.

Soil depth plays a critical role in determining infiltration capacity, water storage, percolation rates, runoff, and, consequently, flood intensity. Shallower soils saturate quickly, increasing the likelihood of flash floods, while deeper soils have a greater capacity to retain water, reducing soil erosion (Anni et al. 2020; Begum et al. 2006). The soil depth layer was created by georeferencing and digitizing the map provided by the Soil Survey of India.

Aspect significantly affects soil moisture, snowmelt, vegetation cover, infiltration, and drainage patterns, all of which influence flood intensity in mountainous regions. Slopes that receive direct sunlight can alter melting rates, soil dryness, vegetation cover, and other factors, thereby increasing the risk of flooding (Haghizadeh et al. 2017; Nachappa et al. 2020). The raster layer depicting aspect was generated using the 12.5-m Alos Palsar DEM and Arc-GIS software.

15.3.2 Flood Conditioning Factor Integration Model—Fuzzy Logic Approach

The present study employs a fuzzy logic approach to identify flood-vulnerable zones, utilizing the fundamental components of fuzzy systems: fuzzy sets, fuzzy production, and fuzzy membership rules (Mahato and Pal 2019). Fuzzy logic revolves around the concept of membership grades, which determine the extent to which objects belong to a particular set. This approach utilizes fuzzy set operators such as union (And), intersection (Or), and product to group two membership functions.

In the context of creating a flood-vulnerable zone map, fuzzy logic methodology has been adopted. Through the integration of thematic layers using fuzzy overlay techniques, flood vulnerability zones have been delineated. Fuzzy logic assigns values to locations on a scale from 0 to 1, similar to Boolean logic, which operates between 0 and 1. Fuzzy logic is often considered closely aligned with human reasoning as it quantifies the degree of truth.

The workflow for implementing fuzzy logic involves several key criteria:

1. Collection of all criteria layers.
2. Assignment of fuzzy membership values to all data layers.
3. Execution of the fuzzy overlay function.
4. Verification and application of results.

Fuzzy Membership

Assigning fuzzy membership values is pivotal for analyzing the impact of different thematic layers. In the context of creating a flood vulnerability zone map, fuzzy membership values are assigned to each layer based on their influence on flooding.

Fuzzy Overlay Functions

Fuzzy overlay functions are critical for combining thematic layers effectively. These functions involve understanding the relative relationships between layers and the degree of interaction among them. Various fuzzy operators, such as fuzzy Or, fuzzy And, fuzzy Product, fuzzy Sum, and fuzzy Gamma are employed for overlaying thematic layers. These operators, rooted in fuzzy set theory, play a crucial role in this process.

Fuzzy AND: logical intersection

AND fuzzy algorithm is;

μ combination = MIN (μA, μB,.........μN)

Fuzzy OR: logical union

μ combination = MAX (μA, μB,.............μN)

The Fuzzy Product is, μ combination = μA* μB*...... *μN

The Fuzzy Algebraic Sum is, μ combination = μA + μB + ... + μN

Another approach, i.e., Fuzzy Gamma is represented as;

μ combination = (Fuzzy Algebraic Sum)γ * (Fuzzy Algebraic Product)$^{(1-\gamma)}$

Here, the γ represents the parameter which ranges between (0 and 1). If the γ value is 1 that means same combination with fuzzy sum and if it is (γ) 0 that means the combination is equal with the Fuzzy algebraic product (Rather and Andrabi 2012).

This study employed a comprehensive array of fifteen parameters, each meticulously chosen to capture the various facets of physiographic, hydrologic, and geological attributes in the study area. These parameters collectively form the basis for creating a flood vulnerability zone map using fuzzy logic methodology to the specific geographical characteristics of Chamoli (Fig. 15.2). Each parameter contributes uniquely to the overall understanding of flood susceptibility, allowing for a more robust and accurate flood vulnerability assessment in this mountainous region.

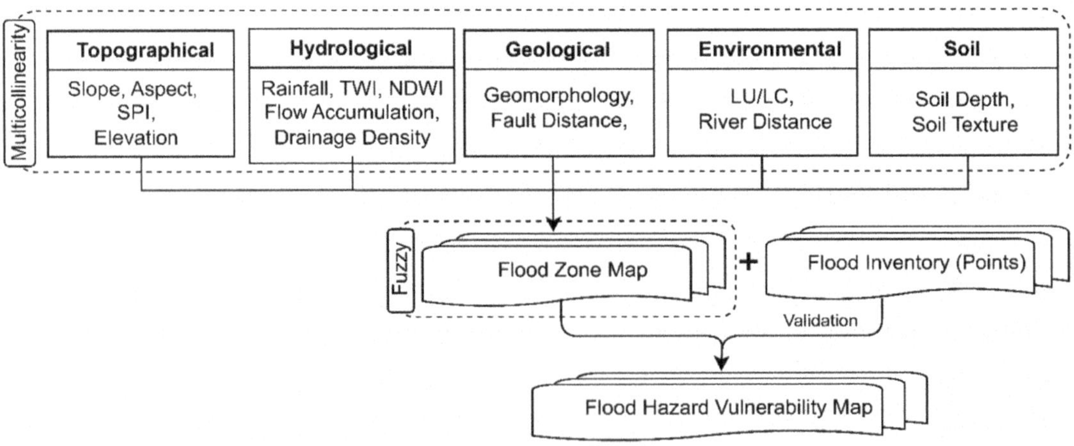

Fig. 15.2 Methodological flow of the study

15.4 Results and Discussion

15.4.1 Flood Conditioning Factors

The results of our study highlight the critical role of several key parameters in assessing flood vulnerability in Chamoli District. Firstly, drainage density and elevation emerged as crucial factors influencing the drainage capability and flood impact. Areas with lower to moderate elevation combined with varying drainage densities were found to be particularly susceptible to flooding. Notably, towns, like Nandprayag, Karna Prayag, Badrinath, Gopeshwar, Gairsain (the summer capital of Uttarakhand), and Narayan Bagar were situated in regions characterized by high drainage density, indicating a higher flood risk. Elevation played a significant role, with lower elevations in proximity to the Alaknanda River indicating a greater vulnerability to flooding, making low-lying areas more prone to waterlogging and river-induced floods (Fig. 15.3).

Slope percentage emerged as another influential factor, with higher percentages indicating a reduced risk of waterlogging and flash floods. Conversely, low-slope regions, especially those near riverbanks and plains, were identified as more vulnerable to flooding. The presence of fault lines within the study area, categorized based on their distance, played a significant role at the local level in shaping hydrological processes. Areas with shorter fault distances were found to be at higher risk of flooding, as represented by the red pixels in our mapping.

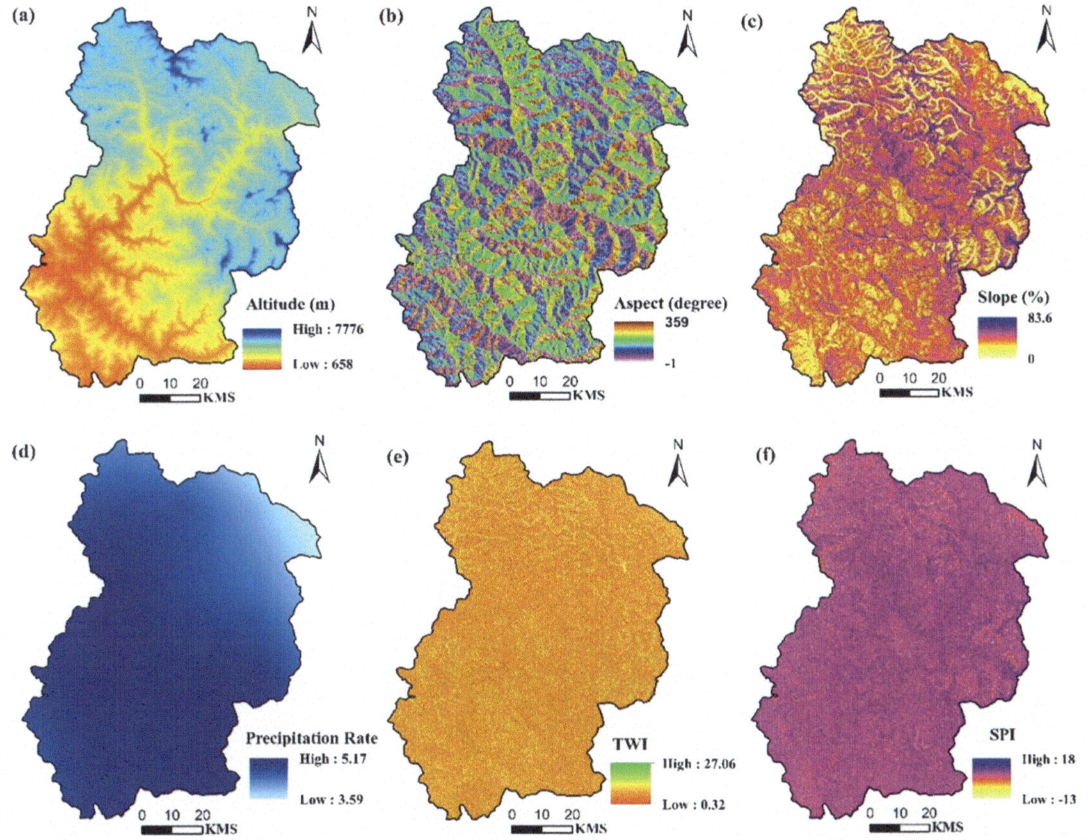

Fig. 15.3 Data layers for flood-vulnerable zones mapping

Moreover, topographic wetness index (TWI) values correlated with surface saturation, suggesting that higher TWI values corresponded to greater flood risk.

Stream power index (SPI) values served as indicators of erosive and sedimentation power within streams and contributed significantly to flood risk mapping. Locations with higher SPI values were identified as having a greater potential for sediment transport and erosion, elevating the risk of flooding. Precipitation rate, a crucial factor in flood susceptibility, varied across the district. Areas with high precipitation rates, such as the southward-facing slopes, were identified as more flood-prone. Geomorphological features, including glacial and riverine elements, played an essential role in determining flood conditions, with heavily dissected hills and valleys dominating the landscape.

Soil texture emerged as a significant factor, categorized into seven types. Sandy and loamy soils exhibited lower flood vulnerability due to their high infiltration capacity, while fine and silty soils were more prone to flooding (Fig. 15.4). Flow accumulation, indicating water concentration, varied across the study area, with areas of low water concentration depicted in dark green and green, while high water concentration regions were represented in red and orange. Soil depth also influenced flood susceptibility, with high-depth soils having greater water storage capacity and reduced vulnerability to flooding. Land use and land cover (LULC) classifications revealed that built-up and agricultural areas were more susceptible to flooding due to their proximity to water bodies.

The normalized difference water index (NDWI) was a valuable tool for identifying

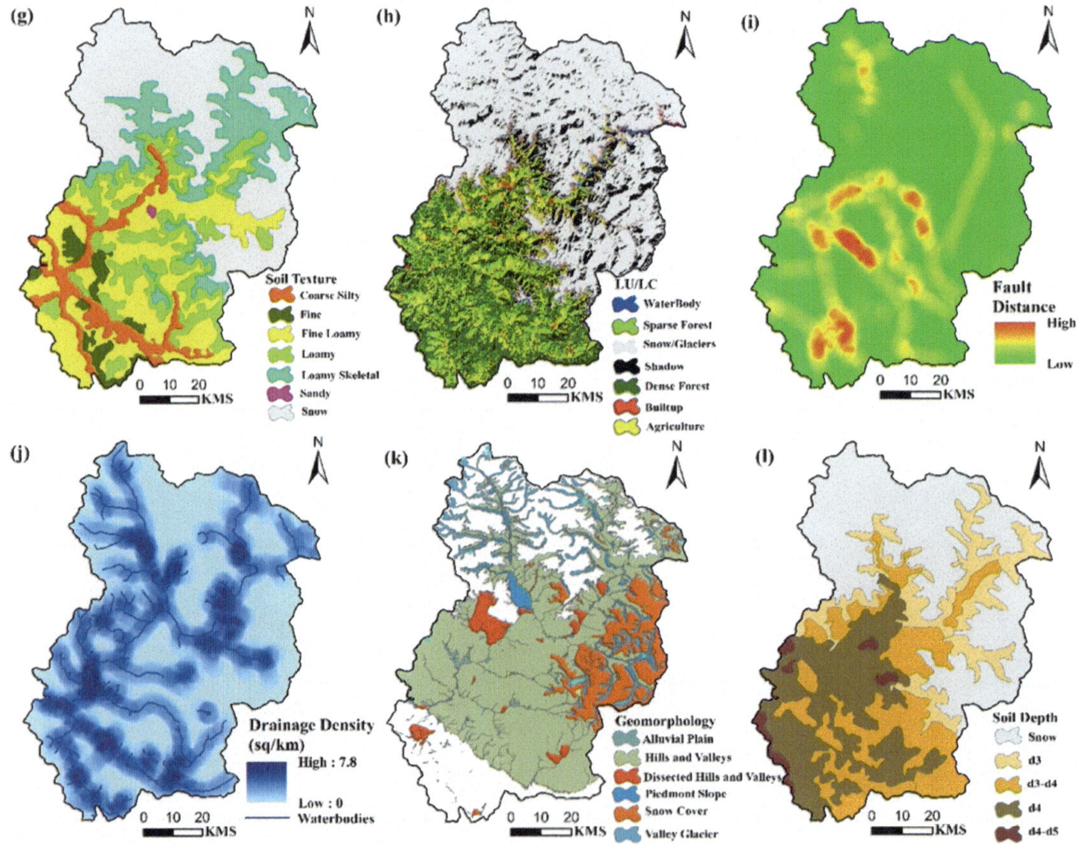

Fig. 15.4 Data layers for flood-vulnerable zones mapping

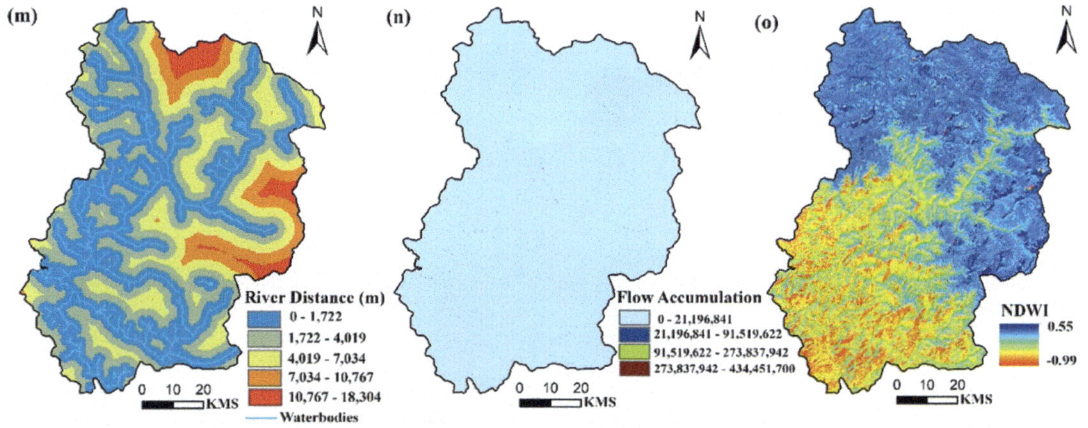

Fig. 15.5 Data layers for flood-vulnerable zones mapping

water-logged areas, with positive values indicating water pixels. Finally, aspect played a significant role, with north-facing aspects more prone to flooding compared to south-facing slopes (Fig. 15.5). The proximity to rivers also influenced flood vulnerability, with areas nearer to rivers at a higher risk of flooding due to factors such as channel capacity and topography.

15.4.2 Flood-Vulnerable Model

The flood vulnerability model categorized Chamoli District into five distinct flood vulnerability zones based on fifteen natural flood conditioning factors. These zones are characterized as follows: very low (21.13%), low (16.86%), moderate (25.05%), high (29.79%), and very high (7.14%) (Fig. 15.6). One significant finding of this study is that seven out of eight previously recorded flood events occurred in the very high and high flood vulnerability zones. These findings underscore the importance of these zones as areas highly prone to flooding. Specifically, the study reveals that nearly one-fifth (21.13%) of the total area falls within the very low flood vulnerability zone, while the low vulnerability zone covers 16.86% of the district. The moderate zone comprises 25.05%, the high zone encompasses 29.79%, and the very high flood vulnerability zone accounts for 7.14% of the total district area. Further analysis using data from Open Street Map indicates that 325 built-up locations are situated within the very high flood vulnerability zones. These locations include 234 hamlets, 12 isolated dwellings, 7 major towns, and 72 villages, emphasizing the significance of addressing flood risks in these areas.

The identification of flood vulnerability zones in Chamoli District provides valuable insights for disaster preparedness and mitigation efforts. It is evident that major towns, such as Karnaprayag, Gopeshwar, Gairsain, Gauchar, Narayan Bagar, Nandprayag, and Badrinath are situated in high and very high flood vulnerability zones. Additionally, numerous villages and religious sites, including Rudraprayag and Badrinath, are also located in these vulnerable areas (Table 15.3). These findings underscore the urgent need for comprehensive flood risk management strategies, including early warning systems, infrastructure development, and community preparedness measures, to mitigate the impact of potential floods and safeguard the lives and property of the residents in these vulnerable zones.

15.4.3 Validation

The flood vulnerability zone map, generated using the eight flood conditioning factors and the fuzzy approach, was subjected to a validation process by comparing it with a flood

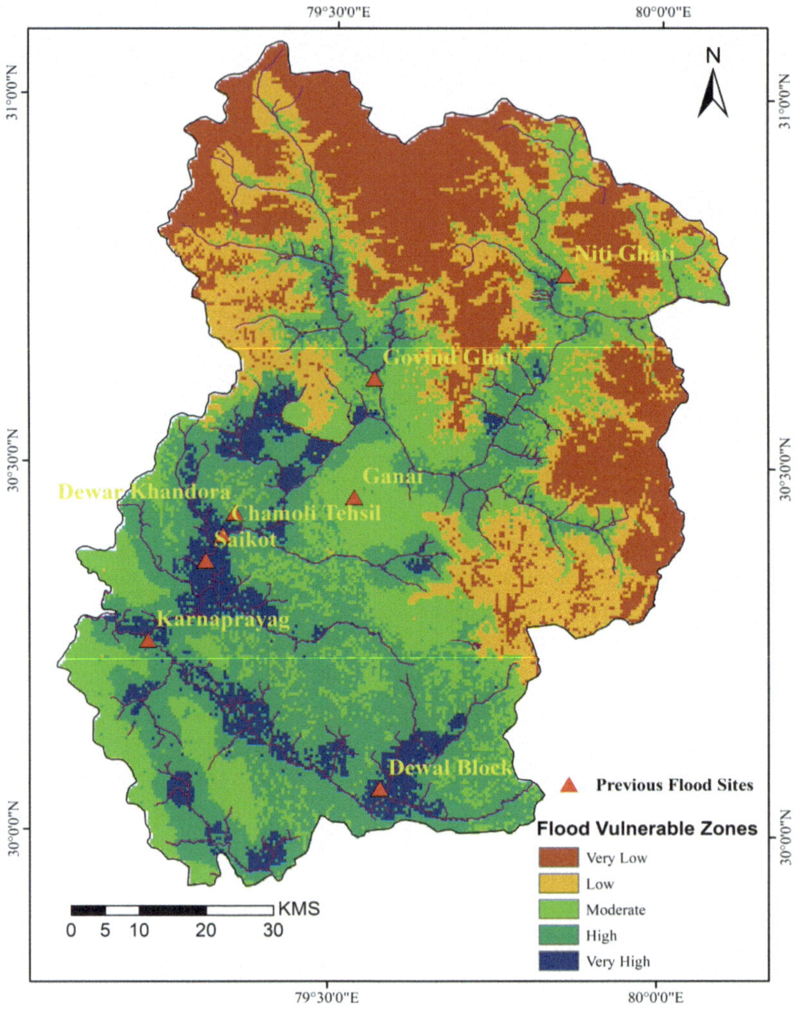

Fig. 15.6 Model of flood-vulnerable zones

Table 15.3 Towns situated in very high flood vulnerability zones

Sl. No.	Town name	Population (Census, 2011)	Faced flood in history
1	Karnaprayag	8297	February 2021
2	Gopeshwar town	21,447	11 August 2006
3	Gairsain	62,412	July 2023
4	Gauchar	8864	June 2013
5	Narayan Bagar	362	June 2014
6	Nandprayag	1641	February 2021
7	Badrinath	2438	2013, 2022, 2023

inventory map created from historical flood records, which were obtained from previous research and government reports. This validation aimed to assess the accuracy and reliability of the vulnerability zone map in representing actual flood events in Chamoli District. The flood inventory map was developed with the assistance of historical data and satellite imagery from sources such as Google Earth, and it documented the locations and dates of previous flood events in the region. The validation process involved overlaying the flood vulnerability zone map with the flood inventory map to determine

Table 15.4 Previous flood locations in Chamoli

S. No.	Location/Place	Date
1	Ganai	20 July 1970
2	Saikot	17 June 1979
3	Dewar Khadora	16 August 1991
4	Govind Ghat	29–30 June 2005
5	Chamoli Tehsil	11 August 2006
6	Rudraprayag	16, 17 June 2013
7	Dewal Block	8–9 August 2019
8	Niti Ghati in Joshimath Tehsil	19 July 2018

the extent to which the identified flood-prone areas corresponded to the historical flood events (Table 15.4).

The validation process revealed compelling results, with a significant correlation between the flood vulnerability zones and the locations of previous flood events. Out of the eight documented flood locations, seven were found to be situated within the high and very high flood vulnerability zones, as identified by the vulnerability zone map. This validation outcome underscores the effectiveness of the developed flood vulnerability model in accurately identifying areas prone to flooding in Chamoli District.

15.4.4 Comparative Analysis with Prior Research

Prior research in Chamoli District predominantly concentrated on flood hazard mapping based on temporal changes in land use and land cover, slope, and aspect (Bhatt et al. 2014), and the examination of cloudburst events and their repercussions on local communities (Khanduri 2021). While these studies undoubtedly provide valuable insights, our research takes a comprehensive approach, encompassing fifteen flood conditioning factors across four distinct categories. Although changes in land use and land cover, particularly vegetation alterations and shifts in rainfall patterns have been recognized as flood triggers in Chamoli District (Verma et al. 2022), it's essential to acknowledge that floods are influenced by an array of factors such as geomorphology, soil type, and proximity to rivers. Some studies have primarily focused on flood hazard zoning (Bhatt et al. 2014), the environmental impact of floods (Sati and Kumar 2022), and flood management challenges (Kansal and Singh 2022). Managing floods in mountainous regions is an intricate process, requiring a blend of indigenous knowledge and advanced technologies, along with regulations on flood catalyst development endeavors like dam construction and hydropower projects.

15.4.5 Degree of Influence

The primary factors contributing to flooding in the study area are natural, as discussed and utilized in this research. Nevertheless, anthropogenic factors, such as the construction of new roads and the widening of existing roads in floodplains, exacerbate overall flood losses. Natural factors may predispose Chamoli District to disasters, but human activities act as catalysts, transforming natural hazards into man-made disasters. Geological conditions, hydro-meteorological processes, land use/land cover changes, developmental activities, dam and hydropower project construction, inadequate infrastructure, deficient management and planning, and encroachments in river floodplain areas are the principal factors enhancing the frequency, intensity, and occurrence of floods. Avalanche occurrences, which can trigger floods, merit inclusion in future flood zonation and mapping studies. The escalating population, while not a flood conditioning factor, is a pivotal concern and a significant contributor to not only floods but also other disasters, as the increased resource demand forces the depletion of natural resources, upsetting the balance in nature. Tourism, though not a flood conditioning factor, indirectly amplifies flood losses by encouraging developmental activities in floodplains, such as the establishment of shops, hotels, and restaurants.

15.4.6 Limitations and Future Prospects

This study predominantly relies on secondary and geospatial data, and as such, the level of accuracy may not match research grounded in primary data collected through field surveys. Variability may occur in specific locations classified as non-very high risk but still prone to future flooding due to anthropogenic factors. Therefore, precise predictions may vary since they hinge on the influence of individual factors with varying weights. For instance, a flood might transpire primarily due to heavy rainfall or a cloudburst, or it could stem from a natural dam breach or lake outburst, as witnessed in the Kedarnath flood of 2013.

Future research should focus on the development of accurate early warning systems, real-time monitoring of all glacial lakes, rainfall prediction, and landslide monitoring, especially concerning landslides leading to outburst lake floods (LLOF). Developmental activities in floodplains, near water bodies, and within high flood vulnerability zones should be promptly restricted. Geologists and geomorphologists should play a pivotal role before and during any developmental activities to ensure environmental and flood risk assessments are adequately addressed.

15.4.7 Insights for Disaster Management

Our research underscores the need for a multifaceted approach to flood risk assessment in Chamoli District, integrating both natural and human-induced factors. By understanding the intricate interplay between geological, hydro-meteorological, and anthropogenic elements, local authorities and stakeholders can formulate more effective disaster management strategies. Moreover, our study reiterates the significance of preserving natural resources and enforcing regulations to safeguard against flood catalyst developmental activities in vulnerable regions. This comprehensive approach, guided by accurate early warning systems and real-time monitoring, is pivotal for enhancing disaster resilience in Chamoli District and other similar mountainous regions.

15.5 Conclusion

In conclusion, this study has significantly advanced our understanding of flood vulnerability in Chamoli District, shedding light on the intricate web of natural and anthropogenic factors that influence flood occurrences. The final flood vulnerability map, based on fifteen comprehensive flood conditioning factors, has been pivotal in categorizing the district into five distinct vulnerability zones, ranging from very low to very high. Notably, approximately 21.13% of Chamoli District's expanse has been identified as a very low flood vulnerability zone, offering a relative safe haven from the perils of flooding. However, our analysis has revealed a stark reality—a striking 7.13% of the region falls within the very high vulnerability zone, encompassing all historical flood events. Alarmingly, a cumulative 40% of the district's territory lies within the high and very high flood hazard zones, emphasizing that Chamoli is profoundly susceptible to future flooding. The flood vulnerability zone (FVZ) map generated in this study carries profound implications for disaster management. It provides crucial insights into the identification of hotspots, areas highly vulnerable to flooding. These findings serve as a cornerstone for the development of emergency preparedness plans, informed land use management, and precise risk assessments. In essence, the comprehensive approach employed in this research, combining natural and human-induced factors, serves as a beacon for disaster resilience. As we move forward, it is imperative to heed the insights offered by this study, adopting proactive management strategies to mitigate the impending threat of floods in Chamoli District and safeguard the lives and livelihoods of its inhabitants.

References

Anni AH, Cohen S, Praskievicz S (2020) Sensitivity of urban flood simulations to stormwater infrastructure and soil infiltration. J Hydrol 588:125028. https://doi.org/10.1016/j.jhydrol.2020.125028

Anthwal A, Joshi V, Sharma A, Anthwal S (2006) Retreat of Himalayan glaciers–indicator of climate change. Nat Sci 4(4):53–59

Arnaud P, Bouvier C, Cisneros L, Dominguez R (2002) Influence of rainfall spatial variability on flood prediction. J Hydrol 260(1–4):216–230. https://doi.org/10.1016/S0022-1694(01)00611-4

Aryal D, Wang L, Adhikari TR, Zhou J, Li X, Shrestha M, Wang Y, Chen D (2020) A model-based flood hazard mapping on the southern slope of Himalaya. Water 12(2):540. https://doi.org/10.3390/w12020540

Badola R, Hussain SA, Dobriyal P, Manral U, Barthwal S, Rastogi A, Gill AK (2018) Institutional arrangements for managing tourism in the Indian Himalayan protected areas. Tour Manage 66:1–12

Begum M, Juraimi AS, Amartalingam R, Bin Man A, Bin Syed Rastans SO (2006) The effects of sowing depth and flooding on the emergence, survival, and growth of Fimbristylis miliacea (L.) Vahl. Weed Biol Manage 6(3):157–164. https://doi.org/10.1111/j.1445-6664.2006.00209.x

Besnard AG, La Jeunesse I, Pays O, Secondi J (2013) Topographic wetness index predicts the occurrence of bird species in floodplains. Divers Distrib 19(8):955–963

Beven K (1987) Towards the use of catchment geomorphology in flood frequency predictions. Earth Surf Proc Land 12(1):69–82. https://doi.org/10.1002/esp.3290120109

Bhatt GD, Sinha K, Deka PK, Kumar A (2014) Flood hazard and risk assessment in Chamoli district, Uttarakhand using satellite remote sensing and GIS techniques. Int J Innov Res Sci, Eng Technol 3(8):9

Das M, Nayak AK, Das B, Verma OP (2018) Groundwater quality assessment and mapping using multivariate statistics and analytic hierarchy process in Bhubaneswar city, Odisha, India. Int J Water 12(3):195–207. https://doi.org/10.1504/IJW.2018.093668

Dash P, Punia M (2019) Governance and disaster: analysis of land use policy with reference to Uttarakhand flood 2013, India. Int J Disaster Risk Reduction 36:101090. https://doi.org/10.1016/j.ijdrr.2019.101090

Dash P, Sar J (2020) Identification and validation of potential flood hazard area using GIS-based multi-criteria analysis and satellite data-derived water index. J Flood Risk Manage 13(3):1–14. https://doi.org/10.1111/jfr3.12620

Diduck A, Sinclair J, Pratap D, Hostetler G (2007) Achieving meaningful public participation in the environmental assessment of hydro development: case studies from Chamoli District, Uttarakhand, India. Impact Assess Proj Appraisal 25(3):219–231. https://doi.org/10.3152/146155107X217299

Gao BC (1996) NDWI—a normalized difference water index for remote sensing of vegetation liquid water from space. Remote Sens Environ 58(3):257–266

Garrote L, Bras RL (1995) A distributed model for real-time flood forecasting using digital elevation models. J Hydrol 167(1–4):279–306. https://doi.org/10.1016/0022-1694(94)02592-Y

Guhathakurta P, Sreejith OP, Menon PA (2011) Impact of climate change on extreme rainfall events and flood risk in India. J Earth Syst Sci 120:359–373. https://doi.org/10.1007/s12040-011-0082-5

Haghizadeh A, Siahkamari S, Haghiabi AH, Rahmati O (2017) Forecasting flood-prone areas using Shannon's entropy model. J Earth Syst Sci 126:1–11. https://doi.org/10.1007/s12040-017-0819-x

Hirabayashi Y, Mahendran R, Koirala S, Konoshima L, Yamazaki D, Watanabe S, Kanae S (2013) Global flood risk under climate change. Nat Clim Chang 3(9):816–821. https://doi.org/10.1038/nclimate1911

Ibarrarán ME, Ruth M, Ahmad S, London M (2009) Climate change and natural disasters: macroeconomic performance and distributional impacts. Environ Dev Sustain 11:549–569. https://doi.org/10.1007/s10668-007-9129-9

Jain SK, Singh RD, Jain MK, Lohani AK (2005) Delineation of flood-prone areas using remote sensing techniques. Water Resour Manage 19:333–347. https://doi.org/10.1007/s11269-005-3281-5

Kabenge M, Elaru J, Wang H, Li F (2017) Characterizing flood hazard risk in data-scarce areas, using a remote sensing and GIS-based flood hazard index. Nat Hazards 89(3):1369–1387. https://doi.org/10.1007/s11069-017-3024-y

Kansal ML, Singh S (2022) Flood management issues in Hilly regions of Uttarakhand (India) under changing climatic conditions. Water 14(12):1879. https://doi.org/10.3390/w14121879

Kazakis N, Kougias I, Patsialis T (2015) Assessment of flood hazard areas at a regional scale using an index-based approach and analytical hierarchy process: application in Rhodope–Evros region, Greece. Sci Total Environ 538:555–563. https://doi.org/10.1016/j.scitotenv.2015.08.055

Khanduri S (2021) Flash flood struck Dhauliganga valley on February 7, 2021: a case study of Chamoli district of Uttarakhand Himalaya in India. Acad Platform J Nat Hazards Disaster Manage 2(1):1–15. https://doi.org/10.52114/apjhad.903387

Kumar R, Kumar A, Shankhwar AK, Vishkarma DK, Sachan A, Singh PV, Jahangeer J, Verma A, Kumar V (2022) Modelling of meteorological drought in the foothills of Central Himalayas: a case study in Uttarakhand State, India. Ain Shams Eng J 13(3):101595. https://doi.org/10.1016/j.asej.2021.09.022

Mafi-Gholami D, Zenner EK, Jaafari A, Bakhtyari HRR, Bui DT (2019) Multi-hazards vulnerability assessment of southern coasts of Iran. J Environ Manage 252:109628. https://doi.org/10.1016/j.jenvman.2019.109628

Mahato S, Pal S (2019) Groundwater potential mapping in a rural river basin by union (OR) and intersection (AND) of four multi-criteria decision-making models. Nat Resour Res 28(2):523–545

Malik ZA, Bhatt AB (2015) Phytosociological analysis of woody species in Kedarnath Wildlife Sanctuary and its adjoining areas in Western Himalaya, India. J for Environ Sci 31(3):149–163

Manfreda S, Di Leo M, Sole A (2011) Detection of flood-prone areas using digital elevation models. J Hydrol Eng 16(10):781–790. https://doi.org/10.1061/(ASCE)HE.1943-5584.0000367

McClean F, Dawson R, Kilsby C (2020) Implications of using global digital elevation models for flood risk analysis in cities. Water Resour Res 56(10): e2020WR028241. https://doi.org/10.1029/2020WR028241

McFeeters SK (1996) The use of the normalized difference water index (NDWI) in the delineation of open water features. Int J Remote Sens 17(7):1425–1432

Mittal H, Wu YM, Lin TL, Legendre CP, Gupta S, Yang BM (2019) Time-dependent shake map for Uttarakhand Himalayas, India, using recorded earthquakes. Acta Geophys 67:753–763. https://doi.org/10.1007/s11600-019-00281-7

Nachappa TG, Piralilou ST, Gholamnia K, Ghorbanzadeh O, Rahmati O, Blaschke T (2020) Flood susceptibility mapping with machine learning, multi-criteria decision analysis and ensemble using Dempster Shafer theory. J Hydrol 590:125275. https://doi.org/10.1016/j.jhydrol.2020.125275

Ogden FL, Raj Pradhan N, Downer CW, Zahner JA (2011) Relative importance of impervious area, drainage density, width function, and subsurface storm drainage on flood runoff from an urbanized catchment. Water Resour Res 47(12). https://doi.org/10.1029/2011WR010550

O'Neill E, Brereton F, Shahumyan H, Clinch JP (2016) The impact of perceived flood exposure on flood-risk perception: the role of distance. Risk Anal 36(11):2158–2186. https://doi.org/10.1111/risa.12597

Pal SC, Chakrabortty R, Saha A, Bozchaloei SK, Pham QB, Linh NTT, ... and Ahmadi K (2022) Evaluation of debris flow and landslide hazards using ensemble framework of Bayesian-and tree-based models. Bull Eng Geol Environ 81:1–25. https://doi.org/10.1007/s10064-021-02546-2

Pande RK (2010) Flash flood disasters in Uttarakhand. Disaster Prev Manage: Inter J 19(5):565–570. https://doi.org/10.1108/09653561011091896

Pandey P, Chauhan P, Bhatt CM, Thakur PK, Kannaujia S, Dhote PR, Aggrawal SP (2021) Cause and process mechanism of rockslide triggered flood event in Rishiganga and Dhauliganga river valleys, Chamoli, Uttarakhand, India using satellite remote sensing and in situ observations. J Indian Soc Remote Sens 49:1011–1024. https://doi.org/10.1007/s12524-021-01360-3

Papaioannou G, Vasiliades L, Loukas A (2015) Multi-criteria analysis framework for potential flood prone areas mapping. Water Resour Manage 29(2):399–418. https://doi.org/10.1007/s11269-014-0817-6

Pham BT, Avand M, Janizadeh S, Phong TV, Al-Ansari N, Ho LS, Prakash I (2020) GIS based hybrid computational approaches for flash flood susceptibility assessment. Water 12(3):683. https://doi.org/10.3390/w12030683

Posthumus H, Hewett CJM, Morris J, Quinn PF (2008) Agricultural land use and flood risk management: engaging with stakeholders in North Yorkshire. Agric Water Manag 95(7):787–798. https://doi.org/10.1016/j.agwat.2008.02.001

Pourali SH, Arrowsmith C, Chrisman N, Matkan AA, Mitchell D (2016) Topography wetness index application in flood-risk-based land use planning. Appl Spat Anal Policy 9:39–54

Ranger N, Hallegatte S, Bhattacharya S, Bachu M, Priya S, Dhore K, Rafique F, Mathur P, Naville N, Henriet F, Herweijer C, Corfee-Morlot J (2011) An assessment of the potential impact of climate change on flood risk in Mumbai. Climatic Change 104:139–167. https://doi.org/10.1007/s10584-010-9979-2

Rather JA, Andrabi ZABR (2012) Fuzzy logic based GIS modeling for identification of ground water potential zones in the Jhagrabaria watershed of Allahabad district, Uttar Pradesh, India. Inter J Adv Remote Sens GIS 1(2):218–233

Saghafian B, Farazjoo H, Bozorgy B, Yazdandoost F (2008) Flood intensification due to changes in land use. Water Resour Manage 22:1051–1067

Sati SP, Gahalaut VK (2013) The fury of the floods in the north-west Himalayan region: the Kedarnath tragedy. Geomat, Nat Hazards Risk 4(3):193–201. https://doi.org/10.1080/19475705.2013.827135

Sati VP, Kumar S (2022) Environmental and economic impact of cloudburst-triggered debris flows and flash floods in Uttarakhand Himalaya: a case study. Geoenviron Disasters 9(1):5. https://doi.org/10.1186/s40677-022-00208-3

Shafapour Tehrany M, Shabani F, Neamah Jebur M, Hong H, Chen W, Xie X (2017) GIS-based spatial prediction of flood prone areas using standalone frequency ratio, logistic regression, weight of evidence and their ensemble techniques. Geomatics, Nat Hazards Risk 8(2):1538–1561. https://doi.org/10.1080/19475705.2017.1362038

Siddique T, Haris PM, Pradhan SP (2022) Unraveling the geological and meteorological interplay during the 2021 Chamoli disaster, India. Nat Hazards Res 2(2):75–83. https://doi.org/10.1016/j.nhres.2022.04.003

Singh G, Pandey A (2021) Flash flood vulnerability assessment and zonation through an integrated approach in the Upper Ganga Basin of the Northwest Himalayan region in Uttarakhand. Int J Disaster Risk Reduction 66:102573. https://doi.org/10.1016/j.ijdrr.2021.102573

Singh S, Kansal ML (2022) Chamoli flash-flood mapping and evaluation with a supervised classifier and NDWI thresholding using Sentinel-2 optical data in

Google earth engine. Earth Sci Inform 15(2):1073–1086. https://doi.org/10.1007/s12145-022-00786-8

Singh RD, Gumber S, Tewari P, Singh SP (2016) Nature of forest fires in Uttarakhand: frequency, size and seasonal patterns in relation to pre-monsoonal environment. Current Sci 398–403. http://www.jstor.org/stable/24908631

Singh VK, Ray PKC, Jeyaseelan APT (2010) Orthorectification and digital elevation model (DEM) generation using Cartosat-1 satellite stereo pair in Himalayan Terrain. J Geogr Inf Syst 2(2):85

Sofia G, Nikolopoulos EI (2020) Floods and rivers: a circular causality perspective. Sci Rep 10(1):5175

Swanson FJ, Johnson SL, Gregory SV, Acker SA (1998) Flood disturbance in a forested mountain landscape: interactions of land use and floods. Bioscience 48(9):681–689. https://doi.org/10.2307/1313331

Thakur PK, Aggarwal SP, Dhote P, Nikam BR, Garg V, Bhatt CM, Chouksey A, Jha A (2019) Hydrometeorological hazards mapping, monitoring and modelling. Remote Sens Northwest Himalayan Ecosyst 139–169

Tiwari JK, Rawat DS (2017) Exploration of some important fodder plants of Joshimath area of Chamoli district of Garhwal, Uttarakhand. Curr Bot 8

Verma S, Sharma A, Yadava PK, Gupta P, Singh J, Payra S (2022) Rapid flash flood calamity in Chamoli, Uttarakhand region during Feb 2021: an analysis based on satellite data. Nat Hazards 112(2):1379–1393. https://doi.org/10.1007/s11069-022-05232-y

Yamazaki D, Kanae S, Kim H, Oki T (2011) A physically based description of floodplain inundation dynamics in a global river routing model. Water Resour Res 47(4)

16 Assessing Land Use/Land Cover Transformation and Wetland Decline in Birbhum District, West Bengal: A Time-Series Analysis

Sk Mohibul, Md Nawaj Sarif, Neha Parveen, Monidip Mondal, Mohammad Itahdur Ali, Basir Ali Karikar, Manirul Mia, and Lubna Siddiqui

Abstract

The examination of changes in land use and land cover (LULC) has become a significant field of study in current research approaches, with the goal of efficiently addressing both urban and rural development. This was achieved by utilizing an unsupervised classification technique and a change detection method. The analysis was carried out throughout the course of three distinct time periods, namely 1977, 1998, and 2018. The primary objective of the present study was to generate LULC maps to assess the changes and impacts on wetlands over a period of forty-one years. The present study undertook a thorough examination of the LULC change (LULCC) in Birbhum District during the years 1977, 1998, and 2018. The results of the study unveiled noteworthy changes in land usage across the specified timeframe. The land area dedicated to agriculture experienced a significant increase throughout the years. In 1977, it accounted for 65.1% of the total land area, which rose to 72.9% in 1998, and further escalated to 74.1% in 2018. On the other hand, there was a decrease observed in aquatic systems, suggesting possible consequences for both water resources and ecosystems. The vegetation exhibited a slight expansion, however there was a significant increase in built-up areas, indicating the process of urbanization and the construction of infrastructure. This study highlights the significance of monitoring LULCC to make well-informed decisions about resource management, urban planning, and environmental preservation. The statement underscores the importance of implementing sustainable land management practices to achieve a harmonious equilibrium between economic progress and the preservation of the environment. The results of this study offer significant contributions to our understanding of the changing LULC patterns in Birbhum District, serving as a basis for future planning and development endeavours.

Keywords

Land use land cover change · Spatio-temporal patterns · Change detection methods · Wetland · Birbhum

16.1 Introduction

Wetlands, defined as areas of land covered by water either temporarily or permanently, manifest enormous diversity in terms of their origin,

S. Mohibul · M. N. Sarif · N. Parveen · M. Mondal · M. I. Ali · B. A. Karikar · M. Mia · L. Siddiqui (✉)
Department of Geography, Faculty of Sciences, Jamia Millia Islamia, New Delhi 110025, India
e-mail: lsiddiqui@jmi.ac.in

S. Talukdar et al. (eds.), *Water Resource Management in Climate Change Scenario*, GIScience and Geo-environmental Modelling, https://doi.org/10.1007/978-3-031-61121-6_16

geographical location, water regime, and chemical composition (Dewan and Yamaguchi 2009). These ecosystems play a crucial role in the hydrological cycle and are among the most productive on Earth (Mohibul et al. 2022). They provide a wide range of services, directly and indirectly supporting millions of people by offering storm and flood control, clean water supply, food production, raw materials, scenic beauty, and recreational and educational benefits (Mohibul et al. 2023). The Millennium Ecosystem Assessment conservatively estimates that wetlands cover approximately seven per cent of the Earth's surface and deliver 45% of the world's natural productivity and ecosystem services (Talukdar et al. 2020a, b). However, the existence of these unique resources is threatened by developmental activities and population pressures (Das et al. 2021). The growing population on Earth's surface has also resulted in the urban agglomeration of regions (Parveen et al. 2022). According to the United Nations' "World Urbanization Prospect 2014" report, the percentage of the global population living in urban areas is expected to reach 66% by 2050, with approximately 54% already residing in urban areas as of 2014 (Parveen et al. 2021). However, the functionality and extent of wetlands rapidly degrade due to activities within their watersheds, primarily agricultural practices such as crop and livestock production. Consequently, long-term planning is required to preserve and conserve these valuable resources. The first step towards achieving this is the development of an updated and accurate database to support research and decision-making. The use of advanced techniques such as satellite remote sensing and geographic information system (GIS) is now essential for the creation of accurate and timely spatial databases covering large areas (Das and Das 2019). The increasing population in developing countries like India has had a vulnerable effect on wetlands, leading to changes in land use and land cover (LULC). Remote sensing techniques offer accurate and valuable data acquisition of the Earth's surface in both time and space (Shahfahad et al. 2023). This study was undertaken to identify changes in the wetlands of Birbhum District and to identify areas that are shrinking and vulnerable due to LULC changes.

On the Earth's surface, there are both natural and human regions. However, the growing population in different countries is affecting nature and leading to a decrease in natural areas as human settlements expand. Wetlands are an important part of the Earth's surface, and understanding their importance and impact on human health is crucial. Wetlands are unique and transitional ecosystems, covering about 6–8% of the Earth's terrestrial surface and accounting for approximately 45% of the total economic value of all global ecosystems (Davidson et al. 2018). LULC mappings are of great importance in current times as they provide valuable information on both natural and human aspects. It facilitates agricultural monitoring, disaster management, soil mapping, wetland mapping, resource extraction, tourism, and urban planning (Nasrin et al. 2023; Sarif et al. 2022).

The sustainable management and wise use of wetlands are crucial for the future well-being of human societies. Wetlands are closely intertwined with human beings, as they provide more than three-quarters of the food required by mankind, including rice, vegetables, fish, and more (Pang et al. 2010). However, these ecosystems face threats from both natural and anthropogenic forces, with wetlands often being altered for other purposes. The primary objective of this study is to examine the shrinking of wetland due to LULC change (LULCC) and conduct a spatial and temporal analysis of the LULCC that occurred in the study area between 1977 and 2018. This study also aims to analyze the reasons behind wetland shrinkage due to the advancement of civilization by creating an LULC map. LULC change have become essential considerations in the management and monitoring of wetlands. With advancements in technology and the use of satellite imagery, it is now crucial to understand the impact of human activities on natural resources over time. In situations of rapid and often unrecorded LULCC, satellite observations provide objective information on human utilization of the landscape.

16.2 Database and Methodology

16.2.1 Study Area

The study area of Birbhum is in the western part of West Bengal, situated between 23° 32′ 30″ and 24° 35′ 0″ north latitude and 87° 5′ 25″ and 88° 1′ 40″ east longitude. It covers an area of approximately 4545 km^2 (1755 sq mi) and has a triangular shape (Fig. 16.1). The southern base of the triangle is formed by the Ajay River, which also serves as the boundary between Birbhum and Burdwan Districts. To the north and west, Birbhum shares its borders with the state of Jharkhand, while to the east, it is bordered by Murshidabad District. Geographically, Birbhum lies at the northeastern end of the Chotonagpur Plateau, where it gradually slopes down and merges with the alluvial plains of the Ganga and Damodar Rivers. The district features a hilly area with natural rock

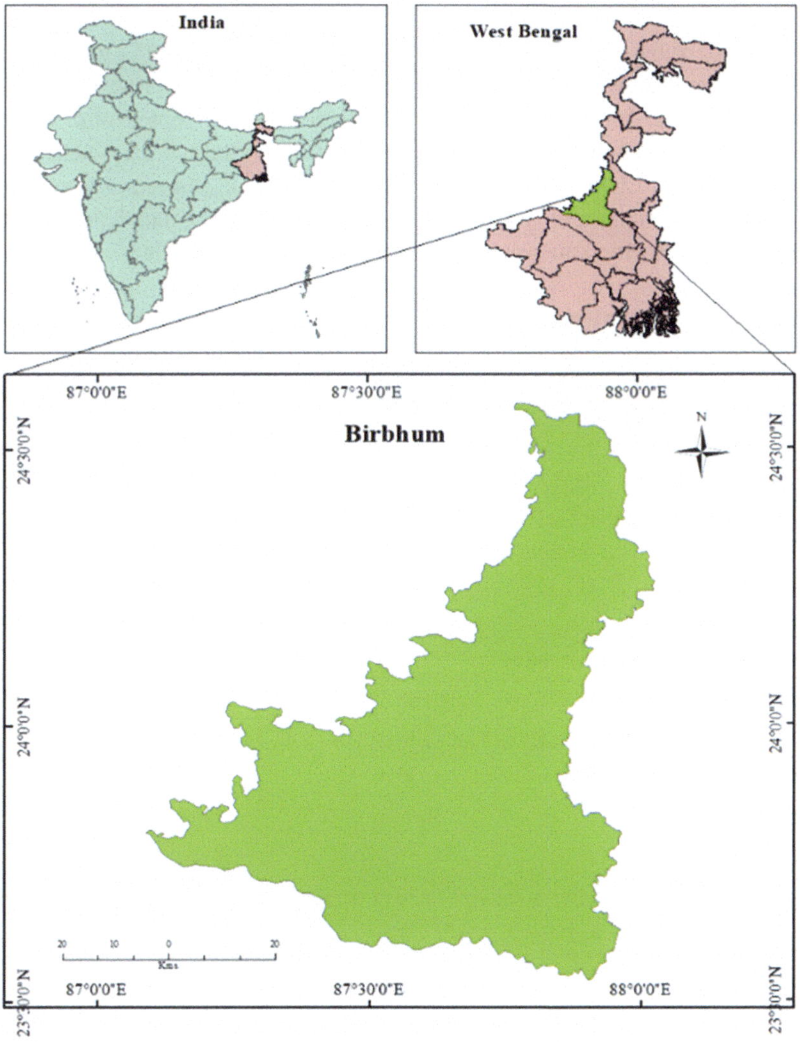

Fig. 16.1 Study area map of Birbhum

formations near the town of Dubrajpur, known as Mama Bhagne Pahar (Hill). The region is traversed by several important rivers, including Ajay, Mayurakshi, Shal, Kopai, Bakreshwar, Brahmani, Hinglo, and Dwarka. Overall, Birbhum's geographical characteristics and its proximity to rivers and hilly areas contribute to its unique landscape and natural diversity.

16.2.2 Methods of Data Analysis

To achieve these objectives, Landsat-7 TM (Thematic Mapper) satellite images were acquired for three time periods: 18 March 1977, 13 December 1998, and 01 October 2018. These images were obtained from the website of the United States Geological Survey (USGS) Earth Explorer (Tables 16.1 and 16.2). The analysis of these satellite images will provide valuable insights into the dynamics of LULC, as well as the changes in the extent and condition of wetlands in Birbhum District over the specified time span.

16.2.2.1 LULC Map Preparation

The ERDAS Imagine 14.0 programme was employed to conduct an unsupervised classification analysis to discern patterns of LULC across the years 1977, 1998, and 2018. The software also facilitated the processing of layer stacking, mosaicking, and a specific subset of the study region (Fig. 16.2). The Landsat image included in the study possesses a spatial resolution of 60 and 30 m. Unsupervised categorization proves to be highly advantageous in scenarios where there is an absence of pre-existing field data or a lack of precise descriptions pertaining to aerial pictures. The user is permitted to assign a significant title to each class, with each title resembling a unique geographical group. In the context of unsupervised categorization, a total of 300 classes were chosen to produce LULC maps for the time periods of 1977, 1998, and 2018. Various classes were identified by the utilization of on-screen visual interpretation keys, including tone, form, texture, size, pattern, and association. In the current study, a categorization of LULC maps into five distinct classes, i.e. agriculture, waterbody, vegetation, built-up, and others. The raw image of the pixel colour discloses distinct representations of several elements: agriculture is depicted in yellow, water bodies in blue, vegetation in green, built-up areas in red, and other features are represented in grey. These colour classifications are utilized for the purpose of sample selection. The options include "Save" and further processing in ERDAS Imaging. The formation of the unsupervised image is followed by the subsequent correction of certain pixels that do not conform to their expected values inside the image. During the process of recoding, the resulting image is displayed, along with the corresponding region for each class.

Table 16.1 Data source and information

Sl. No.	Data type	Date of acquisition	Resolution	Source
1	LANDSAT-2	18 Mar 1977	60 m	USGS-EARTH EXPLORER
2	LANDSAT-6	13 Dec 1998	30 m	USGS-EARTH EXPLORER
3	LANDSAT-8	01 Oct 2018	30 m	USGS-EARTH EXPLORER
4	Shapefile	–	–	Bhukosh, geological survey of India

Table 16.2 Geo-referencing properties of the images of 1977, 1998, and 2018

Data type	File type	Path	Row	Referencing system	Reference units	Spatial resolution	Spectral resolution
Thematic mapper	TIFF	149	043	UTM zone 45N	Metre	30 m and 60 m	4–7, 1–7, 1–9 band data

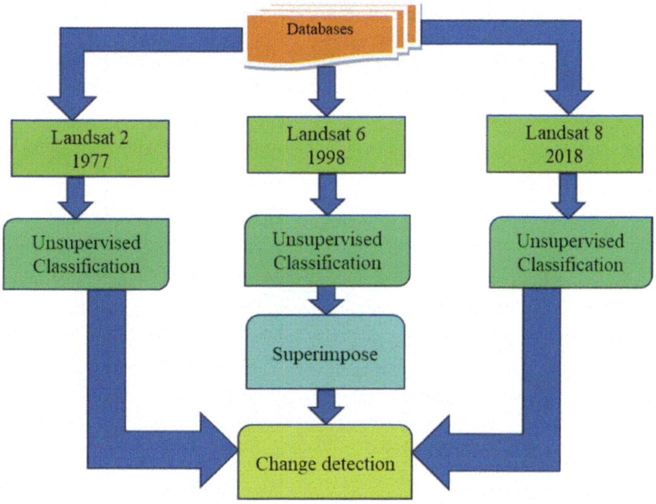

Fig. 16.2 Flowchart of LULC and change detection

16.2.2.2 Change Detection

Change detection is a crucial application of remote sensing technology, allowing the identification and analysis of alterations over time. By comparing data from different time periods, it enables quantitative analysis and provides valuable insights into the characteristics and mechanisms of surface changes. This involves assessing modifications in surface properties, boundary shifts, and patterns before and after changes occur. In the 1980s, scholars began investigating change detection, primarily using multi-band remote sensing images. Change detection methods in multi-band remote sensing images can be broadly categorized into two approaches. The first involves image subtraction, principal component analysis, change vector analysis, and spectral feature variance techniques, focusing on discrepancies between images from distinct time intervals. The second approach begins with picture classification and then analyzes the classified data to detect and characterize changes. Both methods have their strengths and are chosen based on specific objectives and data availability.

Remote sensing technology for change detection facilitates monitoring and understanding the dynamic characteristics and temporal modifications of various features. There are three categories of methods for detecting changes in multispectral image data: spectral type characteristic analysis, vector analysis of spectral variations, and time-series analysis. Spectral type characteristic analysis examines changes in remote sensing images over multiple phases using spectral classification and calculation. Methods include multi-temporal image stacking, algebraic change detection, primary image component change detection, and change detection after classification. Vector analysis of spectral changes aims to determine the magnitude and orientation attributes of alterations based on spectral differences. These methods collectively provide powerful tools for assessing and comprehending changes in remote sensing data over time.

16.3 Result

16.3.1 Land Use Land Cover from 1977 to 2018

16.3.1.1 Land Use Land Cover of Birbhum District, 1977

The LULCC analysis was conducted for the years 1977, 1998, and 2018, as depicted in (Figs. 16.3, 16.4, and 16.5). According to the LULC map of 1977, accompanied by their corresponding areas measured in hectares (ha)

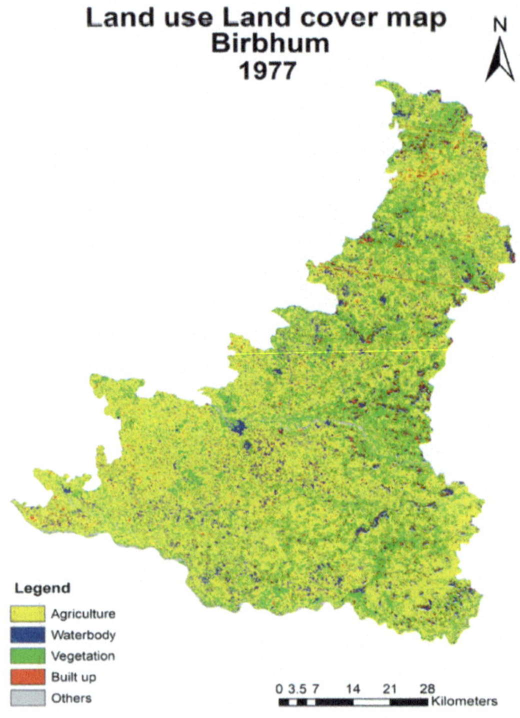

Fig. 16.3 LULC map of Birbhum, 1977

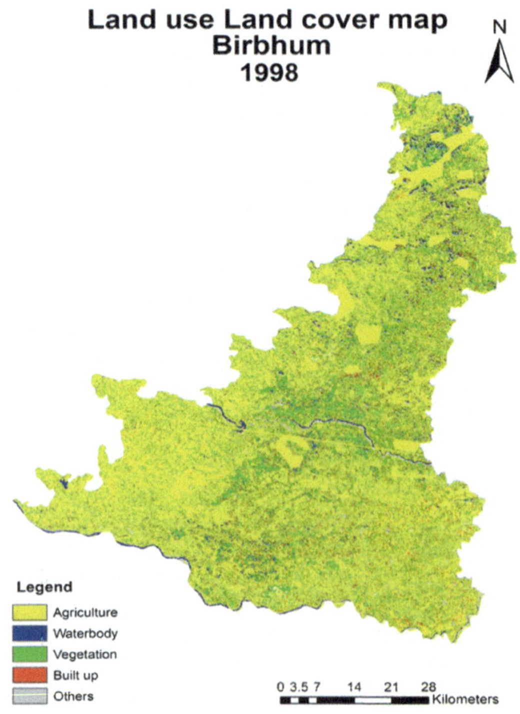

Fig. 16.4 LULC map of Birbhum, 1998

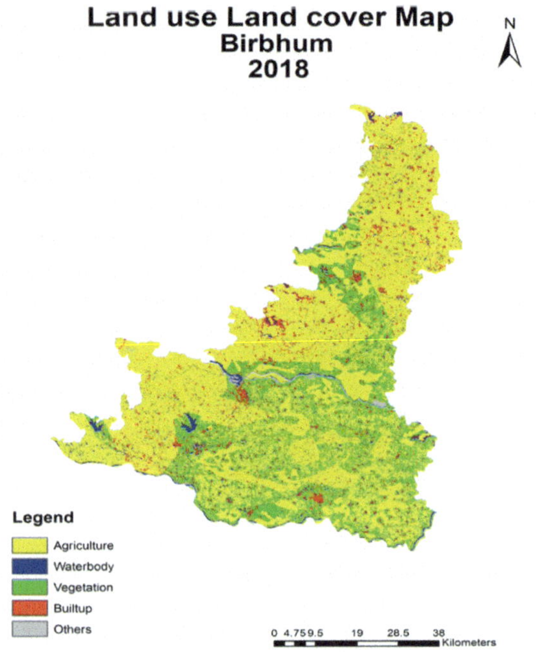

Fig. 16.5 LULC map of Birbhum, 2018

and percentages, the overall extent of the territory under examination amounts to 454,015.4 hectares. Agriculture is the predominant LULC type, surrounding an expansive area of 295,532 hectares, representing around 65.1% of the total land area. This finding suggests that a considerable proportion of the area is allocated to agricultural pursuits. The waterbody class encompasses a land area of 44,967.7 hectares, which accounts for approximately 9.9% of the overall total land area. This comprises various aquatic environments such as lakes, ponds, rivers, and other bodies of water that are found within the geographical area. The vegetation class covers a total area of 72,340.9 hectares, or approximately 15.9% of the overall land area. This contains various ecosystems such as forests, grasslands, and other forms of natural vegetation. The built-up area is comparatively smaller, encompassing 20,229.7 hectares, accounting for 4.5% of the overall land area. This includes metropolitan areas, infrastructure, and human settlements. The "others" category contains a diverse range of land cover types that are not classified within the primary classes. The designated region comprehends a land area of 20,945.1 hectares, constituting approximately 4.6% of the overall total land area (Figs. 16.3, 16.6, and 16.7). In general, the LULC analysis conducted in 1977 indicates a predominant presence of agricultural land, followed by vegetation and water bodies. The fraction of the total area occupied by the built-up area and other land cover types is rather small. The provided information offers valuable insights into the patterns of land utilization and the distribution of various land cover classes within the region during the specified year.

16.3.1.2 Land Use Land Cover of Birbhum District, 1998

The land use and land cover categories for 1998, accompanied by their corresponding areas measured in hectares (ha) and percentages. The cumulative extent of the geographical area under examination amounts to 451,670.92 hectares. In 1998, the dominant LULC class was agriculture, including a total area of 329,204 hectares, representing around 72.9% of the overall land area. This observation suggests a rise in the amount of land dedicated to agricultural purposes when

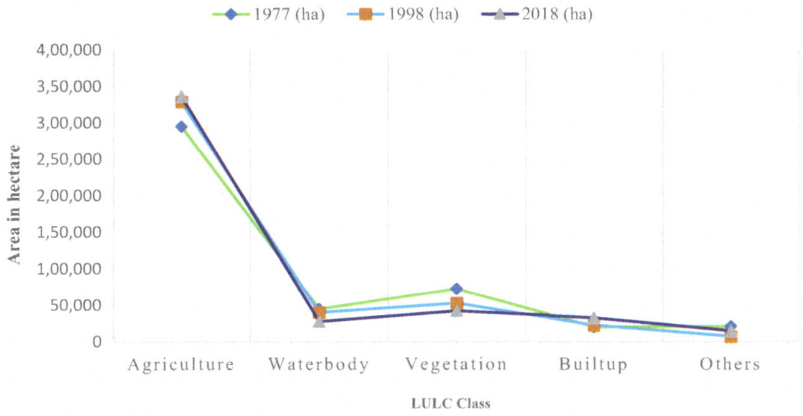

Fig. 16.6 Distribution of LULC types (in hectare-wise) in Birbhum district

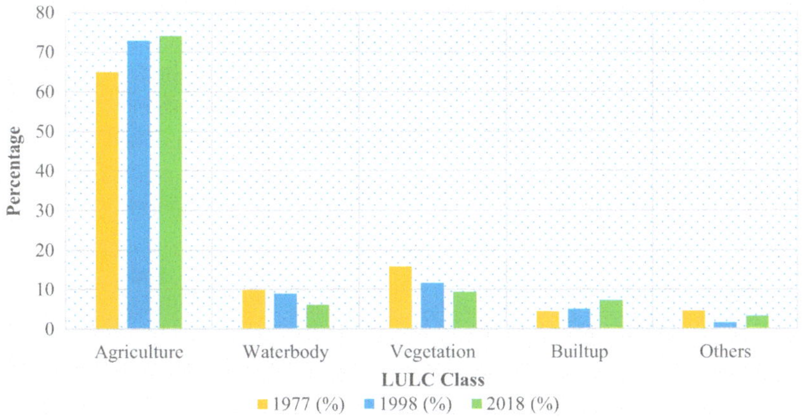

Fig. 16.7 Distribution of LULC types (in percentage-wise) in Birbhum district

compared to the data from 1977. The waterbody class encompasses a land area of 40,056.7 hectares, which accounts for 8.9% of the overall total land area. A marginal decline in the extent of the waterbody is observed when comparing the current data to that of 1977. The vegetation class encompasses a total area of 52,669.1 hectares, or approximately 11.7% of the overall land area. The data from 1977 indicates a marginal expansion in the vegetation area. The total built-up area accounts for 22,518.8 hectares, equivalent to 5.0% of the overall land area. The data from 1977 indicates a modest rise in the built-up area, suggesting the presence of urbanization and infrastructure development. The group referred to as "others" contains a diverse range of land cover types that do not fall within the primary classes previously discussed in (Figs. 16.4, 16.6, and 16.7). The land in question spans a total of 7222.32 hectares, constituting approximately 1.6% of the overall land area. Upon comparing the data from 1998 with that of 1977, it becomes evident that there has been a general upward trend in the expansion of agricultural land and built-up areas. Conversely, there has been a modest decline in the extent of waterbody and vegetation areas. This observation implies alterations in land utilization trends, which may indicate the process of urbanization and the increase of agricultural activities over time. The examination of alterations in LULC provides significant insights into the dynamics of the region, so facilitating informed decision-making in resource management, urban planning, and environmental protection.

16.3.1.3 Land Use Land Cover of Birbhum District, 2018

In 2018, the agriculture class maintained its status as the most extensive LULC category, encompassing a total area of 336,739 hectares. This figure corresponds to around 74.1% of the overall land area. This observation suggests a marginal rise in the amount of agricultural land in relation to the data from 1998. The waterbody class encompasses a land area of 27,745.2 hectares, or 6.1% of the overall total land area. A decline in the surface area of the waterbody is observed when comparing the data from 1977 and 1998, indicating possible alterations in water resources and land utilization practices. The vegetation class incorporates an area of 42,298.5 hectares, which accounts for 9.3% of the overall land area. The data from 1998 indicates a marginal expansion in the vegetation region. The total built-up area covers 32,860.9 hectares, or approximately 7.2% of the overall land area. The data reveals a discernible rise in the extent of developed land when compared to the datasets from 1977 and 1998, suggesting a significant growth of metropolitan areas and the advancement of infrastructure. The group referred to as "others" includes a diverse range of land cover types that do not fall within the primary classes previously discussed. The designated region comprises a landmass of 15,037 hectares, constituting approximately 3.3% of the whole geographical expanse (Figs. 16.5, 16.6, and 16.7). Upon comparing the data from 2018 with that of prior years, it becomes evident that there has been a consistent expansion of agricultural land. Additionally, there is a notable rise in the extent of built-up regions. Nevertheless, a discernible decline in the extent of waterbodies has been observed, indicating possible ramifications for both water resources and ecological systems. The data pertaining to the vegetation class also express a marginal augmentation, suggesting a certain degree of preservation or reestablishment of vegetation. The comprehension of temporal alterations in LULC holds paramount importance in the surveillance and administration of natural resources, urban expansion, and environmental preservation. The data supplied presents valuable information regarding the changing trends in land utilization, emphasizing the necessity of implementing sustainable land management strategies and making well-informed decisions to effectively harmonize economic progress with environmental preservation.

The classified photos were cross-referenced with GPS data, topographical sheets, and existing wetland maps of the research area. This analysis aimed to ascertain the correspondence between the classification of each site observed during ground truthing and its representation on the image. The assessment of accuracies involved the utilization of error matrices, as outlined by Congalton and Green (1999). The categorization accuracies for the respective years of 1977, 1998, and 2018 were determined to be 69.44%, 68.43%, and 70%.

16.3.2 Assess the Shrinking of Wetlands in the Study Area During the Period from 1977 to 2018

The detection of changes in Birbhum is conducted using data from two distinct time periods. The data implementation for this change detection process is sourced from USGS Earth Explorer and Google Earth Image, with specific coordinates being employed. Landsat-2 was launched in March 1977. The study incorporates the utilization of Landsat-6, which was launched in December 1998, and Landsat-8, which was launched in October 2018. An image is captured for the purpose of demarcating the boundary or generating a subset mask for the Birbhum region. The change detection matrices of each class are detailed in the next section.

16.3.2.1 Change Matrix from 1997 to 1998

The agricultural sector has experienced a substantial expansion in terms of land coverage when compared with the year 1977, exhibiting a total area of 3,189,440 hectares. The largest LULC category, with 93.5% of the total area in 1998, is identified. The water body area had a modest reduction to 100,876.5 hectares, which accounts for 2.96% of the overall land area. The vegetation class encloses a land area of 74,583.1 hectares, or 2.19% of the overall total land area. Furthermore, it demonstrates a marginal rise in comparison with the data from 1977. The built-up area experienced a notable expansion, reaching a total of 22,950.8 hectares, which corresponds to around 0.67% of the overall land area. This category encompasses regions characterized by the presence of built infrastructure and inhabited areas. The category labelled "others" encompasses a diverse range of land cover types that do not fall into the classification of the primary groups. The designated region encompasses a landmass measuring 24,732.7 hectares, which accounts for a mere 0.73% of the whole geographical expanse (Fig. 16.8 and Table 16.3). The presented data offers a comprehensive depiction of the distribution of LULC in the year 1998, emphasizing the prevailing prominence of agriculture as the principal category of land utilization. The monitoring of temporal changes in land usage is of paramount importance in facilitating efficient land management and facilitating the planning of sustainable development.

16.3.2.2 Change Matrix from 1998 to 2018

The agricultural class demonstrates a marginal decline in comparison with the data from 1998, with a cumulative land area of 2,987,903 hectares. The largest land use category, comprising 91.6% of the total area in 2018, continues to retain its prominence. The extent of the aquatic body likewise experienced a reduction to 43,994.9 hectares, which accounts for 1.6% of the overall land area. The vegetation class comprises a land area of 52,626.5 hectares, or 1.9% of the overall total land area. There is a marginal rise observed in comparison with the data from 1998. The urbanized land area expanded to 18,506.4 hectares, accounting for 0.7% of the whole geographic expanse. The group labelled "others" surrounding a land area measuring 7197.7 hectares accounts for a mere 0.3% of the overall total land area (Fig. 16.9 and Table 16.4). The presented data offers a comprehensive depiction of the alterations in land utilization from 1998 to 2018. It is noteworthy that agriculture continues to be the predominant land use, albeit with minor reductions in its extent. The observed decline in aquatic bodies and modest rise in vegetation and built-up areas indicate potential alterations in land utilization patterns over a period. The monitoring of these changes holds significance in the context of land management and the formulation of strategies for sustainable development.

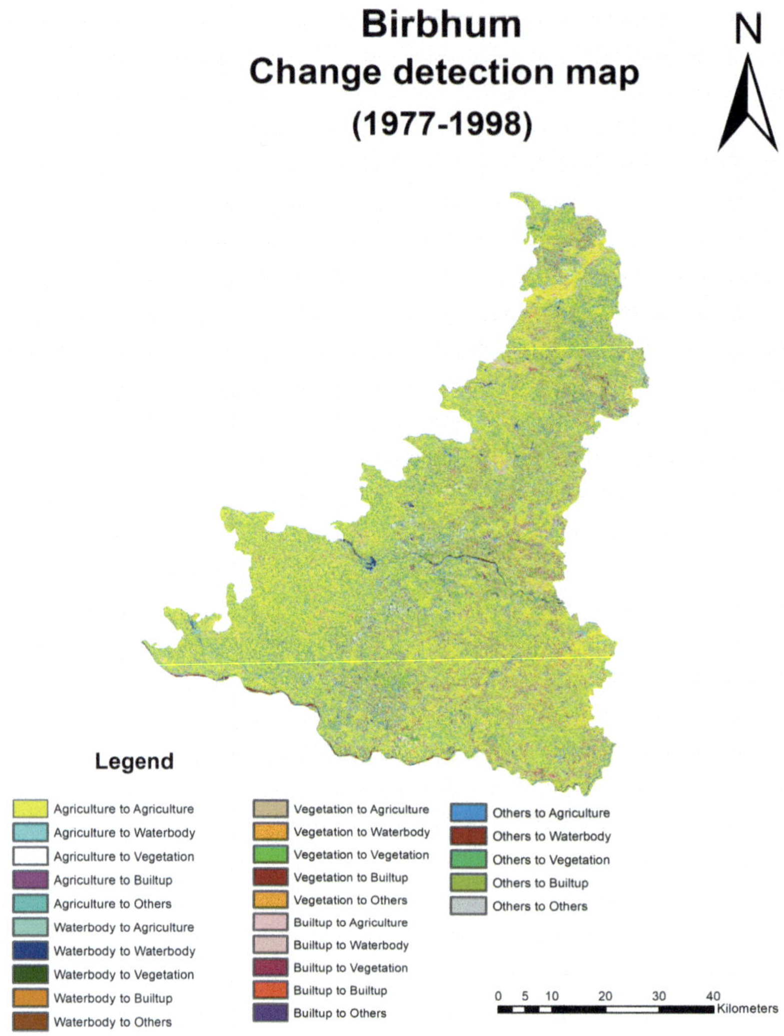

Fig. 16.8 Area under changed along with unchanged area of Birbhum from 1977 to 1998

Table 16.3 Overall change of LULC from 1977 to 1998

1998							
1977	Class	Agriculture	Water body	Vegetation	Built-up	Others	Total
	Agriculture	644,488	650,488	643,488	600,488	650,488	3,189,440
	Water body	36,019.2	4627.17	53,086.48	6599.21	544.41	100,876.5
	Vegetation	48,910.4	8018.55	9781.47	3936.33	3936.33	74,583.1
	Built-up	15,700.8	2746.17	2969.19	1175.94	358.74	22,950.8
	Others	17,209.2	2754.63	2433.24	751.23	1584.36	24,732.7
	Total	762,327.6	668,634.5	670,358.4	657,950.7	656,911.8	3,412,583

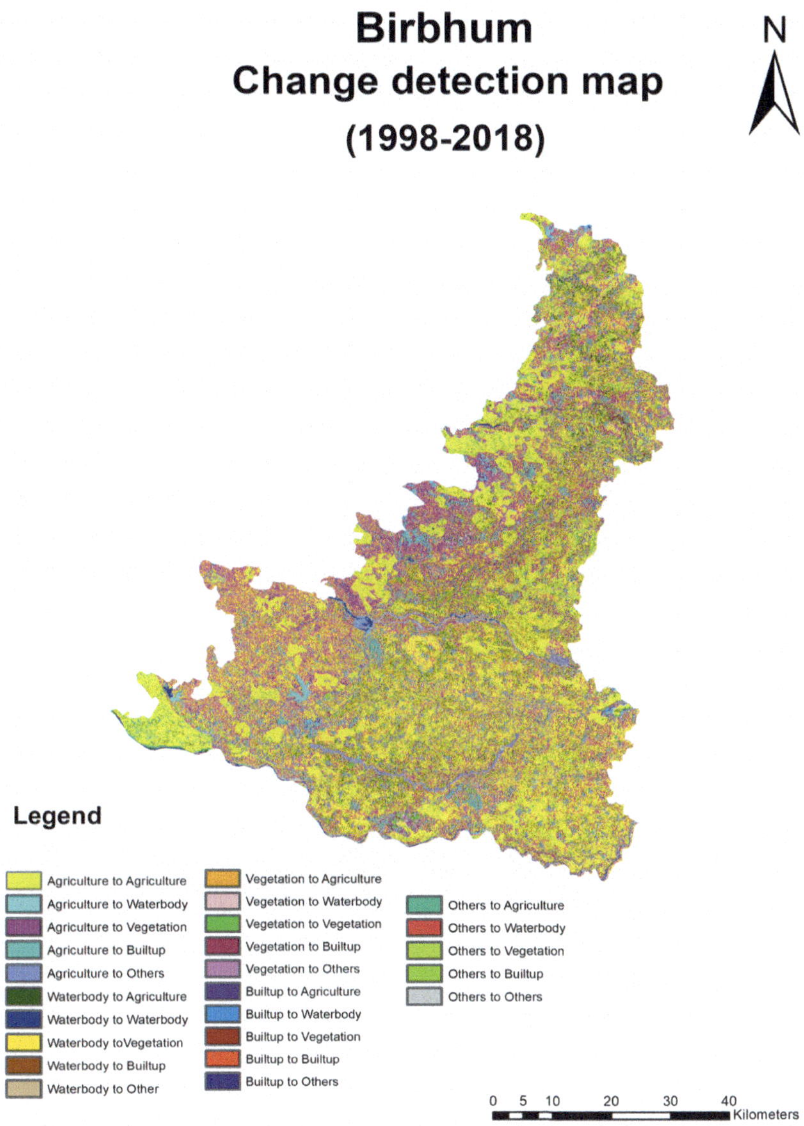

Fig. 16.9 Area under changed along with unchanged area of Birbhum from 1998 to 2018

Table 16.4 Overall change of LULC from 1998 to 2018

2018							
1998	Class	Agriculture	Water body	Vegetation	Built-up	Others	Total
	Agriculture	632,321	628,321	640,589	548,321	538,351	2,987,903
	Water body	25,028.1	4432.43	10,792.7	1622.84	2118.82	43,994.9
	Vegetation	32,083.8	2307.08	15,025.2	1772.42	1437.95	52,626.5
	Built-up	10,921.8	885.127	5629.99	510.255	559.237	18,506.4
	Others	3077.03	872.752	1450.98	415.238	1381.72	7197.7
	Total	719,431.7	656,818.4	681,219.9	652,641.8	653,818.7	2,711,288.7

16.3.2.3 Change Matrix from 1977 to 2018

The agricultural sector experienced a substantial growth in size when compared with the year 1977, as evidenced by a total land area of 290,773.4 hectares. The category holds the distinction of being the most extensive land use classification, comprising 64.2% of the overall land area in the year 2018. The water body area had a growth of 41,841.8 hectares, which accounts for 9.2% of the overall land area. The vegetation class encloses a land area of 72,159.5 hectares, or 15.9% of the overall total land area. There is a significant rise observed in comparison with the year 1977. The built-up area experienced a notable expansion, reaching a total of 23,162.0 hectares, which accounts for approximately 5.1% of the overall land area. The category labelled "others" encompasses a land area measuring 24,798.7 hectares, which accounts for 5.5% of the overall total land area (Fig. 16.10 and Table 16.5). The aforementioned data presents valuable observations regarding the alterations in land utilization from 1977 to 2018. The observed substantial rise in the enrolment of agriculture and vegetation courses suggests the possibility of evolving land use trends over a period. The observed rise in built-up areas is indicative of the processes of urbanization and development. The monitoring of these changes is of utmost importance in the context of sustainable land management and efficient planning.

It is evident that wetlands have been consistently overlooked by human populations for a considerable duration. Between the years 1977 and 1998, a total of 36,019 hectares of water bodies underwent transformed for agricultural use. A total land area measuring 4627 hectares was identified where water bodies exhibited a state of stasis or remained unaltered. The conversion of aquatic bodies to vegetation encompassed an estimated area of 53,086 hectares, while the transition to built-up areas constituted 6599 hectares. Furthermore, a conversion of 544 hectares of aquatic ecosystems occurred, resulting in the transformation of these areas into different land cover categories (Fig. 16.11). Between the years 1998 and 2018, there was a conversion of water bodies into agricultural land, resulting in a total coverage of 25,028 hectares. A total land area measuring 4432 hectares was observed to expose no changes in the presence or characteristics of water bodies. The conversion of water bodies to vegetation encompassed an estimated area of 10,792 hectares, while the transformation into built-up areas occupied 1622 hectares (Fig. 16.11). Furthermore, a total of 2118 hectares of aquatic ecosystems underwent a transformation, resulting in their conversion into alternative land cover classifications. The examination of lakes and their land use/land cover in relation to space and time indicates that the growth in population has resulted in the reduction and modification of wetland systems. According to remote sensing data, it was observed in 1977 that the open water area in Birbhum encompassed a total of 44,967 hectares, representing approximately 9.9% of the overall land area. The region had a continued decline by 1998, and by 2018, it had reached a total of 27,745 hectares, representing a mere 6.0% of the overall land area (Fig. 16.11). Likewise, there has been a decrease in the extent of open water regions in smaller ponds distributed around the region, leading to the complete desiccation of several lakes.

The wetlands have seen considerable alterations due to the progressive encroachment of alluvial land by both small- and large-scale farmers, as well as the associated reclamation efforts. Over the course of time, there have been significant alterations observed in the land use and land cover patterns in various parts of Birbhum. The conversion of agricultural land into urban settlements or industrial facilities has occurred. The implementation of reclamation works inside the lake region has led to a decrease in the initial capacity of the basin. Given the current vulnerability of the lake and its surrounding wetland systems to eutrophication, it is imperative to implement proactive measures aimed at mitigating the progression of this phenomenon.

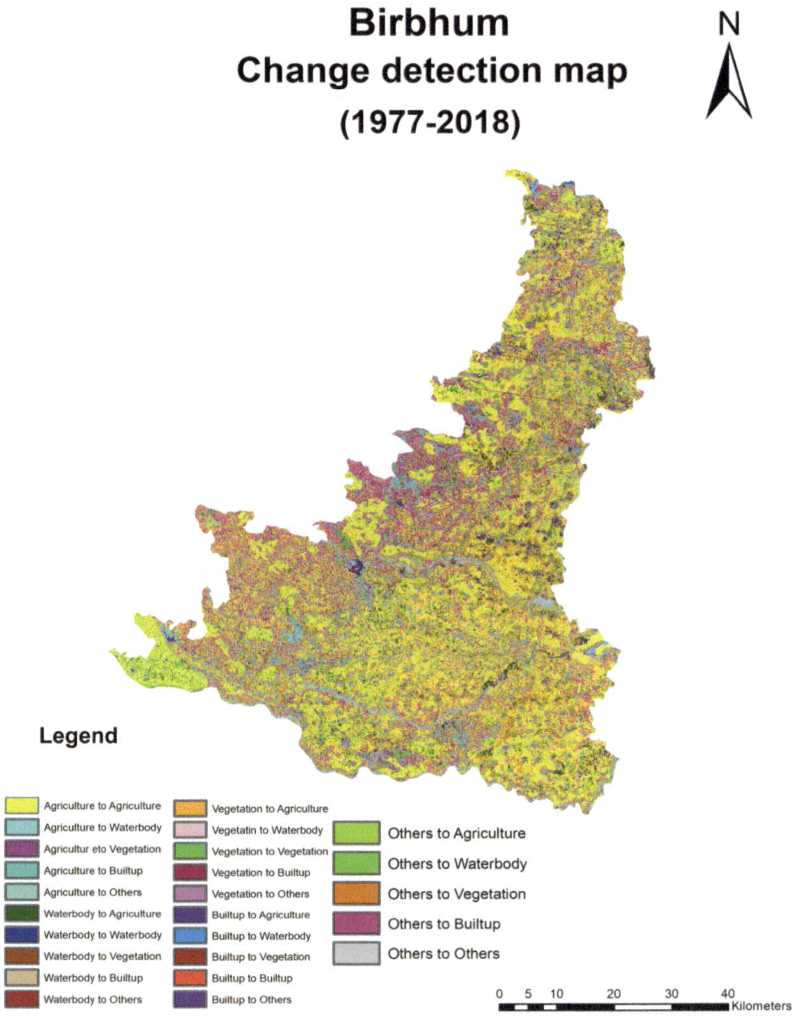

Fig. 16.10 Area under changed along with unchanged area of Birbhum from 1977 to 2018

Table 16.5 Overall change of LULC from 1977 to 2018

2018							
1977	Class	Agriculture	Water body	Vegetation	Built-up	Others	Total
	Agriculture	178,979	15,698.5	76,938.8	10,951.3	8205.84	290,773.4
	Water body	23,496.7	3963.62	10,099.6	2479.75	1802.12	41,841.8
	Vegetation	41,989.8	4420.33	20,737.8	2673.45	2338.15	72,159.5
	Built-up	13,465.5	1774.73	6191.48	1052.55	677.745	23,162.0
	Others	14,577.7	1778.15	5894.32	613.642	1934.93	24,798.7
	Total	272,508.7	27,635.3	119,862	17,770.7	14,958.8	452,735.5

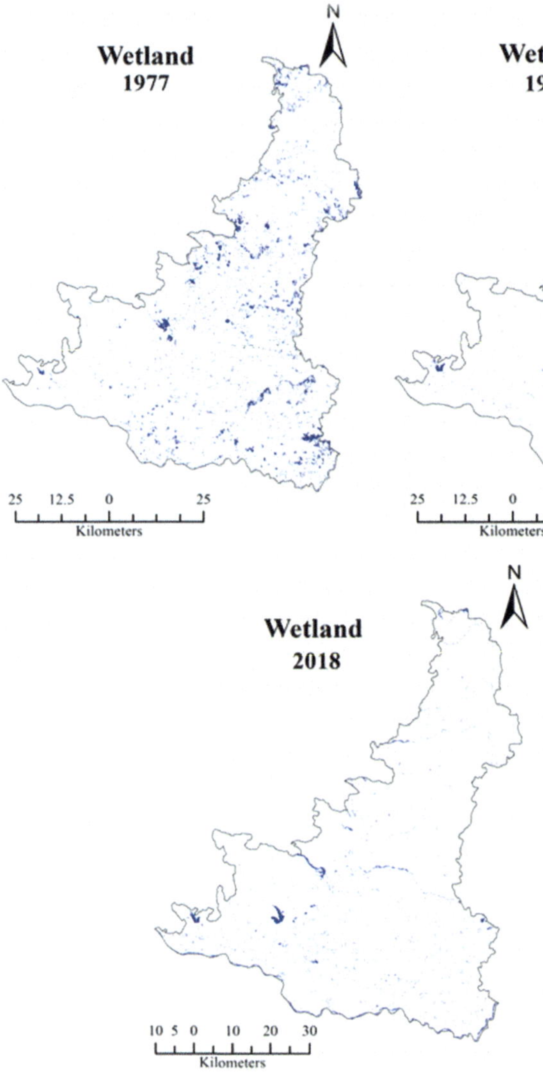

Fig. 16.11 Total wetlands area of Birbhum 1977, 1998, and 2018

16.4 Discussion

The study reveals conspicuous decreasing trends in waterbody or the wetland and vegetation agricultural in the Birbhum District over the period 1977–2018, due to the expansion of agriculture and built-up land (Figs. 16.6 and 16.7). Based on the preceding discourse, it is apparent that the Birbhum District is predominantly characterized by agricultural pursuits, given that a significant proportion of its populace is involved in primary sector occupations. The territory within the district is mostly characterized by the presence of lateritic soil, while certain regions also have alluvial and stony soil. Over the course of several decades, the growth of agricultural fields in plain areas has resulted in a progressive encroachment of built-up areas onto wetlands. The analysis reveals significant temporal trends in land use categories. Agriculture emerges as the predominant land use type, consistently expanding its coverage over the three decades, reaching 74.1% of the total land area by 2018. This substantial increase in agricultural land signifies the region's reliance on agriculture activities.

The current research study demonstrates significant alterations in wetland and vegetation areas, resulting in a net reduction of 17,223 hectares and 30,042 hectares, respectively, over the past forty-one years. The agriculture and built-up area that showed positive change with net increase was 41,207 hectare and 12,631 hectares, respectively. The change matrices from 1997–1998, 1998–2018, and 1977–2018 further illustrate these trends. Agriculture consistently maintains its dominance, with minor fluctuations in extent. Conversely, water bodies show a declining trend, indicating alterations in land utilization practices and their potential ecological consequences. The expansion of built-up areas is evident in all matrices, reflecting urbanization and development. This expansion necessitates careful land use planning to strike a balance between development and environmental preservation. The wetland and vegetation area continuously shrinking due to the population involved in agriculture sector as well as the major urban centre such as Bolpur, Suri, and Rampurhat play an important role for the attraction of employment, education, industries, and various economic activities, hence attracting individuals from towns, cities, blocks districts, as well as neighbouring states (Rahaman et al. 2020; Sinha et al. 2011). Unregulated urban expansion, agricultural practices, industrial development, and other human activities, including construction, have encroached on the wetland environment (Majumdar and Sivaramakrishnan 2015).

According to (Das 2017) demonstrated that about 90% of wetland reduced within 150 years due to different anthropogenic activities. The similar result had been observed that the whole wetland area has experienced a decline of approximately 39.60% because of land reclamation and wetland alteration. The analysis of land use change reveals that the primary factor contributing to wetland degradation is the expansion of agricultural activities, which has experienced a significant increase of 35% (Chatterjee et al. 2022). The numerous past studies confirmed that the LULC area has undergone rapid transformation due to resource management, urbanization and infrastructure development, agricultural expansion, deforestation, climate change impacts, and industrialization (Naikoo et al. 2022; Talukdar et al. 2020a, b). The findings of indicating a shift in land usage patterns between 2008 and 2018 had been substantial changes in the reduction of 19.72 km^2 of vegetation across diverse forest habitats. These changes can be attributed to heightened biotic interaction and over-grazing (Thakur et al. 2020). The survey revealed a constant increase of 21.17% in urban built-up regions in Kolkata, indicating comparable findings. This growth can be attributed to the development of new infrastructure such as roads, flyovers, and settlements. Of contrast, there has been a progressive decline of open terrain, vegetation, agricultural land, and water bodies (Kundu et al. 2020). The specified research region has had a notable transformation in LULC pattern due to the substantial industrial expansion, agricultural activities, and economic growth. The examination of LULCC in Birbhum District across a duration of three decades provides significant perspectives on the developing trends of land usage. The evolving topography of the region is marked by the persistent expansion of agricultural activities, the depletion of water resources, and the escalating urbanization of developed areas. The findings highlight the vitality of implementing sustainable land management practices and making informed decisions to guarantee the future development and environmental preservation of the region.

16.5 Limitations and Future Research

Although the present investigation pertaining to LULCC provides valuable insights, it is subject to certain limitations. The process of categorizing land cover classes has the potential to oversimplify intricate landscapes, hence potentially overlooking crucial nuances. While error matrices are employed for the purpose of evaluating accuracy, this study exhibits a deficiency in providing a comprehensive analysis of potential causes of error and their corresponding effects. Moreover, the lack of future projections restricts its use for long-term strategic planning. The future prospects for study in this field present considerable potential. Examining LULC over extended timeframes can yield a more comprehensive comprehension of enduring patterns. Evaluating the influence of alterations in land use on the local climate and ecosystem services has the potential to expand the environmental outlook. The integration of policy interventions and the utilization of advanced machine learning techniques have the potential to significantly improve the precision and practicality of LULC research. Furthermore, the integration of community viewpoints, the utilization of emerging remote sensing technologies, and the analysis of cross-border influences and urban–rural dynamics have the potential to enhance the research conducted in this field. These many channels provide prospects for addressing the highlighted constraints and making contributions to more complete and knowledgeable investigations of land use.

16.6 Conclusion

This study that investigated a comprehensive examination of LULCC in Birbhum District during the periods of 1977, 1998, and 2018 has unveiled significant transformations in the use of land. The agricultural sector has had a significant increase, asserting its dominance over the landscape with a share of 74.1% in 2018. This trend can be attributed to the shifting economic priorities that have emerged throughout time. On the contrary, there has been a significant decline in the size and quantity of water bodies, which has led to growing apprehension regarding the availability of water resources and the potential impact on ecosystems. The expansion of vegetation has shown a modest increase, however a notable surge in built-up areas signifies a vigorous process of urbanization and infrastructural development. The robustness of our research outcomes is substantiated by the utilization of change matrices and the assessment of categorization accuracies. The monitoring of LULCC is of utmost importance to make well-informed decisions about resource management, urban planning, and environmental preservation. The implementation of sustainable land management practices is crucial in achieving a balance between economic development and the preservation of the environment. The findings establish a crucial basis for future strategic planning in Birbhum District, emphasizing the need for proactive measures to guarantee sustainable development in the face of growing urbanization and evolving land use patterns. This study contributes to the advancement of knowledge regarding LULCC dynamics, providing valuable insights for local authorities and stakeholders to make well-informed decisions that promote the welfare and environmental sustainability of the region.

References

Chatterjee S, Chakraborty K, Mura SNS (2022) Investigating the present status, spatial change, and emerging issues related to riparian wetlands of Bhagirathi-Jalangi floodplain (BJF) in lower deltaic West Bengal, India. Environ Dev Sustain 24(5):7388–7434. https://doi.org/10.1007/s10668-021-01710-8

Congalton R, Green K (1999) Assessing the accuracy of remotely sensed data: principles and practices. CRC/Lewis Press, Boca Raton, FL, 137p

Das N (2017) Human interference and deterioration of the character of wetlands: a micro level study in Labpur block, Birbhum district, West Bengal. IOSR J Humanit Soc Sci 22(01):35–44. https://doi.org/10.9790/0837-2201013544

Das M, Das A (2019) Dynamics of urbanization and its impact on urban ecosystem services (UESs): a study of a medium size town of West Bengal, Eastern India. J Urban Manage 8(3):420–434. https://doi.org/10.1016/j.jum.2019.03.002

Das M, Das A, Pereira P, Mandal A (2021) Exploring the spatio-temporal dynamics of ecosystem health: a study on a rapidly urbanizing metropolitan area of lower Gangetic plain, India. Ecol Indic 125:107584. https://doi.org/10.1016/j.ecolind.2021.107584

Davidson NC, Fluet-Chouinard E, Finlayson CM (2018) Global extent and distribution of wetlands: trends and issues. Mar Freshw Res 69(4):620–627. https://doi.org/10.1071/MF17019

Dewan AM, Yamaguchi Y (2009) Land use and land cover change in Greater Dhaka, Bangladesh: using remote sensing to promote sustainable urbanization. Appl Geogr 29(3):390–401. https://doi.org/10.1016/j.apgeog.2008.12.005

Kundu K, Halder P, Mandal JK (2020) Urban change detection analysis during 1978–2017 at Kolkata, India, using multi-temporal satellite data. J Indian Soc Remote Sens 48(11):1535–1554. https://doi.org/10.1007/s12524-020-01177-6

Majumdar S, Sivaramakrishnan L (2015) Patterns of land use in and around Kolkata city: a spatio-temporal analysis. Indian Cartographer 35(April):218–223

Mohibul S, Siddiqui L, Siddiqui MA, Sarif MN, Parveen N, Islam MS, Khan S, Khanam N, Shariq M, Nasrin T (2022) Spatio-temporal analysis of land use/land cover change using STAR method in Kolkata Urban agglomeration, pp 187–207. https://doi.org/10.1007/978-981-19-3567-1_12

Mohibul S, Sarif MN, Parveen N, Khanam N, Siddiqui MA, Naqvi HR, Nasrin T, Siddiqui L (2023) Wetland health assessment using DPSI framework: a case study in Kolkata metropolitan area. Environ Sci Pollut Res 0123456789. https://doi.org/10.1007/s11356-023-25854-4

Naikoo MW, Rihan M, Shahfahad Peer AH, Talukdar S, Mallick J, Ishtiaq M, Rahman A (2022) Analysis of peri-urban land use/land cover change and its drivers using geospatial techniques and geographically weighted regression. Environ Sci Pollut Res 0123456789. https://doi.org/10.1007/s11356-022-18853-4

Nasrin T, Ramiz M, Sarif N, Hashim M, Siddiqui MA, Siddiqui L, Mohibul S, Mankotia S (2023) Modeling of impact assessment of super cyclone Amphan with machine learning algorithms in Sundarban biosphere. Nat Hazards 117(2):1945–1968. https://doi.org/10.1007/s11069-023-05935-w

Pang A, Li C, Wang X, Hu J (2010) Land use/cover change in response to driving forces of Zoige County, China. Procedia Environ Sci 2(5):1074–1082. https://doi.org/10.1016/j.proenv.2010.10.119

Parveen N, Siddiqui L, Sarif MN, Islam MS, Khanam N, Mohibul S (2021) Industries in Delhi: air pollution versus respiratory morbidities. Process Saf Environ Prot 152:495–512. https://doi.org/10.1016/j.psep.2021.06.027

Parveen N, Siddiqui L, Siddiqui MA, Sarif MN, Islam MS, Khan S, Khanam N, Mohibul S, Shariq M (2022) Monitoring built-up area expansion led by industrial transformation in Delhi using geospatial techniques. Environ Sci Pollut Res 0123456789. https://doi.org/10.1007/s11356-022-23221-3

Rahaman S, Kumar P, Chen R, Meadows ME, Singh RB (2020) Remote sensing assessment of the impact of land use and land cover change on the environment of Barddhaman district, West Bengal, India. Front Environ Sci 8(August):1–15. https://doi.org/10.3389/fenvs.2020.00127

Sarif MN, Siddiqui L, Siddiqui MA, Parveen N, Islam MS, Khan S, Khanam N, Mohibul S, Shariq M, Nasrin T (2022) Household-based approach to assess the impact of river bank erosion on the socio-economic condition of people: a case study of lower Ganga plain 73–101. https://doi.org/10.1007/978-981-19-3567-1_7

Shahfahad, Talukdar S, Naikoo MW, Rahman A, Gagnon AS, Islam ARMT, Mosavi A (2023) Comparative evaluation of operational land imager sensor on board Landsat 8 and Landsat 9 for land use land cover mapping over a heterogeneous landscape. Geocarto Inter 38(1):2152496

Sinha A, Hazra P, Khan TN (2011) Population trends and spatiotemporal changes to the community structure of waterbirds in Birbhum district, West Bengal, India. Proc Zool Soc 64(2):96–108. https://doi.org/10.1007/s12595-011-0018-8

Talukdar S, Pal S, Chakraborty A, Mahato S (2020a) Damming effects on trophic and habitat state of riparian wetlands and their spatial relationship. Ecol Ind 118(July):106757. https://doi.org/10.1016/j.ecolind.2020.106757

Talukdar S, Singha P, Mahato S, Pal S (2020b) Land-use land-cover classification by machine learning classifiers for satellite observations—a review

Thakur TK, Patel DK, Bijalwan A, Dobriyal MJ, Kumar A, Thakur A, Bohra A, Bhat JA (2020) Land use land cover change detection through geospatial analysis in an Indian biosphere reserve. Trees, for People 2(June):100018. https://doi.org/10.1016/j.tfp.2020.100018

Index

S. Talukdar et al. (eds.), *Water Resource Management in Climate Change Scenario*, GIScience and Geo-environmental Modelling, https://doi.org/10.1007/978-3-031-61121-6

The manufacturer's authorised representative in the EU is Springer Nature Customer Service Centre GmbH, Europaplatz 3, 69115 Heidelberg, Germany. If you have any concerns regarding our products, please contact ProductSafety@springernature.com

Printed and bound by CPI Group (UK) Ltd, Croydon, CR0 4YY
01/10/2026
02225965-0019